S0-AGM-791

Technical Mathematics

Third Edition

Jacqueline Austin
Margarita Isern
Miami-Dade Community College

 SAUNDERS COLLEGE PUBLISHING

Philadelphia   New York   Chicago
San Francisco   Montreal   Toronto
London   Sydney   Tokyo   Mexico City
Rio de Janeiro   Madrid

Address orders to:
383 Madison Avenue
New York, NY 10017

Address editorial correspondence to:
West Washington Square
Philadelphia, PA 19105

Text Typeface: Times Roman
Compositor: Monotype Composition Company, Inc.
Acquisitions Editor: Leslie Hawke
Developmental Editor: Jay Freedman
Project Editor: Patrice L. Smith
Copyeditor: Ellen Murray
Managing Editor & Art Director: Richard L. Moore
Design Assistant: Virginia A. Bollard
Text Design: Caliber Design Planning, Inc.
Cover Design: Lawrence R. Didona
Text Artwork: J & R Technical Services, Inc.
Production Manager: Tim Frelick
Assistant Production Manager: Maureen Read

Cover credit: © University Group, Inc., Long Beach, CA

Library of Congress Cataloging in Publication Data

Austin, Jacqueline.
 Technical mathematics.

 Includes index.

 1. Mathematics—1961–   I. Isern,
 Margarita Alejo de Sánchez.   II. Title.
 QA39.2.A89   1983   512′.1   82-60533
 ISBN 0-03-061234-9

TECHNICAL MATHEMATICS

ISBN 0-03-61236-5

678   066   765

CBS COLLEGE PUBLISHING
Saunders College Publishing
Holt, Rinehart and Winston
The Dryden Press

Preface

Technical mathematics courses at a community college face two challenges: to accommodate students with varying degrees of mathematical background, and to teach the mathematical skills required by the various technological departments.

In an attempt to meet these challenges, the authors surveyed the technological departments of Miami-Dade Community College. Each department was asked to identify, from an extensive list of behavioral objectives in mathematics ranging from arithmetic to trigonometry, which skills were needed by the students in their particular area of technology. The departments surveyed were: Aerospace, Architecture and Building Construction, Air Conditioning and Refrigeration, Civil Engineering Technology, Manufacturing Technology, Electronics, and Graphic Arts.

This book evolved from the conclusions of that survey. Every technological department identified the need for the computational skills of arithmetic and a basic knowledge of geometry. This need is created because of the open door policy in the community college, where students of all ages and varying degrees of mathematical background register for technical mathematics courses. Unit 1, Basic Arithmetic Operations, and Unit 2, Applied Geometry, can be taught in full, used as a refresher course, or omitted, depending on the needs of the students enrolled in each individual class.

The topics covered in Units 3 through 8 were mentioned by at least two of the technological departments. However, the departments did emphasize that only a very basic working knowledge of these topics was required.

The book, in its worktext form, emphasizes manipulative skills in mathematics. The rules and examples are stated in a clear and concise manner, making it easier for the technical student to follow. Numerous exercises are given to be used as classwork as well as assignments. Each exercise set is graduated in increasing degree of difficulty, thus making each exercise set a learning tool. Applied problems are included to illustrate applications of the mathematical skills taught. The tests at the end of each unit can be used by the student, as well as the instructor, to diagnose the student's knowledge of the unit.

This book can be used as a two semester course in technical mathematics: Units 1 through 7, the first semester; Units 8 through 13, the second semester.

The book may also be used as a one-semester course. The units on Graphing, Simultaneous Equations, Quadratic Equations, and Logarithms may be omitted without loss of continuity. By studying all the units, except the trigonometry unit, the student will have the necessary background to progress to a full course in trigonometry.

If trigonometry is needed earlier in the course, Sections 13.1 to 13.5 may be taught after Unit 5. Section 13.12 may be taught after Section 11.4 and Section 13.6 after 11.6.

The book may also be used as an individual study course, in which the student progresses at his own speed through the units in the text.

JACQUELINE C. AUSTIN
MARGARITA ISERN

Contents Overview

Contents

Basic Arithmetic Operations

Unit 1 Objectives
1. To add whole numbers.
2. To subtract whole numbers.
3. To multiply whole numbers.
4. To divide whole numbers.
5. To express improper fractions as mixed numbers.
6. To change fractions to equivalent fractions with a given denominator.
7. To reduce fractions to lowest terms.
8. To express numbers in exponential form as factors.
9. To find the prime factorization of whole numbers.
10. To find the lowest common denominator of two or more fractions.
11. To add and subtract fractions.
12. To multiply and divide fractions.
13. To round off decimals and to identify significant figures.
14. To add and subtract decimals.
15. To multiply and divide decimals.
16. To use a calculator to evaluate numerical calculations.
17. To express fractions in decimal form, and to express decimals in fractional form.
18. To change decimals and fractions to percent, and to express percent in decimal and fractional form.
19. To convert a quantity in one unit of measure to an equivalent quantity in another unit of measure.

1.1 Addition of Whole Numbers

In geometry the words point and line are not defined but are words used to define other words. The word "set" is also an intuitive concept and cannot be defined. One method of designating a set is by listing the elements. The set of *natural numbers,* or counting numbers, may be written as N = {1, 2, 3, 4, 5, . . .}. The braces are used to enclose the elements belonging to the set. Capital letters are usually used to name the set. In listing the members of the natural numbers the three dots are used to indicate that the pattern set up by listing the first few elements is to be continued. A set, such as the natural numbers, which cannot be counted, is called an *infinite set.*

If zero is included with the set of natural numbers, the set is called the *whole numbers:* W = {0, 1, 2, 3, 4, . . .}. To indicate that 0 is a member of the set of whole numbers, it is written symbolically as $0 \in W$. Similarly, to show that 0 is not a member of the set of natural numbers we may write $0 \notin N$.

To write any whole number, we use one or more of ten different digits: 0, 1, 2, 3, 4, 5, 6, 7, 8, 9. These digits acquire their value according to their place in the whole number. From right to left the place values are ones, tens, hundreds, thousands, ten thousands, and so on.

The whole number 483 is expressed in expanded notation as follows:

$$483 = 400 \quad + \quad 80 + \quad 3$$
$$= 4 \text{ hundreds} + 8 \text{ tens} + 3 \text{ ones}$$

1

Rule 1: To add two whole numbers, add only those numbers with the same place value, adding the columns from right to left.

Example 1: Add 452 and 317:

$$452 = 4 \text{ hundreds} + 5 \text{ tens} + 2 \text{ ones}$$
$$\underline{+317} = \underline{3 \text{ hundreds} + 1 \text{ ten} + 7 \text{ ones}}$$
$$= 7 \text{ hundreds} + 6 \text{ tens} + 9 \text{ ones}$$
$$= 769$$

Therefore,

$$452$$
$$\underline{+317}$$
$$769$$

Rule 2: If the sum of a column is larger than 9, write the right digit at the bottom of that column and "carry" (add) the left digit to the next column on the left.

Example 2: Add 547 and 316:

$$547 = 5 \text{ hundreds} + 4 \text{ tens} + 7 \text{ ones}$$
$$\underline{+316} = \underline{3 \text{ hundreds} + 1 \text{ ten} + 6 \text{ ones}}$$
$$= 8 \text{ hundreds} + 5 \text{ tens} + 13 \text{ ones}$$
$$= 8 \text{ hundreds} + 5 \text{ tens} + (10 \text{ ones} + 3 \text{ ones})$$
$$= 8 \text{ hundreds} + 5 \text{ tens} + (1 \text{ ten} + 3 \text{ ones})$$
$$= 8 \text{ hundreds} + (5 \text{ tens} + 1 \text{ ten}) + 3 \text{ ones}$$
$$= 8 \text{ hundreds} + 6 \text{ tens} + 3 \text{ ones}$$
$$= 863$$

Therefore,

$$1$$
$$547$$
$$\underline{+316}$$
$$863$$

The number carried (1) is written at the top of the column to which it is being added.

Example 3: Add 4376 and 1258:

$$11$$
$$4376$$
$$\underline{+1258}$$
$$5634$$

Example 4: A central air-conditioning system is installed in a house. The following lengths of 8-inch circular duct are needed: 10 feet for the living room, 5 feet for the dining room, 7 feet for the kitchen, 13 feet for the master bedroom, 3 feet for the master bathroom, 6 feet for the second bedroom, 2 feet for the second bathroom, and 8 feet for the family room. What is the total amount of 8-inch circular duct needed?

$$10$$
$$5$$
$$7$$
$$13$$
$$3$$
$$6$$
$$2$$
$$\underline{+ 8}$$
$$54 \text{ feet}$$

The sum of 22 and 34 is the same as the sum of 34 and 22. $22 + 34 = 56$, and $34 + 22 = 56$. Two whole numbers may be added in any order. This is called the *commutative property* of addition.

To add three whole numbers, $10 + 4 + 6$, group them in order to add two numbers at a time. Parentheses are used to indicate which numbers are added first: $(10 + 4) + 6$ indicates that 10 and 4 are to be added first. The sum, 14, is then added to 6 for a sum of 20. When adding $10 + (4 + 6)$, the parentheses indicate that 4 and 6 are to be added first. The sum, 10, is added to 10 for a sum of 20. Thus, $(10 + 4) + 6 = 10 + (4 + 6)$. Both have a sum of 20. When three numbers are added, the sum of the first two added to the third number is equal to the sum of the last two numbers added to the first number. This is called the *associative property* of addition.

Example 5: Show that $(14 + 21) + 33 = 14 + (21 + 33)$.

$$(14 + 21) + 33 = 14 + (21 + 33)$$
$$35 + 33 = 14 + 54$$
$$68 = 68$$

EXERCISES 1.1

1.
```
  156
+ 233
```

2.
```
  439
+ 250
```

3.
```
  124
+ 467
```

4.
```
  366
+ 539
```

5. $372 + 563 =$

6. $491 + 684 =$

7. $867 + 938 =$

8. $379 + 686 =$

9.
```
  223
  514
+ 647
```

10.
```
  508
  623
+ 781
```

11.
```
   229
  5048
    84
+ 1381
```

12.
```
   719
  3484
   576
+ 2116
```

13. $219 + 5176 + 635 + 4217 =$

14. $74 + 1277 + 752 + 542 =$

15. Add the following lengths: 49 meters, 37 meters, 24 meters, and 19 meters.

16. Add the following weights: 139 kg, 208 kg, 73 kg, and 177 kg.

17. Add the following distances: 1340 km, 4700 km, and 32,000 km.

18. Add the following times: 32 minutes, 58 minutes, and 36 minutes.

19. On a stock room cutter rack, there are 27 angle cutters, 74 saws, and 58 end milling cutters. How many items are on the cutter rack?

20. The electric bill for the Jones family indicated that they used the following kilowatt hours for a period of six months: January, 1475 kw hr; February, 1693 kw hr; March, 1436 kw hr; April, 1876 kw hr; May, 1944 kw hr; June, 2389 kw hr. How many kilowatt hours did the Jones family use during the six-month period?

21. A house purchased for $38,972 appreciated $2618 during a two-year period. What was the value of the house two years after the purchase?

22. How many kilometers did Mr. Black travel during a month if he traveled 1319 km the first week, 643 km the second week, 1045 km the third week, and 482 km the fourth week?

23. A truck driver drove 599 km from Miami to Tampa. The next day he drove 237 km from Tampa to Gainesville. How many km did he drive altogether?

24. A mason laid 334 clay tiles on Monday, 298 clay tiles on Tuesday, 368 clay tiles on Wednesday, 401 clay tiles on Thursday, and 349 clay tiles on Friday. What was the total number of tiles laid that week?

25. During October, a masonry contractor ordered 8500 bricks the first week, 5376 bricks the second week, 4975 bricks the third week, and 3900 bricks the fourth week. How many bricks did the contractor order during October?

26. A mechanical drafting firm has the following equipment: 48 pencils, 9 erasers, 5 protractors, 12 triangles, 17 T-squares, and 4 curves. Find the total pieces of equipment in the inventory.

27. Show that $(62 + 11) + 13 = 62 + (11 + 13)$.

28. Show that $18 + (9 + 6) = (18 + 9) + 6$.

29. Use the commutative property for addition of whole numbers to show that $86 + 54 = 54 + 86$.

30. Use the commutative property to check the addition for $6874 + 2659$.

1.2 Subtraction of Whole Numbers

Subtraction is the inverse of addition. To subtract 6 from 9, find a number which when added to 6 is 9. The 9 is called the *minuend,* and the 6 is called the *subtrahend.* Since 3 added to 6 is 9, 3 is the *difference* between 9 and 6.

Rule 1:   To subtract two whole numbers, find a number that when added to the subtrahend is equal to the minuend.

Example 1:   Subtract 235 from 586:

$$586 = 5 \text{ hundreds} + 8 \text{ tens} + 6 \text{ ones}$$
$$-235 = 2 \text{ hundreds} + 3 \text{ tens} + 5 \text{ ones}$$
$$= 3 \text{ hundreds} + 5 \text{ tens} + 1 \text{ one}$$
$$= 351$$

Therefore,

$$\begin{array}{r} 586 \\ -235 \\ \hline 351 \end{array}$$

Rule 2:   If a number in the subtrahend is larger than the number in the minuend, "borrow" one from the next number to the left in the minuend.

Example 2:   Subtract 368 from 793:

$$793 = 7 \text{ hundreds} + 9 \text{ tens} + 3 \text{ ones}$$
$$-368 = 3 \text{ hundreds} + 6 \text{ tens} + 8 \text{ ones}$$

8 ones cannot be subtracted from 3 ones. Borrow 1 ten from 9 tens, leaving 8 tens. 1 ten plus 3 ones is 13 ones because 1 ten is equal to 10 ones. Next, 8 ones from 13 ones is 5 ones.

$$793 = 7 \text{ hundreds} + 9 \text{ tens} + 3 \text{ ones}$$
$$= 7 \text{ hundreds} + (8 \text{ tens} + 1 \text{ ten}) + 3 \text{ ones}$$
$$= 7 \text{ hundreds} + 8 \text{ tens} + (1 \text{ ten} + 3 \text{ ones})$$
$$= 7 \text{ hundreds} + 8 \text{ tens} + (10 \text{ ones} + 3 \text{ ones})$$
$$= 7 \text{ hundreds} + 8 \text{ tens} + 13 \text{ ones}$$

Thus,

$$793 = 7 \text{ hundreds} + 8 \text{ tens} + 13 \text{ ones}$$
$$-368 = 3 \text{ hundreds} + 6 \text{ tens} + 8 \text{ ones}$$
$$= 4 \text{ hundreds} + 2 \text{ tens} + 5 \text{ ones}$$
$$= 425$$

Therefore,

$$\begin{array}{r} {}^{8\ 13} \\ 7\,\cancel{9}\,\cancel{3} \\ -3\ 6\ 8 \\ \hline 4\ 2\ 5 \end{array}$$

The 1 borrowed is written on the upper left of 3 to represent 13. The 9 is crossed out and 8 is written above it to represent what is left.

Example 3:   Subtract 572 from 930.

$$\begin{array}{r} {}^{8\ 12\ 10} \\ \cancel{9}\,\cancel{3}\,\cancel{0} \\ -5\ 7\ 2 \\ \hline 3\ 5\ 8 \end{array}$$

2 cannot be subtracted from 0. Borrow 1 from the 3 leaving 2 in the tens column and making the ones column a 10. Now 2 can be subtracted from 10 resulting in 8 in the ones column.
In the tens column, 7 cannot be subtracted from 2. Borrow 1 from the 9 in the hundreds column leaving 8 and making the tens column a 12. Now 7 can be subtracted from 12 resulting in 5 in the tens column.
5 can be subtracted from 8, resulting in 3 in the hundreds column.

Example 4: A contractor ordered 6000 feet of oak flooring. He uses 3859 board feet. How many board feet of oak flooring are left over?

$$\begin{array}{r} 6000 \\ -\ 3859 \\ \hline 2141 \text{ feet} \end{array}$$

Rule 3: To do a problem that has more than one addition or subtraction symbol or that requires addition and subtraction in the same problem:

Step 1: Do what is in parentheses.

Step 2: Add or subtract, working left to right.

Example 5:
$$\begin{array}{r} 18 + 6 - 10 \\ 24 - 10 \\ 14 \end{array}$$
Working left to right, addition comes before subtraction.
18 + 6 = 24

Example 6:
$$\begin{array}{r} 18 - 6 + 10 \\ 12 + 10 \\ 22 \end{array}$$
Working left to right, subtraction comes first. 18 − 6 = 12.

Example 7:
$$\begin{array}{r} 18 - (6 + 10) \\ 18 - 16 \\ 2 \end{array}$$
Take care of addition within the parentheses first.

EXERCISES 1.2

1. $\begin{array}{r} 535 \\ -213 \end{array}$

2. $\begin{array}{r} 689 \\ -356 \end{array}$

3. $\begin{array}{r} 735 \\ -216 \end{array}$

4. $\begin{array}{r} 326 \\ -117 \end{array}$

5. $\begin{array}{r} 534 \\ -261 \end{array}$

6. $\begin{array}{r} 408 \\ -123 \end{array}$

7. $\begin{array}{r} 645 \\ -286 \end{array}$

8. $\begin{array}{r} 721 \\ -366 \end{array}$

9. $\begin{array}{r} 705 \\ -199 \end{array}$

10. $\begin{array}{r} 901 \\ -486 \end{array}$

11. $\begin{array}{r} 3001 \\ -1234 \end{array}$

12. $\begin{array}{r} 2005 \\ -1526 \end{array}$

13. $\begin{array}{r} 3000 \\ -1435 \end{array}$

14. $\begin{array}{r} 6000 \\ -1526 \end{array}$

15. $\begin{array}{r} 7000 \\ -4999 \end{array}$

16. $6502 - 2802 =$

17. $4589 - 2398 =$

18. $6835 - 1551 =$

19. A length of pipe measures 25 inches long. If a piece measuring 19 inches is cut from the pipe, how long is the remaining length of pipe?

20. Mr. Harding started out on a 724 km trip by car. After driving 438 km he stopped for lunch. How many km did he have left to travel to complete his trip?

21. If you put $4804 down on a new home, what is the amount of the mortgage (balance) if the home was priced at $36,100?

22. A lumber store stocks 500,000 board feet of pine. If the store sells 136,748 board feet during one given week, how many board feet of pine remain in stock?

23. John weighs 25 kg. His older brother weighs 46 kg. How much heavier is his brother?

24. A piece of plywood measures 15 feet in length. If three pieces of equal lengths, each measuring 2 feet, are sawed off from the board, how long is the piece left?

25. A voltmeter has a resistance of 1098 ohms. How much more resistance does a second voltmeter have if its resistance is 2004 ohms?

26. Mr. Kline noted that the odometer of his car read 4684 kilometers before he left on a trip. When he returned the odometer read 6122 kilometers. How far had he traveled?

27. If there are 200 spark plugs in stock and orders for 12, 6, 18, and 36 spark plugs are filled, how many spark plugs are left in stock?

28. A yearly inventory showed that a hardware store had sold 2389 wooden ladders and 3898 aluminum ladders. How many more aluminum ladders than wooden ladders were sold?

29. $28 + 62 - 31$

30. $67 - 24 + 86$

31. $84 - 20 + 65 - 33$

32. $179 + 224 - 100 - 24$

33. $73 - (22 + 39)$

34. $68 + (87 - 23)$

35. $(5 - 3) + (14 - 2)$

36. $(68 - 21) + (71 - 25)$

1.3   Multiplication of Whole Numbers

Multiplication means repeated addition.

$$3 \times 4 \text{ means} \begin{cases} 3 + 3 + 3 + 3 \\ \quad\quad \text{or} \\ 4 + 4 + 4 \end{cases}$$

In either case, the result is 12. Therefore,

$$3 \times 4 = 12$$

In a multiplication problem, the numbers multiplied together are called *factors*: the result is called the *product*. In the above problem, 3 is a factor, 4 is a factor, and 12 is the product.

Multiplication is also denoted as follows:

3×4 by means of a "times sign" between the factors
$3 \cdot 4$ by means of a dot ($\cdot$) between the factors
$(3)(4)$ by the factors being enclosed in parentheses and adjacent to each other

---

Rule 1:   To multiply any whole number by a one-digit number, multiply each digit of the whole number from right to left by the one-digit number. Then add the products together. The product retains the place value of the digit of the whole number used as a factor.

Example 1:   Multiply 376 by 4.

$$
\begin{aligned}
376 &= 3 \text{ hundreds } + \quad 7 \text{ tens } + \quad 6 \text{ ones} \\
\times 4 &= \times \rule{5cm}{0.4pt} \quad 4 \\
&= 12 \text{ hundreds } + \quad 28 \text{ tens } + \quad 24 \text{ ones} \\
&= 1200 \quad\quad\quad + \quad 280 \quad\quad + \quad 24 \\
&= 1504
\end{aligned}
$$

Therefore,

$$
\begin{array}{r}
{\scriptstyle 3\,2} \\
376 \\
\times 4 \\
\hline
1504
\end{array}
$$

4 times 6 is 24; write 4 and carry 2.
4 times 7 is 28, plus 2 is 30; write 0 and carry 3.
4 times 3 is 12, plus 3 is 15; write 15.

Rule 2: To multiply any two whole numbers, multiply the first factor by each digit of the second factor from right to left. Then add the products together.

Example 2: Multiply 376 by 324.

```
      376
    × 324
     1504  = 4 ones × 376
      752  = 2 tens × 376 (indent to hold place value)
     1128  = 3 hundreds × 376 (indent to hold place value)
   121824
```

Example 3: Multiply 4775 by 302.

```
      4775
    ×  302
      9550  = 2 ones × 4775
         0  = 0 tens × 4775
    14325   = 3 hundreds × 4775
   1442050
```

Very rarely are whole numbers by themselves encountered in technology. The technician will encounter whole numbers with specific units of measure attached, such as 5 kilograms, 7 meters, 12 centimeters, or 100 grams. These numbers, representing a measure, are called *denominate numbers*.

Rule 3: To multiply a denominate number by a number, multiply each unit by the number.

Example 4: Multiply 200 m 4 cm by 3.

```
200 m   4 cm
×          3     Multiply 4 cm by 3; multiply 200 m by 3.
600 m  12 cm
```

Rule 4: To multiply two denominate numbers of like units, multiply the numbers and write the dimensions squared.

Example 5: 3 m × 5 m = 15 sq m

The product is read as "15 square meters."

Rule 5: To multiply three denominate numbers of like units, multiply the numbers and write the dimensions cubed.

Example 6: 25 cm × 4 cm × 3 cm = 300 cu cm

The product is read as "300 cubic centimeters."

Example 7: An apartment building has 24 apartments. If each apartment has 17 electrical outlets, find the total number of electrical outlets in the building.

```
      24
    × 17
     168
      24
     408 electrical outlets
```

Multiplication of whole numbers is commutative. For example, $7 \times 8 = 56$, and $8 \times 7 = 56$. The whole numbers are also associative for multiplication.

$$(6 \times 2) \times 5 = 6 \times (2 \times 5)$$
$$12 \times 5 = 6 \times 10$$
$$60 = 60$$

The *distributive property* of multiplication with respect to addition for whole numbers means that two whole numbers may be added and then multiplied by a whole number, or each number to be added may be multiplied first and the products then added.

$$3 \times (4 + 2) = (3 \times 4) + (3 \times 2)$$
$$3 \times 6 \quad = 12 \quad + 6$$
$$18 = 18$$

Example 8: Multiply the sum of 823 and 684 by 207 and check the answer by using the distributive property.

$$207 \times (823 + 684) = (207 \times 823) + (207 \times 684)$$
$$207 \times 1507 \quad = 170361 + 141588$$
$$311{,}949 = 311{,}949$$

Rule 6: Zero times any whole number is equal to zero. A whole number times zero is equal to zero.

Example 9: $684 \times 0 = 0$, and $0 \times 684 = 0$.

EXERCISES 1.3

Multiply

| 1. $\begin{array}{r} 37 \\ \times\ 4 \end{array}$ | 2. $\begin{array}{r} 26 \\ \times\ 7 \end{array}$ | 3. $\begin{array}{r} 24 \\ \times 52 \end{array}$ | 4. $\begin{array}{r} 68 \\ \times 23 \end{array}$ |

| 5. $\begin{array}{r} 213 \\ \times\ 56 \end{array}$ | 6. $\begin{array}{r} 679 \\ \times\ 31 \end{array}$ | 7. $\begin{array}{r} 207 \\ \times\ 19 \end{array}$ | 8. $\begin{array}{r} 528 \\ \times\ 40 \end{array}$ |

| 9. $\begin{array}{r} 438 \\ \times 253 \end{array}$ | 10. $\begin{array}{r} 6050 \\ \times\ 347 \end{array}$ | 11. $\begin{array}{r} 1003 \\ \times\ 231 \end{array}$ | 12. $\begin{array}{r} 2050 \\ \times\ 306 \end{array}$ |

| 13. $\begin{array}{r} 2504 \\ \times\ 204 \end{array}$ | 14. $\begin{array}{r} 1398 \\ \times\ 652 \end{array}$ | 15. $\begin{array}{r} 6673 \\ \times 4589 \end{array}$ | 16. $\begin{array}{r} 3005 \\ \times 1244 \end{array}$ |

17. 16 years 2 months × 4

18. 240 km × 10

19. 27 kl × 6

20. 28 hr 8 min 3 sec × 5

21. 16 cm × 18 cm

22. 29 km × 36 km

23. 52 cm × 5 cm

24. 29 m × 3 m

25. 9 kg × 3 kg × 7 kg

26. 11 cm × 11 cm × 11 cm

27. How much does it cost to carpet a floor measuring 135 sq meters, if the carpeting costs $6 per sq m?

28. If each bundle of furring contains 12 pieces, how many pieces are there in 197 bundles?

29. A welded support weighs 3 pounds. What is the total weight of 205 supports?

30. 142 bags of sand mix were being loaded onto a truck. If each bag weighed 6 kilograms, how many kilograms of sand mix were on the truck after it was loaded?

31. Counter Industries sold an average of 178 formica counter tops a day. How many counter tops were sold during the month of February if the company was open for business 24 days?

32. A mason buys 43 cubic yards of sand at $7 per cubic yard and 359 cubic yards of concrete at $25 per cubic yard. What was the total cost of his purchase?

Use the distributive property to check your answers to Problems 33 through 40.

33. 20 × (61 + 40)

34. 30 × (45 + 20)

35. (18 + 26) × 14

36. (19 + 31) × 15

37. (21 × 16) + (21 × 43)

38. (23 × 19) + (23 × 41)

39. (401 × 28) + (267 × 28)

40. (510 × 25) + (490 × 25)

1.4 Division of Whole Numbers

Division is the inverse of multiplication. To divide 12 by 3, find a number that when multiplied by 3 is 12. Since 3 × 4 is 12, 12 divided by 3 is 4.

Symbolically, "12 divided by 3" is written:

$$3 \overline{)12}^{\,4}, \quad 12 \div 3 = 4, \quad \frac{12}{3} = 4$$

12 is called the *dividend*, 3 is called the *divisor*, and the result is called the *quotient*.

12 ÷ 3 = 4 because 3 × 4 = 12.
 0 ÷ 3 = 0 because 3 × 0 = 0.
12 ÷ 0 is meaningless because 0 multiplied by any whole number is 0, not 12; therefore, 12 ÷ 0 is said to be *undefined*.

Rule 1: To divide any whole number by a one-digit number, perform the following steps:

Step 1: Divide the left digit of the dividend by the divisor; if not possible, divide the two left-most digits of the dividend by the divisor. Write the result above the digits of the dividend.

Step 2: Multiply the result of Step 1 by the divisor. Write the result under the left digit of the dividend.

Step 3: Subtract the result of Step 2 from the left digit of the dividend.

Step 4: Bring down the rest of the digits of the dividend. Write them immediately to the right of the result of Step 3. Consider this number the "new" dividend.

Step 5: Go back to Step 1, using the "new" dividend. Continue Steps 1 through 5 until there is nothing else to bring down in Step 4. Then the division is completed and the result of the previous Step 3 is called the *remainder*.

Example 1: Divide 753 by 6.

$$
\begin{array}{r}
1 \\
6\,\overline{)\,753} \\
-6 \\
\hline
153
\end{array}
$$

Step 1: Divide 7 by 6. The result is 1. Write it above the 7.

Step 2: Multiply 1 times 6. The result is 6. Write it under the 7.

Step 3: Subtract 6 from 7. The result is 1.

Step 4: Bring down the 5 and the 3, and write them to the right of the 1.

$$
\begin{array}{r}
12 \\
6\,\overline{)\,753} \\
6 \\
\hline
153 \\
-12 \\
\hline
33
\end{array}
$$

Step 1: Dividing 6 into 1 is not possible; hence divide 6 into 15. Write the result (2) to the right of the 1 above the dividend.

Step 2: Multiply 2 times 6. The result is 12. Write it under the 15.

Step 3: Subtract 12 from 15. The result is 3.

Step 4: Bring down the 3 and write it to the right of the 3, the result in the previous subtraction.

$$
\begin{array}{r}
125 \\
6\,\overline{)\,753} \\
6 \\
\hline
153 \\
-12 \\
\hline
33 \\
-30 \\
\hline
3
\end{array}
$$

Step 1: Dividing 6 into 3 is not possible; hence, divide 6 into 33. Write the result (5) above the dividend to the right of the other numbers.

Step 2: Multiply 5 times 6. The result is 30. Write it under the 33.

Step 3: Subtract 30 from 33. The result is 3.

Step 4: There is nothing else to bring down; thus, the remainder is 3.

Therefore,

$$753 \div 6 = 125, \text{ remainder } 3$$

Rule 2:   To divide a whole number by any other smaller whole number, count the number of digits in the divisor. Mark off that same number of digits on the dividend, beginning from the left. Proceed as in Rule 1, modifying Step 1 to read: Divide the marked-off digits of the dividend by the divisor, to get the closest answer possible.

Example 2:   Divide 3470 by 145.

$$
\begin{array}{r}
23 \\
145 \overline{)\ 347\,'0} \quad \textbf{Mark off 3 digits.} \\
-290 \\
\hline
570\,' \\
-435 \\
\hline
135
\end{array}
$$

Therefore,

$$3470 \div 145 = 23, \text{ remainder } 135$$

Rule 3:   To divide a denominate number by a number, divide each unit by the number.

Example 3:   Divide 1287 km by 3.

$$
\begin{array}{r}
429 \text{ km} \\
3 \overline{)\ 1287 \text{ km}} \\
12\phantom{87 \text{ km}} \\
\hline
87 \\
6 \\
\hline
27 \\
27 \\
\hline
\end{array}
$$

Therefore, 1287 km ÷ 3 = 429 km.

Rule 4:   To divide two denominate numbers of the same units, divide the numbers. The result will have no dimension.

Example 4:   10 cm ÷ 5 cm = 2

Because

$$
\begin{array}{r}
2\phantom{0 \text{ cm}} \\
5 \text{ cm} \overline{)\ 10 \text{ cm}} \\
10 \text{ cm} \\
\hline
\end{array}
$$

Rule 5:   To divide two denominate numbers of different units, divide the numbers and write the dimensions as a quotient of the given units.

Example 5:   30 m ÷ 15 sec = 2 m/sec

The result is read "2 meters per second."

Example 6:   In a residential development, 109 identical houses were built. How many feet of tubing were used for each house if a total of 16,132 feet of copper tubing was used in the development?

$$16,132 \text{ feet} \div 109 \text{ houses} = 148 \text{ feet per house}$$

Rule 6: In a problem involving more than one operation (i.e., $+$, $-$, $\times$, $\div$):

Step 1: Do what is in parentheses.

Step 2: Multiply and divide from left to right according to which comes first.

Step 3: Add and subtract from left to right according to which comes first.

Example 7: $26 + 9 \div 3 - 4 \times 7$ **Working left to right, divide 9 by 3, and then multiply**
$26 + \underline{9 \div 3} - \underline{4 \times 7}$ **4×7.**

$26 + 3 \qquad - \quad 28$ **Working left to right, add 26 + 3, and then subtract.**
$\qquad 29 \quad - \quad 28$
$\qquad\qquad 1$

Example 8: $68 + 24 \div (6 - 4) \times 3$ **Subtract within the parentheses first.**
$68 + 24 \div 2 \times 3$ **Working left to right, divide 24 by 2.**
$68 + 12 \times 3$ **Multiple before adding.**
$68 + 36$
104

EXERCISES 1.4

Divide.

1. $7 \div 2$ **2.** $16 \div 4$ **3.** $83 \div 5$ **4.** $73 \div 9$

5. $56 \div 6$ **6.** $48 \div 5$ **7.** $248 \div 6$ **8.** $305 \div 7$

9. $340 \div 5$ **10.** $725 \div 8$ **11.** $25 \div 12$ **12.** $46 \div 15$

13. $675 \div 21$ **14.** $475 \div 37$ **15.** $7660 \div 20$ **16.** $9699 \div 53$

17. $738 \div 248$ **18.** $2346 \div 467$

19. 2432 g ÷ 4

20. 2570 kg ÷ 5

21. 3696 l ÷ 6

22. 1673 km ÷ 7

23. 840 mo ÷ 5 mo

24. 2520 km ÷ 30 km

25. 85 l ÷ 5 l

26. 450 km ÷ 10 hr

27. 39 kl ÷ 3 min

28. A car traveled 686 kilometers on 98 liters of gas. Find the gas consumption in kilometer per liters.

29. A piece of plywood 126 centimeters long is to be cut into 14 equal lengths. What should be the length of each piece?

30. If $1863 is evenly divided among 23 people, how much does each receive?

31. Two cities, 492 km apart, are to be sketched on a map. The scale on the map is 1 cm = 12 km. How many cm apart should the cities be plotted?

32. A mason lays 31 concrete blocks per hour. How many hours will it take him to lay 2666 blocks?

33. The total weight of 46 pieces of pine wood is 276 kg. What is the weight of one piece if the pieces are uniform in weight?

34. $(42 - 36) \times 28$

35. $607 + 625 \div 25$

36. $866 + 34 - 84 \div 4$

37. $164 - 16 \times 8 - 28 \div 7$

38. $129 \div 3 - 7 + 42 \times 61$

39. $428 + 1004 \div 2 - 87$

40. $168 \div 4 + (149 - 69)$

1.5 Fractions

A *fraction* is another way of writing a quotient. "12 divided by 5" is written in fractional form as $\dfrac{12}{5}$ and read, "12 over 5."

The dividend is placed above the line and is called the *numerator*. The divisor is placed below the line and is called the *denominator* of the fraction.

Example 1: Consider $\dfrac{12}{5}$. 12 is the numerator and 5 is the denominator.

A fraction is called a *proper fraction* when the numerator is smaller than the denominator.

A fraction is called an *improper fraction* when the numerator is greater than, or equal to, the denominator.

Example 2: $\dfrac{12}{5}$ and $\dfrac{6}{6}$ are improper fractions.

$\dfrac{3}{7}$ and $\dfrac{1}{10}$ are proper fractions.

Improper fractions may be expressed as a sum of a whole number and a proper fraction. When expressed in this form, they are called *mixed numbers*.

Example 3: $4\dfrac{1}{2}$ and $52\dfrac{3}{4}$ are mixed numbers.

Rule 1:   To convert an improper fraction to a mixed number, divide the numerator by the denominator. The quotient is the whole part of the mixed number. The fractional part consists of the remainder (as the numerator) and the divisor (as the denominator).

Example 4:   Convert $\dfrac{13}{5}$ to a mixed number.

$$5 \overline{)\ 13} \begin{array}{c} 2 \\ \\ \underline{-10} \\ 3 \end{array}$$

Therefore, $\dfrac{13}{5}$ is written as $2\dfrac{3}{5}$.

Rule 2:   To convert a mixed number to an improper fraction, perform the following steps:

Step 1:   Multiply the denominator of the fractional part by the whole part.

Step 2:   Add the result of Step 1 to the numerator of the fractional part.

The numerator of the improper fraction is the result of Step 2. The denominator of the improper fraction is equal to the denominator of the fractional part of the mixed number.

Example 5:   Convert $5\dfrac{3}{8}$ to an improper fraction.

Step 1: $8 \times 5 = 40$

Step 2: $40 + 3 = 43$

Therefore, $5\dfrac{3}{8}$ is written as $\dfrac{43}{8}$.

EXERCISES 1.5

Convert the given improper fractions to mixed numbers:

1. $\dfrac{15}{10}$ 2. $\dfrac{5}{3}$ 3. $\dfrac{22}{9}$ 4. $\dfrac{14}{5}$

5. $\dfrac{17}{6}$ 6. $\dfrac{15}{6}$ 7. $\dfrac{164}{24}$ 8. $\dfrac{305}{25}$

9. $\dfrac{138}{68}$ 10. $\dfrac{5094}{204}$

Convert the given mixed numbers to improper fractions:

11. $2\dfrac{1}{5}$ 12. $3\dfrac{3}{4}$ 13. $2\dfrac{7}{8}$ 14. $5\dfrac{2}{7}$

15. $10\frac{2}{3}$ **16.** $13\frac{5}{11}$ **17.** $25\frac{13}{25}$ **18.** $100\frac{5}{9}$

19. $203\frac{1}{100}$ **20.** $435\frac{25}{481}$

21. If the diameter of a washer is $\frac{103}{32}$ inches, express the diameter as a mixed number.

22. If a chisel weighs $1\frac{3}{4}$ pounds, express its weight as an improper fraction.

23. If the interior of a van measures $5\frac{3}{8}$ feet long by $12\frac{7}{8}$ feet wide, express the dimensions as improper fractions.

24. An axle shaft is $\frac{285}{8}$ inches in length. Express its length as a mixed number.

1.6   Equivalent Fractions

The value of a fraction is not changed when it is multiplied by 1. For example, $\frac{3}{4} \times 1 = \frac{3}{4}.$ Since $1 = \frac{5}{5}, \frac{3}{4}$ may be multiplied by $\frac{5}{5}$ without changing its value:

$$\frac{3}{4} \times \frac{5}{5} = \frac{3 \times 5}{4 \times 5} = \frac{15}{20}.$$

Two fractions are *equivalent* if they indicate the same quantity. The fraction $\frac{3}{4}$ is equivalent to the fraction $\frac{15}{20}.$ Equivalence is denoted by an equal sign:

$$\frac{3}{4} = \frac{15}{20}.$$

Rule 1:   To write an equivalent fraction with a specified denominator, perform the following steps:

Step 1:   Divide the specified denominator by the denominator of the given fraction to obtain a number, n.

Step 2:   Multiply the given fraction by $\dfrac{n}{n}$ or 1.

Example 1:   Write a fraction equivalent to $\dfrac{4}{7}$ with a denominator of 35.

Step 1: $35 \div 7 = 5$

Step 2: $\dfrac{4}{7} = \dfrac{4}{7} \times \dfrac{5}{5} = \dfrac{20}{35}$

Therefore, $\dfrac{4}{7} = \dfrac{20}{35}$.

EXERCISES 1.6

Change the given fraction to an equivalent fraction with the indicated denominator:

1. $\dfrac{1}{3} = \dfrac{}{6}$

2. $\dfrac{1}{5} = \dfrac{}{10}$

3. $\dfrac{2}{3} = \dfrac{}{18}$

4. $\dfrac{5}{6} = \dfrac{}{12}$

5. $\dfrac{3}{4} = \dfrac{}{16}$

6. $\dfrac{1}{3} = \dfrac{}{21}$

7. $\dfrac{31}{54} = \dfrac{}{216}$

8. $\dfrac{18}{13} = \dfrac{}{143}$

9. $\dfrac{303}{400} = \dfrac{}{1200}$

10. $\dfrac{1005}{69} = \dfrac{}{207}$

11. If a motor is rated as having $\dfrac{3}{4}$ horsepower, express the horsepower with a denominator of 16.

12. If insulation is $\dfrac{3}{8}$ inch thick, express the thickness with a denominator of 64.

1.7   Fractions in Lowest Terms

A fraction is in *lowest terms* when the numerator and the denominator have no common factor other than 1.

Example 1: $\dfrac{5}{7}$ is in lowest terms.

$\dfrac{5}{10}$ is not in lowest terms because 5 and 10 are both divisible by 5.

Rule 1:   To reduce a fraction to lowest terms, perform the following steps:

Step 1:   Find the greatest number that divides evenly into both the numerator and the denominator.

Step 2:   Divide the numerator and denominator by the result of Step 1.

The resulting quotients in Step 2 are the numerator and the denominator, respectively, of the equivalent fraction in lowest terms.

Example 2:   Reduce $\dfrac{63}{72}$ to lowest terms.

Step 1:   9 is the greatest number that divides evenly into both 63 and 72.

Step 2: $\dfrac{63}{72} = \dfrac{63 \div 9}{72 \div 9} = \dfrac{7}{8}$

Therefore, $\dfrac{63}{72} = \dfrac{7}{8}$

If difficulty is encountered in finding the greatest number that divides evenly into both the numerator and the denominator (Rule 1, Step 1), the following rule may be used instead of Rule 1.

Rule 2:   To reduce a fraction to lowest terms, perform the following steps:

Step 1:   Find a number that divides evenly into both the numerator and the denominator.

Step 2:   Divide the numerator and the denominator by the result of Step 1. The quotients are the numerator and the denominator, respectively, of an equivalent fraction.

Step 3:   Go back to Step 1, considering the equivalent fraction found in Step 2 as the one to be reduced to lowest terms.

Continue these three steps until the result of Step 1 is 1.

Example 3:   Reduce $\dfrac{60}{90}$ to lowest terms.

$$\dfrac{60}{90} = \dfrac{60 \div 5}{90 \div 5} = \dfrac{12}{18} = \dfrac{12 \div 2}{18 \div 2} = \dfrac{6}{9} = \dfrac{6 \div 3}{9 \div 3} = \dfrac{2}{3} \text{ or } \dfrac{60 \div 10}{90 \div 10} = \dfrac{6}{9} = \dfrac{6 \div 3}{9 \div 3} = \dfrac{2}{3}$$

Therefore,

$$\dfrac{60}{90} = \dfrac{2}{3}$$

EXERCISES 1.7

Reduce the following fractions to lowest terms:

1. $\dfrac{6}{8}$
        2. $\dfrac{4}{12}$
        3. $\dfrac{6}{30}$
        4. $\dfrac{18}{36}$

5. $\dfrac{21}{25}$ **6.** $\dfrac{36}{132}$ **7.** $\dfrac{76}{51}$ **8.** $\dfrac{240}{135}$

9. $\dfrac{242}{422}$ **10.** $\dfrac{220}{43}$

11. If the inside diameter of a lock washer is $\dfrac{75}{192}$ inch, express the diameter in lowest terms.

12. If a piston weighs $\dfrac{18}{24}$ pound, express its weight in lowest terms.

1.8   Exponents

An *exponent* is a number written to the upper right of a non-zero whole number. It represents how many times the whole number is multiplied by itself.

Example 1: $3^2 = 3 \times 3 = 9$

$5^3 = 5 \times 5 \times 5 = 125$

$8^1 = 8$

Conversely, any product in which one or more factors is repeated may be expressed by means of exponents.

Example 2: $6 \times 6 = 6^2$

$3 \times 7 \times 7 \times 7 = 3 \times 7^3$

$4 = 4^1$

Note: 3^2 is read "three to the second power," or "three squared"; 5^3 is read "five to the third power," or "five cubed."

Any non-zero number to the zero power is defined to be one.

Example 3: $7^0 = 1$

$3000^0 = 1$

$10^0 = 1$

Whole numbers may be expressed in expanded notation using powers of ten.

Example 4: $4326 = 4000 + 300 + 20 + 6$

$= 4 \times 1000 + 3 \times 100 + 2 \times 10 + 6 \times 1$

$= 4 \times 10^3 + 3 \times 10^2 + 2 \times 10^1 + 6 \times 10^0$

Write using exponents:

1. 8×8 **2.** $2 \times 2 \times 2$ **3.** $1 \times 1 \times 1 \times 1 \times 1$ **4.** 9

5. $5 \times 5 \times 5 \times 5$ **6.** $4 \times 4 \times 4$

Write without exponents and simplify:

7. 4^3 **8.** 3^0 **9.** 1^7 **10.** 9^2

11. 6^3 **12.** 3^1 **13.** 8^2 **14.** 2^0

15. $2^3 \times 3^2$ **16.** $5^1 \times 4^2$ **17.** $8^3 \times 3^5$ **18.** $6^3 \times 2^2$

Express the following whole numbers in expanded notation using powers of ten:

19. 327   **20.** 46   **21.** 79   **22.** 6954

23. 705   **24.** 254786

1.9   Prime Numbers and Prime Factorization

A *prime* number is a whole number greater than 1, evenly divisible only by itself and 1. The prime numbers are 2, 3, 5, 7, 11, 13, 17, 19, 23, 29, 31, and so on.
    Any whole number greater than 1 is either prime or can be represented as a product of prime numbers in only one way, disregarding the order of the factors.

Example 1: $12 = 2 \times 2 \times 3$
          $10 = 2 \times 5$
          $55 = 5 \times 11$
          $36 = 2 \times 2 \times 3 \times 3$

    When a prime number is repeated as a factor, exponents are used instead of repeating the factor.

Example 2: $12 = 2^2 \times 3$
          $36 = 2^2 \times 3^2$
           $8 = 2^3$

    The *prime factorization* of a whole number is the expression of the number as a product of prime numbers.

Rule 1: To find the prime factorization of a whole number, try to divide the whole number by the first prime number, which is 2. If the whole number is divisible by 2, divide it by 2; then try to divide the quotient by 2; continue to divide each new quotient by 2 until division by 2 is no longer possible; then try the next prime number. With the next prime, follow the same procedure as with the first prime. Continue this process, using consecutive primes, until the new quotient is 1. Then the product of the divisors will be the prime factorization of the whole number.

Example 3:   Find the prime factorization of 36.

$$
\begin{array}{c|l}
2 & \underline{36} \quad \text{Try 2: } 36 \div 2 = 18 \\
2 & \underline{18} \quad \text{Try 2: } 18 \div 2 = 9 \\
  & \quad\quad\; \text{Try 2: } 9 \text{ is not divisible by 2.} \\
3 & \underline{9} \quad\; \text{Try 3: } 9 \div 3 = 3 \\
3 & \underline{3} \quad\; \text{Try 3: } 3 \div 3 = 1 \\
  & \;1
\end{array}
$$

Therefore, $36 = 2 \times 2 \times 3 \times 3 = 2^2 \times 3^2$

Example 4:   Find the prime factorization of 147.

$$
\begin{array}{c|l}
  & \quad\quad\; \text{Try 2: } 147 \text{ is not divisible by 2.} \\
3 & \underline{147} \quad \text{Try 3: } 147 \div 3 = 49 \\
  & \quad\quad\quad \text{Try 3: } 49 \text{ is not divisible by 3.} \\
  & \quad\quad\quad \text{Try 5: } 49 \text{ is not divisible by 5.} \\
7 & \underline{49} \quad \text{Try 7: } 49 \div 7 = 7 \\
7 & \underline{7} \quad\; \text{Try 7: } 7 \div 7 = 1 \\
  & \;1
\end{array}
$$

Therefore, $147 = 3 \times 7 \times 7 = 3 \times 7^2$

Prime factorization can be used to reduce fractions to lowest terms. The fraction $\dfrac{10}{15}$ can be reduced to lowest terms by expressing 10 and 15 in factored form and dividing the numerator and the denominator by the factors that they have in common. For example,

$$
\frac{10}{15} = \frac{2 \cdot \cancel{5}^{\,1}}{3 \cdot \underset{1}{\cancel{5}}} = \frac{2}{3}
$$

Rule 2:   To reduce a fraction to lowest terms using factoring, perform the following steps:

Step 1:   Find the prime factorization of the numerator and the denominator.

Step 2:   Divide the numerator and the denominator by the factors that they have in common.

Step 3:   Multiply the remaining factors in the numerator. Do the same for the denominator.

The resulting fraction is the equivalent in lowest terms.

Example 5:   Reduce $\dfrac{63}{72}$ to lowest terms using factoring.

Step 1: $\dfrac{63}{72} = \dfrac{3 \cdot 3 \cdot 7}{2 \cdot 2 \cdot 2 \cdot 3 \cdot 3}$ **Find the prime factorization of the numerator and the denominator.**

Step 2: $= \dfrac{\cancel{3}^{\,1} \cdot \cancel{3}^{\,1} \cdot 7}{2 \cdot 2 \cdot 2 \cdot \cancel{3} \cdot \cancel{3}}$ **Divide by the common factors.**

Step 3: $= \dfrac{7}{8}$ **Multiply the remaining factors.**

Therefore,

$$\frac{63}{72} = \frac{7}{8}$$

EXERCISES 1.9

1. List the prime numbers less than 50 in increasing order.

Find the prime factorization of the following and list factors in exponential form:

2. 12   **3.** 18   **4.** 45   **5.** 70

6. 605   **7.** 1000   **8.** 143   **9.** 375

Reduce the following fractions to lowest terms using factoring:

10. $\dfrac{6}{8}$ **11.** $\dfrac{4}{12}$ **12.** $\dfrac{6}{30}$ **13.** $\dfrac{18}{36}$

14. $\dfrac{21}{25}$ **15.** $\dfrac{36}{132}$ **16.** $\dfrac{76}{51}$ **17.** $\dfrac{240}{135}$

18. $\dfrac{242}{422}$ **19.** $\dfrac{220}{43}$

20. A brick wall has $\dfrac{90}{240}$ inch mortar joints between the bricks. Reduce the thickness of the mortar joints to lowest terms using factoring.

21. A steel block is $\frac{162}{384}$ inch thick. Express the thickness in lowest terms using factoring.

22. An eraser is $\frac{175}{560}$ inch thick. Express the thickness in lowest terms using factoring.

1.10 Lowest Common Denominator

The *lowest common denominator* (L.C.D.) of two or more fractions is the smallest common multiple of the denominators of the fractions.

Rule 1: To find the lowest common denominator of 2 or more fractions, perform the following steps:

Step 1: Find the prime factorization of each of the denominators of the fractions.

Step 2: Take all the different primes that occur in the factorizations of Step 1 and write them as a product.

Step 3: Raise each of the factors of the product of Step 2 to the highest exponent at which they occur in Step 1.

The L.C.D. is the result of Step 3.

Example 1: Find the L.C.D. of $\frac{5}{12}$ and $\frac{7}{18}$.

Step 1: $12 = 2^2 \times 3$
$18 = 2 \times 3^2$

Step 2: 2×3

Step 3: $2^2 \times 3^2$

Therefore, the L.C.D. $= 2^2 \times 3^2 = 4 \times 9 = 36$.

Example 2: Find the L.C.D. of $\frac{8}{45}$ and $\frac{11}{70}$.

Step 1: $45 = 3^2 \times 5$
$70 = 2 \times 5 \times 7$

Step 2; $2 \times 3 \times 5 \times 7$

Step 3: $2 \times 3^2 \times 5 \times 7$

Therefore, the L.C.D. $= 2 \times 3^2 \times 5 \times 7 = 630$.

To convert fractions to a common denominator means to find equivalent fractions with denominators equal to the lowest common denominator.

Rule 2: To convert two or more fractions to a common denominator, perform the following steps:

Step 1: Find the L.C.D. of the fractions.

Step 2: Write an equivalent fraction with denominator equal to the L.C.D. for each of the given fractions.

Example 3: Convert $\dfrac{5}{12}$ and $\dfrac{7}{18}$ to a common denominator.

> **Step 1:** L.C.D. = 36
>
> **Step 2:** $\dfrac{5}{12} = \dfrac{5 \times 3}{12 \times 3} = \dfrac{15}{36}$
>
> $\dfrac{7}{18} = \dfrac{7 \times 2}{18 \times 2} = \dfrac{14}{36}$
>
> Therefore, $\dfrac{15}{36}$ and $\dfrac{14}{36}$ are equivalent fractions with a common denominator.

Example 4: Convert $\dfrac{8}{45}$ and $\dfrac{11}{70}$ to a common denominator.

> **Step 1:** L.C.D. = 630
>
> **Step 2:** $\dfrac{8}{45} = \dfrac{8 \times 14}{45 \times 14} = \dfrac{112}{630}$
>
> $\dfrac{11}{70} = \dfrac{11 \times 9}{70 \times 9} = \dfrac{99}{630}$
>
> Therefore, $\dfrac{112}{630}$ and $\dfrac{99}{630}$ are equivalent fractions, respectively, to the originals and have a common denominator.

EXERCISES 1.10

Find the lowest common denominators of the given fractions and then convert them to fractions with a common denominator.

1. $\dfrac{1}{4}, \dfrac{1}{2}$　　2. $\dfrac{1}{5}, \dfrac{2}{10}$　　3. $\dfrac{3}{5}, \dfrac{1}{3}$　　4. $\dfrac{2}{3}, \dfrac{1}{2}$

5. $\dfrac{1}{4}, \dfrac{5}{6}, \dfrac{1}{2}$　　6. $\dfrac{7}{9}, \dfrac{2}{3}, \dfrac{1}{6}$　　7. $\dfrac{1}{2}, \dfrac{3}{4}, \dfrac{5}{8}$　　8. $\dfrac{3}{8}, \dfrac{3}{4}, \dfrac{5}{6}$

9. $\dfrac{3}{12}, \dfrac{5}{18}, \dfrac{1}{4}$　　10. $\dfrac{3}{6}, \dfrac{5}{9}, \dfrac{10}{14}$　　11. $\dfrac{22}{35}, \dfrac{8}{15}$　　12. $\dfrac{9}{42}, \dfrac{7}{84}$

13. If a washer has a thickness of $\frac{3}{16}$ inch and an inside diameter of $\frac{7}{8}$ inch, express both dimensions with the same denominator.

14. If the pitch of a thread of a 1-inch pipe is $\frac{1}{24}$ of an inch and the pitch of a thread of a 2-inch pipe is $\frac{1}{18}$ of an inch, express both pitches with a common denominator.

1.11   Addition and Subtraction of Fractions

Rule 1:   To add or subtract two fractions with the same denominator, perform the following steps:

Step 1:   Add or subtract the numerators.

Step 2:   Form a fraction with the result of Step 1 as a numerator, and the common denominator of the given fractions as a denominator.

Step 3:   If possible, reduce the fraction obtained in Step 2 to lowest terms.

Example 1: $\dfrac{1}{8} + \dfrac{3}{8} = \dfrac{1+3}{8} = \dfrac{4}{8} = \dfrac{1}{2}$

$\dfrac{7}{9} - \dfrac{2}{9} = \dfrac{7-2}{9} = \dfrac{5}{9}$

Rule 2:   To add or subtract two fractions with different denominators, perform the following steps:

Step 1:   Convert the fractions to a common denominator.

Step 2:   Add or subtract the fractions using Rule 1.

Example 2: $\dfrac{5}{12} + \dfrac{2}{15} =$

Step 1:

$$12 \qquad = 2^2 \times 3$$
$$15 \qquad = 3 \times 5$$
$$\text{L.C.D.} = 2^2 \times 3 \times 5 = 60$$
$$\frac{5}{12} \qquad = \frac{5 \times 5}{12 \times 5} = \frac{25}{60}$$
$$\frac{2}{15} \qquad = \frac{2 \times 4}{15 \times 4} = \frac{8}{60}$$

First, find L.C.D.

Then, convert the fractions to a denominator equal to L.C.D.

Step 2: $\dfrac{25}{60} + \dfrac{8}{60} = \dfrac{25+8}{60} = \dfrac{33}{60} = \dfrac{11}{20}$

Rule 3: To add two mixed numbers, perform the following steps:

Step 1: Add the fractional parts using Rule 1 or Rule 2. Reduce the result to lowest terms.

Step 2: If the result of Step 1 is a proper fraction, go to Step 3. If the result of Step 1 is an improper fraction, convert the improper fraction to a mixed number.

Step 3: Add the whole parts of the given mixed numbers. Add to this sum the whole part of the mixed number obtained in Step 2, if applicable.

The result is a mixed number with the result of Step 3 as the whole part and the result of Step 2 as the fractional part.

Example 3: Add $9\frac{1}{2}$ and $6\frac{1}{4}$.

Step 1: $\frac{1}{2} + \frac{1}{4} = \frac{2}{4} + \frac{1}{4} = \frac{2+1}{4} = \frac{3}{4}$

Step 2: Not applicable because Step 1 is a proper fraction.

Step 3: $9 + 6 = 15$

Therefore, $9\frac{1}{2} + 6\frac{1}{4} = 15\frac{3}{4}$.

Example 4: Add $2\frac{2}{3} + 1\frac{2}{3}$

Step 1: $\frac{2}{3} + \frac{2}{3} = \frac{4}{3}$

Step 2: $\frac{4}{3} = 1\frac{1}{3}$

Step 3: $2 + 1 = 3$
$3 + 1 = 4$

Therefore, $2\frac{2}{3} + 1\frac{2}{3} = 4\frac{1}{3}$

Rule 4: To subtract two mixed numbers, perform the following steps:

Step 1: Subtract the fractional part of the second number from the fractional part of the first number. If the fractional part of the first number is smaller than the fractional part of the second, borrow 1 from the whole part of the first and add it to the fractional part in order to perform the subtraction.

Step 2: Subtract the whole part of the second number from the whole part of the first number, if no borrowing was performed in Step 1. Subtract the whole part of the second number from one less than the whole part of the first, if borrowing was performed in Step 1.

The result is the mixed number with whole part equal to the result of Step 2 and fractional part equal to the result of Step 1.

Example 5: $4\frac{2}{3} - 1\frac{1}{3} =$

Step 1: $\frac{2}{3} - \frac{1}{3} = \frac{1}{3}$

Step 2: $4 - 1 = 3$

Therefore, $4\frac{2}{3} - 1\frac{1}{3} = 3\frac{1}{3}$

Example 6: $6\frac{2}{5} - 2\frac{3}{5} =$

Step 1: $\frac{2}{5} - \frac{3}{5}$ Since $\frac{2}{5}$ is smaller than $\frac{3}{5}$, borrowing is necessary. Borrow 1 from 6, leaving 5.

$$1 + \frac{2}{5} = \frac{5}{5} + \frac{2}{5} = \frac{7}{5}$$

$$\frac{7}{5} - \frac{3}{5} = \frac{4}{5}$$

Step 2: $5 - 2 = 3$

Therefore, $6\frac{2}{5} - 2\frac{3}{5} = 5\frac{7}{5} - 2\frac{3}{5} = 3\frac{4}{5}$

Example 7: $5\frac{1}{3} - 1\frac{1}{6} =$

Step 1: $\frac{1}{3} - \frac{1}{6} = \frac{2}{6} - \frac{1}{6} = \frac{1}{6}$

Step 2: $5 - 1 = 4$

Therefore, $5\frac{1}{3} - 1\frac{1}{6} = 5\frac{2}{6} - 1\frac{1}{6} = 4\frac{1}{6}$

Example 8: $5\frac{1}{2} - 1\frac{3}{4} =$

Step 1: $\frac{1}{2} - \frac{3}{4} = \frac{2}{4} - \frac{3}{4}$ Since $\frac{2}{4}$ is smaller than $\frac{3}{4}$, borrow 1 from 5, leaving 4.

$$1 + \frac{2}{4} = \frac{4}{4} + \frac{2}{4} = \frac{6}{4}$$

Step 2: $4 - 1 = 3$

Therefore, $5\frac{1}{2} - 1\frac{3}{4} = 4\frac{6}{4} - 1\frac{3}{4} = 3\frac{3}{4}$

Example 9: $6 - 2\frac{2}{3} =$

Step 1:   Since the first number has no fractional part, borrow 1 from 6.

$$6 - 1 = 5$$

$$1 = \frac{3}{3}$$

$$1 - \frac{2}{3} = \frac{3}{3} - \frac{2}{3} = \frac{1}{3}$$

Step 2: $5 - 2 = 3$

Therefore, $6 - 2\frac{2}{3} = 5\frac{3}{3} - 2\frac{2}{3} = 3\frac{1}{3}$

Example 10: The cubic yards of mortar that a mason estimates needing for a job are $12\frac{3}{4}$, $8\frac{1}{2}$, and $9\frac{2}{3}$. Find the total amount of mortar needed.

$$12\frac{3}{4} + 8\frac{1}{2} + 9\frac{2}{3} = 12\frac{9}{12} + 8\frac{6}{12} + 9\frac{8}{12}$$

$$= 29\frac{23}{12}$$

$$= 30\frac{11}{12} \text{ cubic yards}$$

EXERCISES 1.11

Perform the indicated operations and reduce all answers to lowest terms:

1. $\dfrac{1}{5} + \dfrac{3}{5}$ **2.** $\dfrac{2}{6} + \dfrac{1}{6}$ **3.** $\dfrac{7}{10} + \dfrac{1}{10}$ **4.** $\dfrac{5}{8} + \dfrac{5}{8}$

5. $\dfrac{3}{4} + \dfrac{3}{4} + \dfrac{1}{4}$ **6.** $\dfrac{2}{9} + \dfrac{5}{9} + \dfrac{7}{9}$ **7.** $\dfrac{7}{3} - \dfrac{2}{3}$ **8.** $\dfrac{8}{9} - \dfrac{3}{9}$

9. $\dfrac{10}{3} - \dfrac{7}{3}$ **10.** $\dfrac{11}{12} - \dfrac{6}{12}$ **11.** $\dfrac{1}{4} + \dfrac{1}{2}$ **12.** $\dfrac{1}{5} + \dfrac{2}{10}$

13. $\dfrac{3}{5} + \dfrac{1}{3}$ **14.** $\dfrac{2}{3} + \dfrac{1}{2}$ **15.** $\dfrac{1}{4} + \dfrac{5}{6} + \dfrac{1}{2}$ **16.** $\dfrac{7}{9} + \dfrac{2}{3} + \dfrac{1}{6}$

17. Show that $\left(\dfrac{3}{12} + \dfrac{5}{18}\right) + \dfrac{1}{4} = \dfrac{3}{12} + \left(\dfrac{5}{18} + \dfrac{1}{4}\right)$

18. Show that $\left(\dfrac{3}{7} + \dfrac{5}{9}\right) + \dfrac{10}{14} = \dfrac{3}{7} + \left(\dfrac{5}{9} + \dfrac{10}{14}\right)$

19. $\dfrac{9}{10} - \dfrac{2}{5}$ **20.** $\dfrac{8}{9} - \dfrac{12}{15}$ **21.** $\dfrac{2}{5} - \dfrac{6}{15}$ **22.** $\dfrac{2}{3} - \dfrac{1}{2}$

23. $2\dfrac{2}{5} + 1\dfrac{1}{5}$ **24.** $4\dfrac{3}{8} + 2\dfrac{3}{8}$ **25.** $7\dfrac{2}{3} + 3\dfrac{2}{3} + 4\dfrac{2}{3}$ **26.** $3\dfrac{2}{5} - 1\dfrac{3}{5}$

27. $2\dfrac{2}{3} + 5\dfrac{1}{12}$ **28.** $3\dfrac{5}{6} + 1\dfrac{1}{5}$ **29.** $2\dfrac{1}{4} + \dfrac{5}{6} + 1\dfrac{1}{2}$ **30.** $2\dfrac{5}{8} + \dfrac{15}{16} + 5\dfrac{5}{6}$

31. What is the difference in the lengths of two couplings if one is $4\dfrac{2}{3}$ cm long and the other measures $2\dfrac{1}{4}$ cm in length?

32. A counter top $\dfrac{7}{8}$ in. thick is covered with $\dfrac{1}{16}$ in. veneer. What is the total thickness in inches?

33. Jack mails packages weighing $4\dfrac{1}{3}$ kg, $2\dfrac{5}{6}$ kg, $2\dfrac{2}{3}$ kg, and 1 kg. What is the total weight of the packages?

34. A gasoline tank holds $60\dfrac{1}{2}$ liters of gasoline. If the tank has $23\dfrac{2}{7}$ liters in it, how much gasoline is needed to fill the tank?

35. Find the difference in diameter between two rods measuring $\frac{3}{4}$ cm and $\frac{41}{64}$ cm in diameter.

36. From a steel bar $16\frac{1}{2}$ cm in length, cut three pieces of length $1\frac{5}{8}$ cm, $3\frac{1}{4}$ cm, $3\frac{5}{32}$ cm. Find the final length of the board if each cut wastes $\frac{1}{16}$ cm.

37. A partition has a thickness of $4\frac{3}{4}$ in. If the drywall on each side is $\frac{1}{2}$ in. thick, how thick is the stud in between?

38. A board $1\frac{1}{16}$ in. thick was planed $\frac{1}{8}$ in. on one side. What was the original thickness of the board?

39. If three pieces of steel bar measuring $12\frac{3}{4}$ in., $43\frac{1}{16}$ in., and $6\frac{7}{8}$ in. are welded together, what is the total length of the resulting steel bar?

40. If a piece of angle iron $2\frac{5}{8}$ in. is cut from a piece 12 in. long, how much is the remaining piece?

1.12   Multiplication and Division of Fractions

Rule 1:   To multiply two fractions, perform the following steps:

Step 1:   Multiply the numerators of the fractions.

Step 2:   Multiply the denominators of the fractions. The product is the fraction with numerator equal to the result of Step 1 and denominator equal to the result of Step 2. This product should be reduced to lowest terms, if possible.

Example 1:   Multiply $\frac{3}{4}$ and $\frac{2}{9}$.

Step 1: $3 \times 2 = 6$

Step 2: $4 \times 9 = 36$

Therefore, $\frac{3}{4} \times \frac{2}{9} = \frac{6}{36} = \frac{1}{6}$

Note: It is possible to divide the numerator of one fraction and the denominator of the other fraction by the same number in order to simplify the multiplication of the numerators and the denominators.

Example 2:   Multiply $\frac{3}{4}$ and $\frac{2}{9}$.

Divide the numerator of the first and the denominator of the second by 3.
Divide the denominator of the first and the numerator of the second by 2. Thus,

$$\frac{\overset{1}{\cancel{3}}}{\underset{2}{\cancel{4}}} \times \frac{\overset{1}{\cancel{2}}}{\underset{3}{\cancel{9}}} = \frac{1 \times 1}{2 \times 3} = \frac{1}{6}$$

Rule 2:   To multiply mixed numbers and fractions, convert the mixed numbers to improper fractions and apply Rule 1.

Example 3:   Multiply $1\frac{1}{3}$ and $\frac{5}{8}$.

$$1\frac{1}{3} \times \frac{5}{8} = \frac{4}{3} \times \frac{5}{8}$$

Now divide the numerator of the first and the denominator of the second by 4. Thus,

$$\frac{\overset{1}{\cancel{4}}}{3} \times \frac{5}{\underset{2}{\cancel{8}}} = \frac{1 \times 5}{3 \times 2} = \frac{5}{6}$$

To divide two fractions, the quotient can also be written as a fraction, with the dividend in the numerator and the divisor in the denominator.

For example, $\frac{3}{4} \div \frac{7}{8}$ can be written as $\dfrac{\frac{3}{4}}{\frac{7}{8}}$ If the numerator and the denominator

are multiplied by $\frac{8}{7}$, $\left(\dfrac{\frac{8}{7}}{\frac{8}{7}} = 1 \right)$

$$\frac{3}{4} \div \frac{7}{8} = \dfrac{\frac{3}{4}}{\frac{7}{8}} = \dfrac{\frac{3}{4} \times \frac{8}{7}}{\frac{7}{8} \times \frac{8}{7}} = \dfrac{\frac{3}{4} \times \frac{8}{7}}{\frac{\cancel{7}}{\cancel{8}} \times \frac{\cancel{8}}{\cancel{7}}} = \dfrac{\frac{3}{4} \times \frac{8}{7}}{1} = \frac{3}{4} \times \frac{8}{7}$$

then the fraction is ultimately reduced to the product of two fractions in the numerator and 1 in the denominator. More precisely, the fraction is reduced to the product of the dividend and the reciprocal of the divisor. The reciprocal of $\frac{7}{8}$ is $\frac{8}{7}$.

Rule 3: To divide two fractions, multiply the dividend by the reciprocal of the divisor.

Example 4: Divide $\frac{3}{8}$ by $\frac{6}{7}$.

$$\frac{3}{8} \div \frac{6}{7} = \frac{3}{8} \times \frac{7}{6} = \frac{\cancel{3}^1}{8} \times \frac{7}{\cancel{6}_2} = \frac{1 \times 7}{8 \times 2} = \frac{7}{16}$$

Rule 4: To divide two mixed numbers or a mixed number and a fraction, convert the mixed number(s) to improper fractions and divide using Rule 3.

Example 5: Divide $4\frac{1}{16}$ by $8\frac{1}{4}$.

$$4\frac{1}{16} \div 8\frac{1}{4} = \frac{65}{16} \div \frac{33}{4}$$
$$= \frac{65}{16} \times \frac{4}{33}$$
$$= \frac{65}{\cancel{16}_4} \times \frac{\cancel{4}^1}{33}$$
$$= \frac{65 \times 1}{4 \times 33}$$
$$= \frac{65}{132}$$

Example 6: A piece of rigid conduit measures $7\frac{1}{2}$ ft. in length. How many pieces measuring $1\frac{1}{4}$ ft. each can be cut?

$$7\frac{1}{2} \div 1\frac{1}{4} = \frac{15}{2} \div \frac{5}{4}$$
$$= \frac{15}{2} \cdot \frac{4}{5}$$
$$= \frac{\cancel{15}^3}{\cancel{2}_1} \cdot \frac{\cancel{4}^2}{\cancel{5}_1}$$
$$= 6$$

EXERCISES 1.12

Perform the indicated operations and express the result in lowest terms.

1. $\frac{2}{3} \times \frac{6}{8}$ 2. $\frac{1}{3} \times \frac{4}{5}$ 3. $\frac{3}{4} \times \frac{5}{8}$ 4. $\frac{5}{6} \times \frac{1}{3}$

5. $\dfrac{9}{10} \times \dfrac{2}{3} \times \dfrac{5}{8}$ 6. $\dfrac{2}{3} \times \dfrac{3}{4} \times \dfrac{1}{5}$ 7. $16 \times \dfrac{5}{8}$ 8. $\dfrac{9}{10} \times 30$

9. $6 \times \dfrac{3}{4}$ 10. $\dfrac{5}{6} \times 9$ 11. $\dfrac{4}{5} \times 2\dfrac{1}{2}$ 12. $2\dfrac{1}{3} \times \dfrac{3}{7}$

13. $3\dfrac{3}{4} \times 2\dfrac{1}{2}$ 14. $\dfrac{3}{4} \times \dfrac{8}{9} \times 2\dfrac{1}{3}$ 15. $\dfrac{2}{3} \div \dfrac{2}{5}$ 16. $\dfrac{9}{10} \div \dfrac{5}{8}$

17. $\dfrac{3}{4} \div \dfrac{11}{12}$ 18. $\dfrac{7}{8} \div 4$ 19. $27 \div \dfrac{3}{5}$ 20. $\dfrac{5}{6} \div 2$

21. $2\dfrac{1}{4} \div 5$ 22. $6 \div 1\dfrac{1}{2}$ 23. $5\dfrac{1}{3} \div 2\dfrac{2}{5}$ 24. $1\dfrac{2}{3} \div 6\dfrac{2}{3}$

25. What is the total voltage of 20 batteries if each one is $1\dfrac{1}{2}$ volts?

26. How many pieces of pipe $4\dfrac{1}{4}$ in. long can be cut from a pipe $29\dfrac{3}{4}$ in. in length?

27. Frank saved $\dfrac{3}{5}$ of the money needed to buy a new lawn mower. If the mower cost $80, how much has he saved?

28. A recipe calls for $\frac{3}{8}$ of a liter of cream. If you are making $\frac{1}{2}$ the recipe, how much cream should you use?

29. A strip of formica $17\frac{1}{2}$ ft long is to be fastened with screws. The screws are to be placed $1\frac{1}{4}$ ft apart with a screw placed at either end. How many screws will be needed?

30. 48 liters of a liquid is $\frac{5}{6}$ water. How much of the liquid is water?

31. The total weight of 8 boxes is $16\frac{5}{8}$ kg. What is the weight of one box if all boxes are of uniform weight?

32. Mark has 160 basketball schedules to distribute. Bill offered to distribute $\frac{3}{8}$ of the schedules for him. How many schedules did Mark distribute?

33. If one container holds $\frac{5}{8}$ liter of milk, how many liters of milk do you have if you have 7 containers?

34. If you travel $402\frac{1}{5}$ km in 5 hours, how fast are you traveling?

35. On a map 2 cm represent 20 km. If two cities are $4\frac{3}{5}$ cm apart on the map, how many km apart are the two cities?

36. If a machine produces $66\frac{2}{3}$ articles every fifteen minutes, how many articles will it produce in an hour and fifteen minutes?

37. How many pieces of copper wire measuring $1\frac{3}{4}$ meters can be cut from a piece 28 meters in length?

38. A truck traveled 424 kilometers in $4\frac{3}{4}$ hours. What was the average speed of the truck?

39. $\left(\dfrac{1}{2} + \dfrac{1}{3}\right) \times \dfrac{3}{5}$

40. $\dfrac{13}{14} - \dfrac{3}{4} \times \dfrac{4}{7}$

41. $3\dfrac{1}{5} - \dfrac{7}{8} \div \dfrac{3}{8}$

42. $\dfrac{1}{2} + \dfrac{3}{4} \times \dfrac{2}{5} + \dfrac{1}{2} - \dfrac{5}{4}$

43. $1\dfrac{1}{4} + \dfrac{3}{16} \div \dfrac{3}{4}$

44. $\dfrac{3}{4} - \dfrac{3}{20} - \dfrac{3}{5} + \dfrac{1}{2} \times \dfrac{1}{3}$

1.13   Decimals

In the previous sections quantities other than whole numbers were expressed as fractions or mixed numbers. Fractions and mixed numbers may also be written in *decimal notation* in which a point (called a *decimal point*) is placed to the right of the whole number. Each place to the right of the decimal point represents one tenth the value of the one to its left.

In order to write a number in decimal notation the place value concept of whole numbers is extended. From left to right, the place values are tenths, hundredths, thousandths, ten-thousandths, and so on.

Millions	Hundred-thousands	Ten-thousands	Thousands	Hundreds	Tens	Units	Decimal Point	Tenths	Hundredths	Thousandths	Ten-thousandths
1,000,000	100,000	10,000	1000	100	10	1	.	$\dfrac{1}{10}$	$\dfrac{1}{100}$	$\dfrac{1}{1,000}$	$\dfrac{1}{10,000}$

The first digit to the right of the decimal point represents the number of one tenths $\left(\dfrac{1}{10}\right)$, the second digit to the right of the decimal point represents the number of one hundredths $\left(\dfrac{1}{100}\right)$, and so on.

Consider the decimal .589:

$$.589 = 5 \text{ tenths} + 8 \text{ hundredths} + 9 \text{ thousandths}$$
$$= 5 \times \dfrac{1}{10} + 8 \times \dfrac{1}{100} + 9 \times \dfrac{1}{1000}$$
$$= \dfrac{5}{10} + \dfrac{8}{100} + \dfrac{9}{1000}$$
$$= \dfrac{5}{10} \cdot \dfrac{100}{100} + \dfrac{8}{100} \cdot \dfrac{10}{10} + \dfrac{9}{1000}$$
$$= \dfrac{500}{1000} + \dfrac{80}{1000} + \dfrac{9}{1000}$$
$$= \dfrac{589}{1000}$$

The fraction $\frac{589}{1000}$ is written in words as "five hundred eighty-nine thousandths." Therefore, the decimal .589 which is equal to $\frac{589}{1000}$ is written as "five hundred eighty-nine thousandths."

Rule 1:   To write a decimal in words:

Step 1:   Write the whole number in words.

Step 2:   Write "and" for the decimal point.

Step 3:   Write in words the decimal part to the right of the decimal as a whole number followed by the name of the place value of the last digit to the right of the decimal.

Example 1:   Write the decimal 68.72 in words

68	.	72

sixty-eight and seventy-two hundredths

2 is the last digit to the right of the decimal. It is in the hundredths place.

Example 2:   Write one hundred forty-two and two hundred twenty-six thousandths in decimal notation.

one hundred forty-two and two hundred twenty-six thousandths
142 . 226

The *significant figures* in a measurement of a physical quantity are those digits that are known to be reliable. The reliability of the measurement depends on the sensitivity of the measuring instruments.

By way of example, 103 kiloliters has three significant figures. 0.0042 cm has two significant figures; 54,000,000 kilometers has two significant figures.

Rule 2:   The following digits in a measurement are considered significant figures:
1. All non-zero digits.
2. All zeros between significant figures.
3. When a decimal point is indicated: all zero digits which lie to the right of the last non-zero digit.

Example 3:   3,080 has 3 significant figures
110,000 has 2 significant figures
0.00320 has 3 significant figures
50.04   has 4 significant figures
540(has 2 significant figures
540.   has 3 significant figures
540.0   has 4 significant figures

In science and technology it often occurs that all of the decimal places indicated are not necessary. The decimal number is then approximated by another decimal with fewer places. This approximation is called *rounding off*.

For example, 3.1416 has four decimal places (five significant figures). It can be approximated by a number with two decimal places by rounding off to the hundredths digit, 3.14. Therefore, 3.14 is a two decimal approximation of 3.1416. It is written,

3.1416 $\doteq$ 3.14
$\doteq$ means "is approximately"

A decimal is rounded off when only a lesser number of decimal places need be considered and any digits in smaller place values are disregarded. A rounded-off decimal is an approximation of the original decimal.

Rule 3: To round off a decimal to the nearest indicated place value, consider the first digit to the right of the indicated place value and follow the rules of the case to which it applies:

Case 1: If it is a 0, 1, 2, 3, or 4, retain the digit of the indicated place value and change all the digits to the right of the indicated place value to zero.

Case 2: If it is a 5 followed by non-zero digits to the right, or if it is a 6, 7, 8, or 9, add one to the digit of the indicated place value and change all the digits to the right of the indicated place value digit to zero.

Case 3: If it is a 5 followed by zeros to the right, consider the digit at the indicated place value and apply any one of the following subcases:

Sub-case A: If it is an even number, retain it and change all the digits to the right of it to zero.

Sub-case B: If it is an odd number, add one to it and change all the digits to the right of it to zero.

Example 4: Round off 2.57834 to the nearest hundredths. The hundredths digit is 7.

The first digit to the right is 8; apply Case 2. Thus, add 1 to 7 and change 8, 3, and 4 to zero. Therefore, the rounded off decimal is approximately ($\doteq$) 2.58000.

$$2.57834 \doteq 2.58000 \text{ or } 2.58$$

Example 5: Round off 0.0031 to the nearest thousandths. 1 is the first digit to the right of the indicated place value; apply Case 1. Retain the 3 and change the 1 to 0. Therefore, 0.0031 is approximately 0.003. It is denoted by: $0.0031 \doteq 0.003$

Example 6: Round off 37.50 to the nearest unit. 5 is the first digit to the right of the indicated place value and it is followed by zeros; apply Case 3, Sub-case B. Therefore, $37.50 \doteq 38$

Solutions to problems in technology usually are carried to more significant figures than the data from which they are derived, and thus appear to have a greater reliability than they actually have.

EXERCISE 1.13

Write the following decimals in words:

1. 52.74   2. 6.816   3. 0.8423   4. 629.004

5. 78.08   6. 9.408   7. 4568.86   8. 300.01

9. 6894.275   10. 68.20

Write the following in decimal notation:

11. sixty-one and three tenths

12. four hundred three and two hundredths

13. five and one hundred seventy-two thousandths

14. sixty-five hundredths

15. four million and three ten-thousandths

16. eight thousand twenty-four and six hundred twenty-three thousandths

17. four hundred and five tenths

18. sixty-nine and five hundredths

19. seven thousand and one thousand two hundred eighty-six ten-thousandths

20. fifty-four and five hundred one thousandths

In problems 21 through 32, indicate the number of significant figures:

21. 38.03 _____4_____   **27.** 1,280 _____3_____

22. 70 _____1_____   **28.** 0.05 _____2_____

23. 78,000 _____2_____   **29.** 7.500 _____2_____

24. 1,250 _____3_____   **30.** 1,002 _____4_____

25. 5.03 _____3_____   **31.** 0.0070 _____1_____

26. 903 _____3_____   **32.** 0.0396 _____3_____

In problems 33 through 45, round off each decimal to the nearest indicated place value:

33. 0.025 to tenths   *Thousandths*

34. 550.000 to hundreds _____

35. 77.298 to hundredths _____   *77.300*

36. 48.5 to tens   _____   *48.6*

37. 247.32 to hundreds   _____

38. 27.802 to hundreds _____

39. 7.153 to tenths   _____

40. 3.5 to units   _____

41. 5.555 to tenths   _____

42. 107.35 to tens   _____

1.14   Addition and Subtraction of Decimals

Rule 1:   To add or subtract decimals, perform the following steps:

Step 1:   Write the numbers to be added or subtracted in a column in such a way that the decimal points fall underneath each other.

Step 2 (Optional):   Fill in the missing zeros so that all the rows have the same number of decimal places.

Step 3:   Add or subtract as if they were whole numbers, disregarding the decimal point.

Step 4:   Place a decimal point in the result, so that it is lined up with the other decimal points.

Step 5:   If the numbers are approximations, round off the result to the least number of decimal places occurring in the original data.

Example 1:   Add 23.5, 6.836, 300.07

Step 1:   23.5   **Align the decimal points.**
             6.836
           +300.07

Step 2:   23.500
             6.836   **Complete the rows by inserting zeros.**
           +300.070

Step 3:   330 406   **Add.**

Step 4:   330.406   **Insert decimal point lined up with decimal points above.**

Step 5:   If the numbers are exact, the solution is 330.406. If the numbers are approximations, round off as indicated; in this case, to one decimal place, because 23.5 is the number with the least number of decimal places. Therefore, $330.406 \doteq 330.4$

Example 2:   5.397 subtracted from 23.7.

Step 1:   23.7
             − 5.397

Step 2:   23.700
             − 5.397

Step 3:   18 303

Step 4:   18.303

Step 5:   If the numbers are exact, the solution is 18.303; if the numbers are approximations, the solution is 18.3.

Example 3:   In one week a drafter worked 8.5 hours, 9 hours, 7.75 hours, 9.25 hours, and 10 hours. What was the total hours worked that week?

$$
\begin{array}{ccc}
8.5 & & 8.50 \\
9 & & 9.00 \\
7.75 & \text{or} & 7.75 \\
9.25 & & 9.25 \\
+10 & & +10.00 \\
\hline
& & 44.50 \text{ hours}
\end{array}
$$

EXERCISES 1.14

Perform the indicated operation, assuming that the given data are approximations:

1. .7 + .4

.11

2. .8 + .8

.16

3. .74 + .88

.0162

4. 3.76 + 8.26 + 8.87

5. 9.5 + 42.78 + 342.08

6. 23.7 + .63 + 8 + 231.75

7. .4 + 4 + 6.328 + 12.53

8. .8 − .4

9. 27.54 − 7.89

10. 4 − .672

11. 16.02 − 6.5

12. 17 − .302

13. 4.006 − 2.5

14. 17 − 3.02

15. .03 − .0006

16. 5.3 − .667

17. 9.3 − 2.054

18. 27.8 − 15.839

19. A carpenter is purchasing a piece of molding needed for his work. He wants to cut four pieces measuring lengths of 14.5 cm, 12.4 cm, 8.7 cm, and 16.5 cm. If each cut uses .5 cm, what is the length of the piece of molding he should purchase?

20. On a blueprint, an estimator measures lengths of 4.25 cm, 0.7 cm, 2.8 cm, and 3.02 cm. What is the overall length measured?

21. Tickets for a football game are $6.25 for adults and $2.35 for children. How much change will Mr. Sims receive from a ten-dollar bill if he purchases a ticket for himself and his son?

22. Three steel beams are being loaded onto a truck. Find the total weight of the beams loaded if their individual weights are 146.8 kg, 267.3 kg, and 289.7 kg.

23. Mrs. Dixon bought 4.14 kg of meat, 4.41 kg of potatoes, and 2.66 kg of tomatoes. What was the weight of her three purchases?

24. The temperature at 8 a.m. was 12.8°. By noon the temperature was 21.9°. What was the rise in temperature during the four hours?

25. The atomic weight of platinum is 195.1, and the atomic weight of tin is 118.7. What is the difference of the atomic weights of the two elements?

26. A fluorescent light which regularly sold for $16.87 was on sale for $13.99. A work bench was also marked down from $29.95 to $19.99. How much could be saved by buying both the light and the bench on sale?

27. At 0°C the speed of sound in sea water is 1404 m/sec, and in air it is 332 m/sec. How much faster is the speed of sound in sea water than in air?

28. Mr. Carl used his car to make deliveries for his job. During one week he drove the following distances: 74.7 km, 120.7 km, 84.2 km, 68.3 km, and 74.6 km. What was the total number of kilometers traveled?

1.15   Multiplication and Division of Decimals

Rule 1:   To multiply two numbers in decimal notation, perform the following steps:

Step 1:   Multiply as if they were whole numbers, disregarding the decimal point.

Step 2:   Count the number of places to the right of the decimal point in each of the factors and add these two counts.

Step 3:   Place the decimal point in the result of Step 1 so that there are as many digits to the right as the result of Step 2.

Step 4:   If the numbers are approximations, round off the result to the number of significant figures equal to the least number of significant figures occurring in the original data.

Example 1:   Multiply 3.4 by 2.37.

Step 1: $34 \times 237 = 8058$ **Multiply, disregarding decimal.**

Step 2:   1 decimal place in the first factor.
2 decimal places in the second factor.
$1 + 2 = 3$ **Add the number of decimal places in the factors.**

Step 3:   8.058   **Indicate three decimal places.**

Step 4:   If the numbers are approximations, round off to two significant figures; i.e., 8.1

Example 2:   Multiply 0.34 by 100,000

Step 1: $34 \times 100,000 = 3,400,000$

Step 2:   2 decimal places in the first factor.
0 decimal places in the second factor.
$2 + 0 = 2$

Step 3:   34,000.00 or 34,000

Step 4:   If the original numbers are approximations, round off the result to one significant figure because 100,000 has one significant figure. Therefore, the result is 30,000

The whole number 3,000,000 is equal to $3. \times 1,000,000$ or $3. \times 10^6$. A number is said to be in *scientific notation* if it is expressed as the product of a number between 1 and 10 and a power of 10. $3. \times 10^6$ is an example of a number in scientific notation.

Rule 2: If the decimal point is moved *n* places to the left in order to have one non-zero digit to the left of the decimal, multiply by 10^n.

Example 3: The following numbers are expressed as equivalent numbers in scientific notation:

$$64. = 6.4 \times 10^1$$
$$798. = 7.98 \times 10^2$$
$$5286. = 5.286 \times 10^3$$
$$32.6 = 3.26 \times 10^1$$
$$422.7 = 4.227 \times 10^2$$

Rule 3: To divide two numbers in decimal notation, perform the following steps:

Step 1: Count the number of decimal places in the divisor.

Step 2: Move the decimal point to the right in both the divisor and the dividend as the result of Step 1; that is, the number of decimal places required to have the decimal point after the last digit in the divisor.

Step 3: Write the divisor and the dividend in long division form and divide as if they were whole numbers.

Step 4: The decimal point in the result should be aligned with the decimal point of the dividend.

Step 5: If the numbers are approximations, round off to the least number of significant figures occurring in the original data.

Example 4: Divide 6.356 by 2.8.

Step 1: Note that there is one decimal place in the divisor.

Step 2: Move decimal to the right one decimal place: 63.56 and 28.

Steps 3 and 4:
```
        2.27
   28)63.56
       56
        7 5
        5 6
        1 96
        1 96
```

Step 5: If the numbers are approximations, round off to two significant figures. Here, then, the result is 2.3.

Example 5: Divide 35.72 by .001.

Step 1: Note that there are three decimal places in the divisor.

Step 2: Move decimal three places to the right: 35720. and 001.

Step 3: $35720 \div 1 = 35720$

Step 4: 35720

Step 5: If the numbers are approximations, round off to one significant figure. The result, then, is 40,000.

Rule 4: To divide two decimals when the divisor is larger than the dividend, divide disregarding the decimal point, adding as many zeros to the right of the dividend as are needed to obtain the result with one more significant digit than is needed. Then the result is rounded off to the correct number of significant digits.

Example 6:   Divide .408 by 7.5.

Step 1:   There is one decimal place in the divisor.

Step 2: 4.08 and $75.$

Step 3:
$$
\begin{array}{r}
05 \\
75\overline{)4.08} \\
3\ 75 \\
\hline
33
\end{array}
$$

Two significant digits are needed in the result. Therefore add zeros to the dividend until three significant digits are obtained.

Step 4:
$$
\begin{array}{r}
.0544 \\
75\overline{)4.0800} \\
3\ 75 \\
\hline
330 \\
300 \\
\hline
300 \\
300 \\
\hline
\end{array}
$$

Step 5:   The result is then rounded off to two significant digits, and the result is 0.054.

Rule 5:   If the decimal point is moved n places to the right, in order to have one non-zero digit to the left of the decimal, multiply by $\dfrac{1}{10^n}$.

Example 7:   The following numbers are expressed as equivalent numbers in scientific notation:

$$.28 = 2.8 \times \frac{1}{10^1}$$

$$.051 = 5.1 \times \frac{1}{10^2}$$

$$.0065 = 6.5 \times \frac{1}{10^3}$$

EXERCISES 1.15

Perform the indicated operations, assuming that the given data are approximations.

1. $.4 \times .9$ **2.** $14.3 \times .6$ **3.** 31.5×4 **4.** $.34 \times 1.6$

5. 2.5×1.8 **6.** $.82 \times .96$ **7.** 8.47×4.6 **8.** $.573 \times .032$

9. 4.5×10 **10.** 4.5×100 **11.** 4.5×1000 **12.** 2.56×0.01

13. 5.4 ÷ 2   **14.** .63 ÷ 3   **15.** 52.655 ÷ 5   **16.** .32 ÷ 5

17. 4.57 ÷ 2   **18.** .36 ÷ .6   **19.** 0.145 ÷ 0.11   **20.** 2.46 ÷ .12

21. 15.8 ÷ 0.015   **22.** 23.4 ÷ 0.036   **23.** 2.47 ÷ 10   **24.** 64.98 ÷ 100

25. .016 ÷ 1000

26. If a gross (144) of light bulbs cost $24.48, what is the cost per light bulb?

27. A carpenter bought four drills for $99.44. What was the price of each drill?

28. From a sale of sixty-five tickets for a school play there was a return of $152.75. What was the price of each ticket?

29. The weight of one steel beam is 221.42 kilograms. What is the weight of fourteen steel beams of the same weight?

30. A mason worked 39.5 hours one week, 34 hours another week, 23.5 hours the third week, and 37 hours the fourth week. What was the average number of hours worked per week? (To find the average, add the total hours worked and divide the result by the number of weeks.)

31. A part-time technician earns $3.57 an hour. What is his salary if he works 27 hours?

32. If resistors cost $0.06 each, what is the cost of 250 resistors?

33. If 126 light bulbs cost $59.22, what is the price of one light bulb?

34. A worker earns $208.00 per week. If he works an even 40 hours, what is his salary per hour?

35. Light bulbs that regularly sold for $.30 each were on sale four for $.99. How much could be saved by buying two dozen bulbs at the sale price?

36. Mr. Burns purchased a TV antenna for $14.99. He also bought antenna wire for $3.49 and an indoor antenna for $7.99. How much change did he receive if he gave the clerk two twenty-dollar bills?

Express the following as equivalent numbers in scientific notation:

37. 998.   **38.** 60,000.   **39.** 5,486,000.   **40.** 7,820.

41. If the latent heat of fusion for water is 244 BTU per pound, express the amount of BTU per pound in scientific notation.

42. A compressor runs at a speed of 1800 revolutions per minute. Express the speed in scientific notation.

Perform the indicated operations and give the exact answers:

43. 6.24 + .006 × 2.1

44. 8.74 − 2.211 ÷ 1.1

45. .86 − .09 ÷ .4 + .003

46. 4.71 + (2.006 ÷ 1.7) − 1.77

47. 18.64 + (2.1 + 6.8) ÷ .002

48. 5.7 + 6.24 × 3.86 − 1.04

1.16 Hand-Held Calculators

The operations of addition, subtraction, multiplication, and division of whole numbers, fractions, and decimals, and the use of exponents have been discussed in Sections 1.1 to 1.15. To work the exercises in these sections and throughout the text, a scientific calculator can be used.

The scientific calculator you purchase will have an instruction manual with it. Read the manual carefully and get to know your calculator as you progress through this text.

Example 1:   Evaluate $25.79 - 6.8 + 3.4$ on the calculator. To enter 25.79: press 2; press 5; press the decimal point; press 7; press 9.

Enter	Press	Enter	Press	Enter	Press	Display
25.79	$\boxed{-}$	6.8	$\boxed{+}$	3.4	$\boxed{=}$	$22.39 = 22.4$

If your calculator does not have the $=$ key, consult your calculator manual.

Example 2:   Evaluate $4 + 16 \times 2$ on the calculator.

Enter	Press	Enter	Press	Enter	Press	Display
4	$\boxed{+}$	16	$\boxed{\times}$	2	$\boxed{=}$	36

If your calculator displays 40 instead of 36, consult your manual. The rule that tells you to multiply 16 and 2 before adding will have to be followed as you enter the numbers into the calculator.

Example 3:   Evaluate $25.79 - (6.8 + 3.4)$ on the calculator.

The operation within each parentheses must be taken care of first.

$$25.79 - (6.8 + 3.4)$$

Enter	Press	Enter	Press	Press
6.8	$\boxed{+}$	3.4	$\boxed{=}$	$\boxed{STO}$

$$= \quad 25.79 - 10.2$$

Enter	Press	Press	Press	Display
25.79	$\boxed{-}$	$\boxed{RCL}$	$\boxed{=}$	15.59

$$= \quad 15.59 \doteq 15.6$$

Example 4:   Evaluate 614.7^3 on the calculator.

Enter	Press	Enter	Press	Display
614.7	$\boxed{y^x}$	3	$\boxed{=}$	232268138.5 or 2.3227 08 in scientific notation

Example 5:   Evaluate $\dfrac{5907 - 4683}{12 + 4 \times 48}$ on the calculator.

To calculate, enter the denominator first and store.

Enter	Press	Enter	Press	Enter	Press	Press
12	$\boxed{+}$	4	$\boxed{\times}$	48	$\boxed{=}$	$\boxed{STO}$

Enter	Press	Enter	Press	Press	Press	Press	Display
5907	$\boxed{-}$	4683	$\boxed{=}$	$\boxed{\div}$	$\boxed{RCL}$	$\boxed{=}$	6

The *square root* of a number is one of two equal factors. It is denoted by the radical sign, $\sqrt{}$.

Example 6:   Find the square roots of 25, 36, and 49.

$$\sqrt{25} = 5 \text{ because } 5 \cdot 5 = 25$$
$$\sqrt{36} = 6 \text{ because } 6 \cdot 6 = 36$$
$$\sqrt{49} = 7 \text{ because } 7 \cdot 7 = 49$$

The square root of a number between 36 and 49 is a non-repeating, non-terminating decimal. The scientific calculator will give the approximation.

Example 7:   Find the square root of 41 approximated to four significant figures on the calculator.

Enter	Press	Display
41	$\boxed{\sqrt{x}}$	6.4031242

Therefore, $\sqrt{41} \doteq 6.403$

EXERCISE 1.16

Perform the indicated operations on the calculator:

1. $168.92 - 4.60 + 18.75$

2. $2169 + 5378 - 2005$

3. $68 + 92 \times 74$

4. $168 + 75 \times 32$

5. $9876 - 48 \times 16$

6. $18,768 - 40,095 \div 27$

7. $3.14 \,(20 + 50)(60 - 30)$

8. $70 \times (4 + 65) \times 0.667$

9. $\dfrac{0.333 \times 38 \times 57}{27}$

10. $\dfrac{(30 + 50) \times 42}{2}$

11. 68^4

12. 624^3

13. $8.6^2 + 3.27^3$

14. $16^3 + 24^2 \div 4 + 6^4 \times 15$

15. $628 \div 4^2 + 16^3 \times 20$

16. $\dfrac{8^2 + 92^2}{8519 - 7986}$

17. $\dfrac{6981 - 9^4}{3^4 + 24}$

18. Find $\sqrt{168}$ approximated to four decimal places.

19. Find $\sqrt{423}$ approximated to five decimal places.

20. Find $\sqrt{1648}$ approximated to five decimal places.

21. Find the exact value of $\sqrt{35,721} + \sqrt{974,169}$

22. Find the exact value of $\sqrt{52,441} + 87^2$

23. Find the exact value of $\sqrt{9216} - \sqrt{7056}$

24. Find $\dfrac{\sqrt{14.2}}{\sqrt{8.7}}$ approximated to three significant figures.

1.17   Fractions and Decimal Equivalences

Rule 1:   To convert a decimal to a fraction, perform the following steps:

Step 1:   Consider the number without the decimal point and write it as the numerator of the fraction.

Step 2:   Consider the largest number of places to the right of the decimal point. The denominator of the fraction is 10 raised to the above-mentioned number.

Example 1:   Convert .483 to fractional notation.

Step 1:   483 is the numerator of the fraction.

Step 2:   There are three places to the right of the decimal point.

10^3 is the denominator of the fraction.

Therefore, $.483 = \dfrac{483}{1000}$

Example 2:   Convert 26.09 to fractional notation.

Step 1:   2609 is the numerator of the fraction.

Step 2:   There are two places to the right of the decimal point.

$10^2 = 100$ is the denominator.

Therefore, $26.09 = \dfrac{2609}{100}$

A common fraction whose denominator is equal to a power of 10 can be expressed as a decimal number. For example: $\dfrac{4}{10} = .4; \dfrac{61}{100} = .61; \dfrac{893}{1000} = .893$.

To write $\dfrac{1}{25}$ as a decimal, express $\dfrac{1}{25}$ as an equivalent fraction with a denominator of 100. $\dfrac{1}{25} \times \dfrac{4}{4} = \dfrac{4}{100} = .04$. $\dfrac{1}{25}$ may also be expressed as a decimal by dividing the denominator, 25, into the numerator, 1:

$$\begin{array}{r} .04 \\ 25\overline{)1.00} \\ \underline{1.00} \end{array}$$

Rule 2:   To convert a fraction to decimal notation, divide the numerator of the fraction by the denominator of the fraction.

Example 3:   Convert $\dfrac{1}{5}$ to decimal form.

$$1 \div 5 = ?$$

$$\begin{array}{r} 0.2 \\ 5\overline{)1.0} \\ \underline{1.0} \end{array}$$

Therefore, $1 \div 5 = 0.2$

That is, $\dfrac{1}{5} = 0.2$

When the denominator divided into the numerator has a remainder of zero, the decimal is called a *terminating decimal*.

Example 4:   Convert $\dfrac{2}{9}$ to decimal form.

$$\begin{array}{r} 0.2222 \\ 9\overline{)2.0000} \\ \underline{1\ 8} \\ 20 \\ \underline{18} \\ 20 \\ \underline{18} \\ 20 \\ \underline{18} \\ 2 \end{array}$$

Notice that if, in this example, more zeros were added, the division would never be completed. This type of decimal is called a *non-terminating repeating decimal*. It is denoted by writing several of the repeating digit(s) and drawing a line above the last repeated digit.

Therefore, $\dfrac{2}{9} = 0.222\overline{2}$ or $0.\overline{2}$

On the calculator:

Enter	Press	Enter	Press	Display
2	÷	9	=	.22222222

EXERCISES 1.17

Convert from decimals to fractions:

1. .6   **2.** .8   **3.** .36   **4.** .95

5. .06   **6.** .2   **7.** 5.10   **8.** 3.01

9. 28.583   **10.** 41.379   **11.** .042   **12.** .058

13. 90.005   **14.** 420.05   **15.** 420.5

Convert the following fractions to decimals:

16. $\dfrac{1}{4}$ **17.** $\dfrac{1}{2}$ **18.** $\dfrac{5}{8}$ **19.** $\dfrac{2}{3}$

20. $\dfrac{6}{10}$ **21.** $\dfrac{3}{5}$

22. Ventilation standards require department stores to have a minimum of 0.05 cubic foot per minute of ventilation for each square foot of floor. Express the minimum as a fraction.

23. A gear has pitch 0.0625 inch. Express the pitch as a fraction.

24. The diameter of a bolt is $1\frac{4}{5}$ inches. Express the diameter as a decimal.

25. The thickness of a nail is $\frac{9}{32}$ inch. Express the thickness as a decimal.

1.18 Percent

The word *percent* means "per hundred." One percent of a number represents one one-hundredth $\left(\frac{1}{100}\right)$ of the number; twenty-five percent of a number represents twenty-five one-hundredths $\left(\frac{25}{100} = \frac{1}{4}\right)$ of the number.

Percent is denoted by the symbol %. One percent is written 1%; 25 percent is written 25%.

The symbol % is equivalent to two decimal places. 1% is equivalent to 0.01; 25% is equivalent to 0.25; 7% is equivalent to 0.07; 0.5% is equivalent to 0.005.

Rule 1: To change a whole number or a decimal number to percent, multiply by 100 (move the decimal point two places to the right) and add a percent symbol.

Example 1: Change 0.48 to percent.

Multiplying by 100 moves the decimal point two places to the right.

Therefore, 0.48 = 48%

Example 2: Change 83.1 to percent.

Multiplying by 100 moves the decimal point two places to the right; it is necessary to insert a zero before the decimal point to hold the place value.

Therefore, 83.1 = 8310%

Rule 2: To change a fraction to a percent, perform the following steps:

Step 1: Convert the fraction to a decimal, rounding off to ten thousandths if necessary.

Step 2: Use Rule 1 to change the result of Step 1 to percent.

Example 3: Change $\frac{2}{5}$ to percent.

Step 1: $\frac{2}{5} = 0.4$

Step 2: 0.4 = 40%

Example 4: Change $\frac{2}{3}$ to percent.

Step 1: $\frac{2}{3} = 0.66666\overline{6} \doteq 0.6667$

Step 2: $0.6667 = 66.67\%$

Rule 3: To change a percent to a decimal, drop the percent symbol and divide by 100, which moves the decimal point two places to the left.

Example 5: Change 52% to a decimal.

52% is understood to have a decimal point at 52.0% Dropping the percent symbol and dividing by 100 moves the decimal point two places to the left. Therefore, 52% = 0.52.

Example 6: Change 3% to a decimal.

To move the decimal two places to the left it is necessary to insert zeros. Therefore, 3% = 0.03.

Rule 4: To change a percent to a fraction, perform the following steps:

Step 1; Change the percent to a decimal.

Step 2: Change the decimal in the result of Step 1 to a fraction, and reduce it to lowest terms if necessary.

Example 7: Change 20% to a fraction.

Step 1: $20\% = 0.20 = 0.2$

Step 2: $0.2 = \frac{2}{10} = \frac{1}{5}$

Therefore, $20\% = \frac{1}{5}$

Example 8: Change 0.5% to a fraction.

Step 1: $0.5\% = 0.005$

Step 2: $0.005 = \frac{5}{1000} = \frac{1}{200}$

Therefore, $0.5\% = \frac{1}{200}$

EXERCISES 1.18

Change the following decimals to percent:

1. 0.75   **2.** 0.47   **3.** 0.478   **4.** 0.732

5. 0.013   **6.** 0.4367   **7.** 0.008   **8.** 0.0005

9. 3.4   **10.** 4

Change the following fractions to percent:

11. $\frac{1}{4}$ **12.** $\frac{3}{5}$ **13.** $\frac{5}{6}$ **14.** $\frac{1}{3}$

15. $\frac{7}{8}$ **16.** $\frac{1}{6}$ **17.** $2\frac{3}{8}$ **18.** $6\frac{2}{3}$

Change the following percents to decimals:

19. 23% **20.** 57% **21.** 2% **22.** 0.7%

23. 2.5% **24.** 43.7% **25.** 84.3% **26.** $5\frac{4}{5}\%$

27. $3\frac{1}{4}\%$ **28.** $6\frac{3}{8}\%$ **29.** 123% **30.** 445%

Change the following percents to fractions:

31. 30% **32.** 25% **33.** $15\frac{1}{2}\%$ **34.** $64\frac{3}{4}\%$

35. 3.21% **36.** 7.4% **37.** 75% **38.** 100%

39. 825% **40.** 350%

1.19   Measures

In technical problems, numbers represent a measurement or calculations derived from measurements. The number 60 has no physical meaning to a technician without a unit of measurement; 60 meters, however, denotes a specific length.

Length, volume, mass, time, temperature, and electric current are measurable quantities.

There are two systems of measurement: the metric system and the British system. In metric measure the standard unit of length is the meter; the standard unit of weight is the gram; and the standard unit of liquid volume is the liter. Table 1 shows how the meter, the gram, and the liter are subdivided. Each unit is related to the next unit by a factor of ten. Prefixes for the meter, the gram, and the liter are milli, 0.001; centi, 0.01; deci, 0.1; deka, 10; hecto, 100; and kilo, 1000.

TABLE 1. Metric System

L E N G T H	1 millimeter (mm)	=	0.001	meter
	1 centimeter (cm)	=	0.01	meter
	1 decimeter (dm)	=	0.1	meter
	1 meter (m)	=	1	meter
	1 dekameter (dam)	=	10	meters
	1 hectometer (hm)	=	100	meters
	1 kilometer (km)	=	1000	meters
W E I G H T	1 milligram (mg)	=	0.001	gram
	1 centigram (cg)	=	0.01	gram
	1 decigram (dg)	=	0.1	gram
	1 gram (g)	=	1	gram
	1 dekagram (dag)	=	10	grams
	1 hectogram (hg)	=	100	grams
	1 kilogram (kg)	=	1000	grams
V O L U M E	1 milliliter (ml)	=	0.001	liter
	1 centiliter (cl)	=	0.01	liter
	1 deciliter (dl)	=	0.1	liter
	1 liter (l)	=	1	liter
	1 dekaliter (dal)	=	10	liters
	1 hectoliter (hl)	=	100	liters
	1 kiloliter (kl)	=	1000	liters

Rule 1:   To change a unit of length, weight, or liquid measure, within the metric system, to the next larger unit, multiply by 0.1.

Rule 2:   To change a unit of length, weight, or liquid measure, within the metric system, to the next smaller unit, multiply by 10.

Example 1:   Change 23.7 meters to dekameters.

$23.7 \times 0.1 = 2.37$ **Rule 1**

23.7 meters = 2.37 dekameters

Example 2:   Change 468 meters to kilometers.

$468 \times 0.001 = 0.468$ **Rule 1**
(m to dam to hm to km = $0.1 \times 0.1 \times 0.1 = 0.001$)

468 meters = 0.468 kilometer

Example 3:   Change 0.56 grams to decigrams.

$0.56 \times 10 = 5.6$ **Rule 2**

0.56 gram = 5.6 decigrams

Example 4:   Change 0.655 hectoliters to liters.

$0.655 \times 100 = 65.5$ **Rule 2**
(hl to dal to 1 = $10 \times 10 = 100$)

0.655 hectoliter = 65.5 liters

Table 2 shows how units of area are related. Each unit is related to the next unit by a factor of 100. This table also shows that units of volume are related to one another by factors of 1000.

TABLE 2. **Metric** **System**					
	A R E A	1 square millimeter (mm²)	=	0.000001	square meter
		1 square centimeter (cm²)	=	0.0001	square meter
		1 square decimeter (dm²)	=	0.01	square meter
		1 square meter (m²)	=	1	square meter
		1 square dekameter (dam²)	=	100	square meters = 1 are
		1 square hectometer (hm²)	=	10,000	square meters = 1 hectare
		1 square kilometer (km²)	=	1,000,000	square meters
	V O L U M E	1 cubic millimeter (mm³)	=	0.000000001	cubic meter
		1 cubic centimeter (cm³)	=	0.000001	cubic meter
		1 cubic decimeter (dm³)	=	0.001	cubic meter
		1 cubic meter (m³)	=	1	cubic meter
		1 cubic dekameter (dam³)	=	1,000	cubic meters
		1 cubic hectometer (hm³)	=	1,000,000	cubic meters
		1 cubic kilometer (km³)	=	1,000,000,000	cubic meters

Rule 3: To change a unit of square measure, within the metric system, to the next larger unit of square measure, multiply by 0.01.

Rule 4: To change a unit of square measure, within the metric system, to the next smaller unit of square measure, multiply by 100.

Rule 5: To change a unit of cubic measure, within the metric system, to the next larger unit of cubic measure, multiply by 0.001.

Rule 6: To change a unit of cubic measure, within the metric system, to the next smaller unit of cubic measure, multiply by 1000.

Example 5: Change 0.386 cubic centimeter to cubic millimeters.

$0.386 \times 1000 = 386$ **Rule 6**

0.386 cubic centimeter = 386 cubic millimeters

Example 6: Change 9640 square centimeters to square meters.

$9640 \times 0.0001 = 0.9640$ **Rule 3**
(cm² to dm² to m² $= 0.01 \times 0.01 = 0.0001$)

9640 square centimeters = 0.9640 square meter.

It is possible to convert a given unit of measurement in the metric system to an equivalent unit of measurement in the British system, or from a unit of measurement in the British system to a unit of measurement in the metric system. Table 3 shows the equivalences between the two systems of measurement and gives an equivalence factor. An *equivalence factor* is a fraction that is equal to one. For example, from Table 3, 1 inch = 2.54 centimeters. Therefore, $1 = \dfrac{1 \text{ in.}}{2.54 \text{ cm}}$, or $1 = \dfrac{2.54 \text{ cm}}{1 \text{ in.}}$ are equivalence factors.

Rule 7: To convert one quantity in a unit of measure to an equivalent quantity in another unit of measure:

Step 1: Find the equivalence in the table that involves the two measurements.

Step 2: Express the equivalence in a fractional form equal to one, that is, the equivalence factor. The numerator of the fraction should have the unit of measure to be obtained, the denominator of the fraction should have the measure of the unit to be changed.

Step 3:   Multiply the unit of measure to be changed by the conversion factor.

Example 7:   Change 422 feet to meters.

Step 1:   From Table 3, 3 feet $\doteq$ .9144 meters.

Step 2:   The measure to be changed is feet, and the measure to be obtained is meters. Therefore, the conversion factor is $\dfrac{.9144 \text{ m}}{3 \text{ ft}}$.

Step 3: $422 \not{\text{ft}} \times \dfrac{.9144 \text{ m}}{3 \not{\text{ft}}} \doteq 128.62$ meters

Therefore, 422 feet $\doteq$ 129 meters.

Example 8:   Change 24 liters to quarts.

Step 1:   From Table 3, 1 quart $\doteq$ 0.9463 liter.

Step 2:   The measure to be changed is liters, and the measure to be obtained is quarts. Therefore, the conversion factor is $\dfrac{1 \text{ qt}}{0.9463 \, l}$.

Step 3: $24 \not{l} \times \dfrac{1 \text{ qt}}{0.9463 \not{l}} \doteq 25.36$ qt.

Therefore, 24 liters $\doteq$ 25 quarts.

TABLE 3. British– Metric Equivalents		British System	Metric Equivalent
L		1 inch (in.)	= 2.54 cm
I	12 in.	= 1 foot (ft)	= 30.48 cm
N	3 ft	= 1 yard (yd)	= .9144 m
E	$5\frac{1}{2}$ yd	= 1 rod (rd)	= 5.0292 m
A			
R	5280 ft	= 1 mile (mi)	= 1.6093 km
S		1 square inch (sq in.)	= 6.4516 cm²
Q	144 sq in.	= 1 square foot (sq ft)	= 0929 m²
U	9 sq ft	= 1 square yard (sq yd)	= .8316 m²
A	$30\frac{1}{4}$ sq yd	= 1 square rod (sq rd)	= 25.293 m²
R			
E	160 sq rd	= 1 acre (A)	= 4047.9 m²
	640 A	= 1 square mile (sq mi)	= 2.5899 km²
C		1 cubic inch (cu in)	= 16.387 cm³
U	1.728 cu in	= 1 cubic foot (cu ft)	= .0283 m²
B	27 cu ft	= 1 cubic yard (cu yd)	= .7646 m³
I			
C			
L		1 pint (pt	= .4732 liter = 473.176 cc
I	2 pt	= 1 quart (qt)	= .9463 liter
Q	4 qt	= 1 gallon (gal)	= 3.7853 liters = .0038 cu m
U			
I			
D			
W		1 ounce (oz)	= 28.35 g
E	16 oz	= 1 pound (lb)	= 453.59 g
I	2000 lb	= 1 short ton (st)	
G			
H			
T			

Example 9: Change 684,000 meters to miles.

Table 3 does not show a relation between meters and miles, but a relation between kilometers and miles. Use Rule 1 to change meters to kilometers.

$$684,000 \times 0.001 = 684 \quad \textbf{Rule 1}$$

684,000 meters = 684 kilometers.

Step 1: From Table 3, 1.6093 kilometers $\doteq$ 1 mile.

Step 2: Since kilometers is the unit of measure to be changed, and miles is the unit of measure to be obtained, the conversion factor is $\dfrac{1 \text{ mi}}{1.6093 \text{ km}}$.

Step 3: $684 \text{ km} \times \dfrac{1 \text{ mi}}{1.6093 \text{ km}} \doteq 425.03 \text{ mi}$

Therefore, 684,000 meters $\doteq$ 425 miles.

Example 10: Change 98 feet 8 inches to feet using three significant figures.

$$\begin{aligned}
98 \text{ feet } 8 \text{ inches} &= 98 \text{ feet } + 8 \text{ inches} \\
&= 98 \text{ ft } + \left(8 \text{ in.} \times \frac{1 \text{ ft}}{12 \text{ in.}} \right) \\
&= 98 \text{ ft } + 0.667 \text{ ft} \\
&= 98.667 \text{ ft} \\
&= 98.7 \text{ ft}
\end{aligned}$$

EXERCISES 1.19

1. Change 48 centimeters to millimeters.

2. Change 37 deciliters to liters.

3. Change 45.38 milligrams to grams.

4. Change 88.3 centimeters to kilometers.

5. Change 1.683 kilograms to decigrams.

6. Change 281.6 grams to milligrams.

7. Change 30 kiloliters to liters.

8. Change 4.5 meters to centimeters.

9. Change 0.0031 milliliter to liters.

10. Change 44 centigrams to grams.

11. Change 932 square decimeters to square meters.

12. Change 337.1 square centimeters to square millimeters.

13. Change 0.454 square dekameter to square decimeters.

14. Change 515 square centimeters to square meters.

15. Change 222 cubic decimeters to cubic meters.

16. Change 17.17 cubic dekameters to cubic meters.

17. Change 4083 cubic centimeters to cubic meters.

18. Change 4.11 meters to inches.

19. Change 5.222 pounds to grams.

20. Change 3.1 quarts to liters.

21. Change 8 centimeters to inches.

22. Change 2.6 feet to centimeters.

23. Change 5.1 liters to pints.

24. Change 142 kilometers to miles.

25. Change 8.5 milligrams to ounces.

Change to feet and give the answer to three significant figures:

26. 1 ft 4 in.

27. 112 ft 9 in.

28. 9 ft $3\frac{1}{2}$ in.

29. 2 ft $8\frac{1}{4}$ in.

30. 9 in.

31. $\frac{3}{8}$ in.

32. 21 ft 10 in. $\times$ 8 in.

33. 3 ft 5 in. × 2 ft 4 in.

34. 5 ft 2 in. × 8 ft $2\frac{1}{2}$ in. × 18 in.

35. 35 ft 3 in. × 28 ft. × 7 ft 9 in.

Unit 1 Self-Evaluation

Perform the indicated operations:

1. 5168 ÷ 76

2. 3.1 + 710.4 + 0.034

3. 8.2 − 5.132

4. 5.32 × 2.7

5. $83.2 \div 12.1$

6. $6.8 + 14.7 \div .7 - 6.8 \times .01$

7. $\dfrac{5}{8} + \dfrac{7}{12}$

8. $4\dfrac{1}{2} - 1\dfrac{4}{5}$

9. $\dfrac{15}{36} \cdot \dfrac{24}{40}$

10. $10\dfrac{2}{3} \div 8\dfrac{5}{6}$

11. $7\dfrac{1}{3} - 2\dfrac{2}{3} \times \dfrac{3}{4} + \dfrac{1}{16} \div \dfrac{1}{8}$

12. Change $\dfrac{7}{22}$ to a decimal.

13. Change 0.032 to a fraction.

14. Change 76.6 to percent.

15. Round off 65.278 to hundredths.

16. Write $2^2 \times 3^3$ without exponents and multiply.

17. Convert 500 miles to kilometers.

18. Change 0.068 kilogram to grams.

19. Find the prime factorization of 2475.

20. Write the decimal six hundred four and three hundred seventy-three thousandths.

21. Write 6.0007 in words.

22. Use the associative property to evaluate and check: $6.08 \times (18.72 + 24.09)$

23. Evaluate $7^3 + 16^2 \div 2 + 18 \times 12^2 - 2592$

24. Find the cost of covering a floor with tile if the tile costs $14.35 per square meter, and the floor to be covered measures 23 square meters.

25. Eight pieces of plywood, each measuring $20\frac{1}{3}$ cm in length, were cut from a 500 cm piece. If each cut wasted $\frac{1}{16}$ cm, what was the length of the piece of plywood left after the cuttings?

Applied Geometry

Unit 2 Objectives
1. To distinguish between line segments, rays, and lines; to determine whether lines are parallel, intersecting, or coincident.
2. To determine whether angles are acute, obtuse, or right; to recognize vertical angles, adjacent angles, alternate interior and exterior angles, and corresponding angles.
3. To find the area and perimeter of triangles; to use the Pythagorean Theorem.
4. To find the area and perimeter of polygons.
5. To find the area and perimeter of a circle; to convert degree measure to radian measure, and radian measure to degree measure.
6. To sketch and find the surface area and volume of solids.

2.1   Points and Lines

Geometry is the branch of mathematics that studies points, lines, angles, and figures in space. It studies their shapes and properties, as well as the relationship of one figure to another.

Technicians, surveyors, designers, engineers, and draftsmen will often encounter geometric problems in their daily work. The geometry to be considered in this unit will deal with the properties of geometric objects, with emphasis on practical applications.

In geometry concepts that are not defined are called *undefined terms*. Point and line are undefined terms; they underlie the definitions of all other geometric terms.

A *point* is represented by a dot. A capital letter is placed by the dot to name the point or to distinguish it from another point.

The above dots represent point *A*, point *B*, and point *C*.

A *line,* represented by a straight mark, can be thought of as a sequence of points extending indefinitely in both directions along the straight mark. Lines are named by using any two points on the line or by using one script small letter.

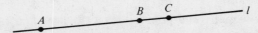

A, B, and *C* are points on the given line. The line above may be named *AB, AC, BC,* or simply, *l*.

73

A *line segment* is the sequence of points between two given points on the line. The line segment has two end-points. It is named by the two end-points with a horizontal line above them.

Line segments $\overline{PQ}$ and $\overline{AB}$ are represented as follows:

Line segments can be measured by a ruler. Technicians and scientists measure line segments with a caliper for greater accuracy.

A *ray* is a sequence of points extending infinitely in only one direction and having an end-point in the other direction. A ray is named on the end-point followed by any other point on the ray with an arrow above.

Rays $\overrightarrow{RS}$ and $\overrightarrow{MN}$ are represented as follows:

Two lines are related to each other in several ways. They may be classified as parallel, intersecting, or coincident lines, according to the number of points that they have in common. *Parallel lines* have no points in common.

Lines PQ and RS are parallel. This is denoted by writing $PQ \parallel RS$.

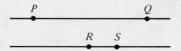

Intersecting lines have one point in common.

Lines AB and CD are intersecting lines because they meet at E. E is said to be the *point of intersection*.

Coincident lines have all points in common.

Lines KL and MN are coincident lines because they are two names for the same line. This is denoted $KL = MN$. Note that $KL = MN = KM = LM = LN$. Any pair of points is sufficient to determine a line.

EXERCISES 2.1

1. How many end-points does a line have?

2. How many end-points does a ray have?

3. How many end-points does a line segment have?

Consider the figure below, given $l \parallel m$ and answer questions 4 through 9.

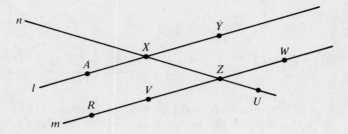

4. Name two rays in line l.

5. Name three line segments in line n.

6. Name $\overrightarrow{XZ}$ in another way.

7. Name a line parallel to XY.

8. Name a line intersecting VW at Z.

9. Name a line coincident with WZ.

10. How many points in common do coincident lines have?

11. How many points in common do parallel lines have?

12. How many points in common do intersecting lines have?

Given:

13. Name l in three different ways.

14. Name three different lines segments on l.

15. Name four different rays on l.

2.2   Angles

An angle is generated by rotating a half-line about its end point from some initial position to some terminal position. The measurement of the angle is the amount of the rotation.

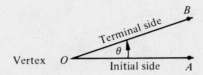

There are three ways of naming an angle: (1) by using three capital letters with the letter of the vertex placed in the middle, such as $\angle AOB$ or $\angle BOA$; (2) by using only the capital letter of the vertex, such as O; (3) by using a small letter in the interior of the angle (usually a Greek letter), such as $\angle \theta$ (theta).

If the rotation of the half-line is *counterclockwise,* the angle is considered to be *positive.*

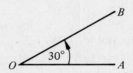

The most common unit of angular measure is the degree. A *degree* can be defined as $\frac{1}{360}$ th of a complete revolution. Each degree has 60 minutes and each minute has 60 seconds. The number of degrees in an angle can be measured by using an instrument called the *protractor.*

Angles having the same initial sides and the same terminal sides are called *coterminal angles.* 30° and 390° are examples of coterminal angles.

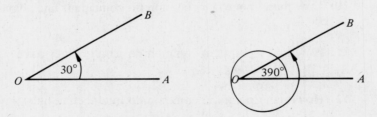

Example 1:   Use a protractor to draw the following angle, and indicate one positive coterminal angle:

160°   360° + 160° = 520°
               160° and 520° are coterminal

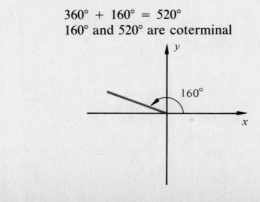

Complementary angles are two angles whose sum is 90°. For example, given that $A = 44°$, B, the complement of A, is equal to $90° - 44°$ or $46°$. A and B are said to be complementary angles.

Example 2: Find the complement of 47°.

$90° - 47° = 43°$

47° and 43° are complementary angles. $47° + 43° = 90°$.

When two lines intersect to form four equal angles, each angle is said to be a *right angle*. A right angle has a measure of 90°. In the figures, right angles will be indicated by the symbol "⌐."

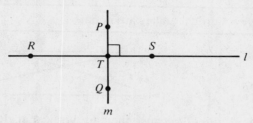

∠PTS, ∠STQ, ∠QTR, and ∠RTP are right angles.

Intersecting lines meeting at right angles are called *perpendicular lines*. Lines *l* and *m* above are perpendicular: this is denoted $l \perp m$.

Angles that measure less than a right angle are called *acute angles*. Angles that measure more than a right angle are called *obtuse angles*.

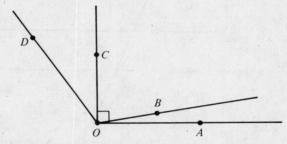

∠*AOC* is a right angle; ∠*AOB* is an acute angle because it measures less than ∠*AOC*; and ∠*AOD* is an obtuse angle because it measures more than ∠*AOC*.

When both sides of an angle lie in opposite directions along the same line, the angle is called a *straight angle*. A straight angle has a measure of 180°.

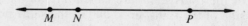

∠*PNM* is a straight angle because $\overrightarrow{NP}$ and $\overrightarrow{NM}$ both lie on the same line but in opposite directions. ∠PNM = 180°.

Two angles with the same vertex and a common side between them are called *adjacent angles*.

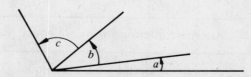

∠*a* and ∠*b* are adjacent angles because they have the same vertex and one side in common. ∠*a* and ∠*c* are not adjacent angles because they do not have a common side although they have a common vertex.

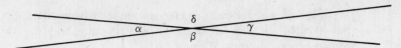

When two lines intersect, they form four angles. The pairs of angles with a common vertex, but no common sides, are called *vertical angles*.

$\angle\alpha$ and $\angle\gamma$ are vertical angles because they are formed by two intersecting lines and have no common sides. $\angle\beta$ and $\angle\delta$ are also vertical angles.

$\angle\alpha$ and $\angle\delta$ jointly form a straight angle. $\angle\gamma$ and $\angle\delta$ also jointly form a straight angle. Therefore, $\angle\alpha$ must measure the same as $\angle\gamma$. This conclusion can be formulated into a rule.

Rule 1:   Vertical angles are equal.

Example 3:

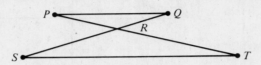

$\angle PRS = \angle QRT$ because they are vertical angles.
$\angle PRQ = \angle SRT$ because they are vertical angles.

When a line intersects a pair of parallel lines, eight angles are formed. The intersecting line is called the *transversal*. The four angles between the parallel lines are called *interior angles*. The four outer angles are called *exterior angles*.

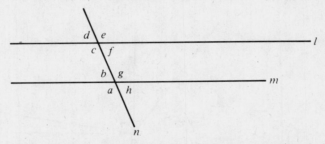

l and m are parallel lines intersected by the transversal line, n. Eight angles are formed: $\angle b$, $\angle c$, $\angle f$, and $\angle g$ are interior angles and $\angle a$, $\angle d$, $\angle e$, and $\angle h$ are exterior angles.

Two non-adjacent interior angles on opposite sides of the transversal are called *alternate interior angles*. Two non-adjacent exterior angles on opposite sides of the transversal are called *alternate exterior angles*. Two angles on the same side of the transversal and in the same position with respect to each of the parallel lines are called *corresponding angles*.

$\angle b$ and $\angle f$ are alternate interior angles, as are $\angle c$ and $\angle g$. $\angle a$ and $\angle e$ are alternate exterior angles, as are $\angle d$ and $\angle h$. $\angle e$ and $\angle g$ are corresponding angles because they are to the right of the transversal and each one is above a parallel line. Other pairs of corresponding angles are: $\angle a$ and $\angle c$, $\angle b$ and $\angle d$, and $\angle f$ and $\angle h$.

Rule 2:   Alternate interior angles are equal.

Rule 3:   Alternate exterior angles are equal.

Rule 4:   Corresponding angles are equal.

Example 4: Name the equal angles in the following figure, given *AF* ‖ *BD*.

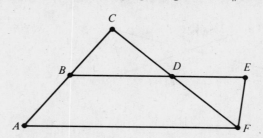

∠*FAB* = ∠*DBC* (corresponding angles)
∠*AFD* = ∠*BDC* (corresponding angles)
∠*BDC* = ∠*EDF* (vertical angles)
∠*BDF* = ∠*CDE* (vertical angles)
∠*AFD* = ∠*EDF* (alternate interior angles)

EXERCISE 2.2

Use a protractor to draw the following angles:

1. 60°   **2.** 140°   **3.** 315°   **4.** 15°

5. 620°   **6.** 390°

Find the complement of the following angles:

7. 32°   **8.** 23°   **9.** 53°   **10.** 61°

11. 45°   **12.** 8°

State whether the following angles are acute or obtuse:

13. 2°   **14.** 89°   **15.** 92°   **16.** 119°

17. 99°   **18.** 135°   **19.** 140°   **20.** 160°

Name a coterminal angle to the given angles:

21. 33°   **22.** 89°   **23.** 6°   **24.** 50°

25. 125°   **26.** 253°   **27.** 380°   **28.** 400°

Determine the degree measure of the indicated angles:

29.

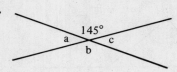

30.

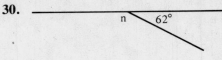

31.

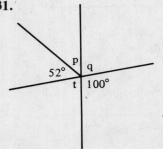

32.

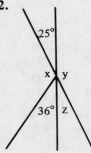

33. Given: $l \parallel n$

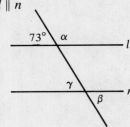

34. Given: $l \parallel n$

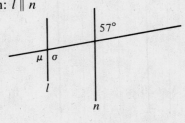

Consider the following figure and find the measure in degrees of the given angles in problems 35 through 42:

Given: KB $\parallel$ JG , LI $\perp$ KB , $\angle$B = 35°, $\angle$DCE = 40° , $\angle$BAC = 30°

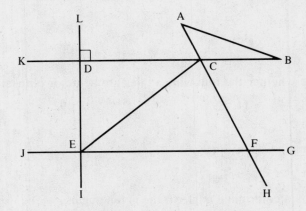

35. $\angle$ACB

36. $\angle$ACD

37. $\angle$DEC

38. $\angle$JEF

39. $\angle$KDL

40. $\angle$ECF

41. $\angle$ACE

42. $\angle$ECB

Consider the figure below and do problems 43 through 46.

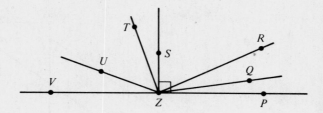

43. Name one right angle.

44. Name two acute angles.

45. Name two obtuse angles.

46. Name one straight angle.

Consider the figure below and do problems 47 through 49.

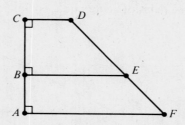

47. Name four right angles.

48. Name two obtuse angles.

49. Name two acute angles.

Consider the figure below and do problems 50 and 51.

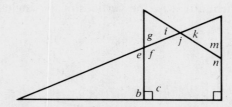

50. Name two pairs of vertical angles.

51. Name four pairs of adjacent angles.

Consider the figure below, given $PV \parallel QW$, and do problems 52 through 54.

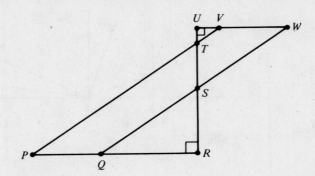

52. Name five pairs of alternate interior angles.

53. Name two pairs of alternate exterior angles.

54. Name five pairs of corresponding angles.

55. Name five pairs of acute equal angles in the following figure, given $AB \parallel CE$:

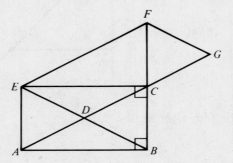

2.3   Triangles

A triangle is a closed figure formed by three line segments and having three angles. It is named by the three letters corresponding to the end-points of the line segments; that is, the vertices of the angles.

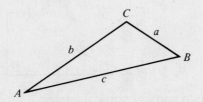

The above triangle is named $\triangle ABC$, where the three letters can be written in any order.

Triangles may be classified according to the relative lengths of their sides.

An *equilateral triangle* has three equal sides. An *isosceles triangle* has two equal sides. A *scalene triangle* has no equal sides.

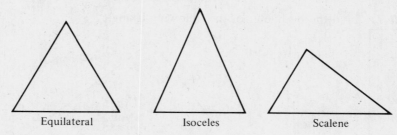

Equilateral   Isoceles   Scalene

Triangles may also be classified according to their angles. A *right triangle* contains a right angle, an *obtuse triangle* contains an obtuse angle, and an *acute triangle* contains three acute angles.

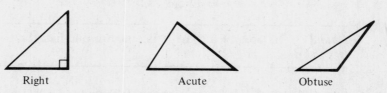

Right   Acute   Obtuse

If the three angles of a triangle are placed adjacent to each other, they will form a straight angle.

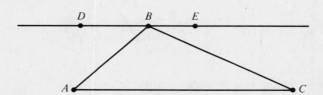

DE, which passes through *B*, is parallel to *AC*. ∠*DBA,* and ∠*A* are alternate interior angles and therefore equal. ∠*EBC* and ∠*C* are alternate interior angles and therefore equal. Since ∠*DBA*, ∠*ABC*, and ∠*EBC* jointly form a straight line, then so will the three angles of the triangle.

$$\angle A + \angle B + \angle C = 180°$$

Rule 1:   The sum of the measures of the three angles of a triangle equals the measure of a straight angle, 180°.

The *base* of a triangle is usually identified as the horizontal side of the triangle, although any side may be used as the base. The *altitude* of a triangle is the length of the line segment drawn perpendicular to the base or its extension from the opposite vertex.

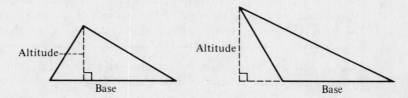

The *perimeter* of a triangle, or of any closed figure, is the sum of the lengths of its sides. The *area* of a triangle, or of any closed figure, is the measure of the surface enclosed by the perimeter.

Rule 2: The perimeter of a triangle is the sum of the lengths of its three sides.

Example 1: Find the perimeter of the following triangle.

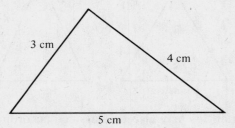

The perimeter = 3 cm + 4 cm + 5 cm
             = 12 cm

Rule 3: The area of a triangle is equal to one-half the product of the base and the altitude.

Example 2: Find the area of a triangle whose base is 8 cm and whose altitude is 5 cm.

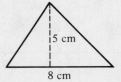

Area = $\frac{1}{2}$ × base × altitude

     = $\frac{1}{2}$ × 8 cm × 5 cm

     = 20 cm²

In a right triangle, the side opposite the right angle is called the *hypotenuse*. The other two sides are referred to as the *legs* of the right triangle. In the figure below, side *c* is the hypotenuse and sides *a* and *b* are the legs:

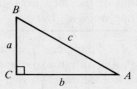

The *Pythagorean Theorem* states that in a right triangle, the square of the hypotenuse is equal to the sum of the squares of the two legs. The Pythagorean Theorem is also used to find either of the two legs of a right triangle.

Rule 4: In a right triangle with legs equal to *a* and *b*, and hypotenuse, *c*, the Pythagorean Theorem can be stated in the following three ways:

$$c^2 = a^2 + b^2$$
$$a^2 = c^2 - b^2$$
$$b^2 = c^2 - a^2$$

Example 3: One leg of a right triangle is 5, and the other leg is 12. Find the length of the hypotenuse.

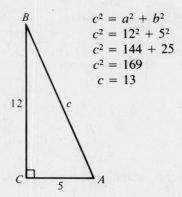

$$c^2 = a^2 + b^2$$
$$c^2 = 12^2 + 5^2$$
$$c^2 = 144 + 25$$
$$c^2 = 169$$
$$c = 13$$

EXERCISES 2.3

Consider the following figure and do problems 1 through 3.

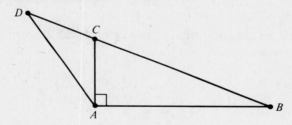

1. Name one right angle.

2. Name one obtuse angle.

3. Name one acute angle.

Consider the figure below, given $\overline{CE} = \overline{DE}$, $\overline{CD} = \overline{ED}$, $\overline{BC} = \overline{FE}$, and do problems 4 through 6.

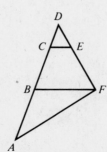

4. Name the equilateral triangle.

5. Name the isosceles triangles.

6. Name the scalene triangles.

7. Find the perimeter and area of

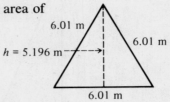

8. Find the perimeter and area of

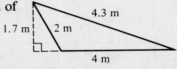

9. Find the perimeter and area of

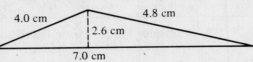

10. The area of a triangle is 30 cm². If the altitude is 6 cm, find the length of the base.

11. The area of a triangle is 54 m². If the base is 12 m, find the length of the altitude.

Given a right triangle *ABC* with side *c* the hypotenuse, find the unknown in problems 12 through 17.

12. $a = 15$
$c = 17$
$b =$

13. $b = 7$
$c = 25$
$a =$

14. $a = 10$
$c = 2\sqrt{41}$
$b =$

15. $a = 5$
$b = 7$
$c =$

16. $a = 80$
$b = 20$
$c =$

17. $b = 12$
$a = 5$
$c =$

18. Find the height of a building if a point 30 m from the foot of the building is 50 m from the top of the building.

19. A rectangle is 14 cm by 22 cm. Find the length of the diagonal.

20. A point B is 2.0 km directly south of point A. If a point, C, is measured to be 3.2 km directly east of A, find the distance from the point B to the point C.

21. A cable, anchored 8 meters from the foot of a tower, stretches 17 meters to the top of the tower. Find the height of the tower.

22. A 20-meter ladder leans against a building. How far up the side of the building does it reach if the foot of the ladder is 5 meters from the base of the building?

23. A marble table is shaped as an equilateral triangle with each side measuring 2 ft 3 in. Find the perimeter of the table.

24. A bathroom tile shaped as an isosceles triangle measures 6 inches at the base and 4 inches at each of the other sides. Find the perimeter of the tile.

25. An antenna is 100 ft high. A guy wire is run from 10 feet below the top antenna and fastened to a point 35 feet from the base of the antenna. Find the length of the guy wire.

26. A square bracket measures 3 ft $5\frac{1}{2}$ in. and 2 ft 4 in. on each of the sides. Find the length of the diagonal in feet.

27. Find the length of the diagonal of a T-square, if the other sides measure 13 inches and 9 inches.

28. A vacant lot is triangular in shape with a base of 95 ft and an altitude of 74 ft. Find the area of the lot.

29. A gable roof end (triangular shape) has a rise (height) of 4 ft 8 in. and a span (base) of 30 ft 4 in. Find the area.

30. A triangle-shaped tile measures 9 inches at the base and 8 inches in height. Find the area of one tile in square feet.

2.4 Polygons

A *polygon* is a closed figure named according to the number of sides it has.

Name	Number of Sides
Triangle	3
Quadrilateral	4
Pentagon	5
Hexagon	6
Heptagon	7
Octagon	8
Nonagon	9
Decagon	10

Polygons that have all equal sides are called *equilateral*. If all angles are equal, they are called *equiangular*. Polygons that are both equilateral and equiangular are called *regular polygons*.

A four-sided polygon is a *quadrilateral*, which is indicated by the symbol △. It is named by the four letters at the vertices.

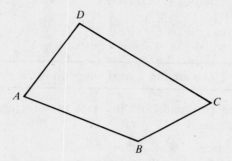

Quadrilaterals may be classified according to the relationship of their sides. An *irregular quadrilateral* has no parallel sides. △ *ABCD* is an irregular quadrilateral.

If a quadrilateral has one and only one pair of parallel sides, it is called a *trapezoid*.

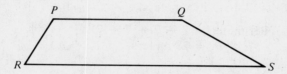

△ *PQRS* is a trapezoid because *PQ ∥ RS* but *PR* is not parallel to *QS*. The parallel sides are called the *bases of the trapezoid*. The perpendicular distance between the two bases is the *altitude of the trapezoid*.

If a quadrilateral has two pairs of parallel sides it it called a *parallelogram*. A parallelogram with equal sides and equal angles is a *square*. A parallelogram with equal angles but unequal sides is a *rectangle*. A parallelogram with equal sides but unequal angles is a *rhombus*.

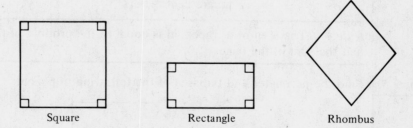

Square   Rectangle   Rhombus

In both a square and a rectangle the sides are perpendicular; thus, all four angles are right angles. A square is merely an equilateral rectangle. In the case of the square and the rectangle, the *base* is one of the sides, usually the horizontal one; the *altitude* is a side perpendicular to the base.

The *base of the rhombus* is one of the sides, and the *altitude of the rhombus* is the perpendicular distance from the base to the side parallel to the base.

Rule 1: The perimeter of a polygon is the sum of the lengths of its sides.

Rule 2; The area of a square is the square of the length of one side.

Example 1:   Find the perimeter and the area of the square where each side measures 3 meters.

Perimeter = 3 + 3 + 3 + 3
= 12 meters

Area   = $(3)^2$
= 9 square meters

Rule 3:   The area of a parallelogram is the product of the base and the altitude.

Example 2:   Find the perimeter and the area of a rectangle measuring 4 centimeters by 5 centimeters.

Perimeter = 4 + 5 + 4 + 5
= 18 centimeters

Area   = base × altitude
= 5 × 4
= 20 square centimeters

Example 3:   Find the perimeter and the area of a rhombus where each side measures 7 meters and its altitude is 5 meters.

Perimeter = 7 + 7 + 7 + 7
= 28 meters

Area   = 7 × 5
= 35 square meters

Example 4:   Find the area of a parallelogram where two sides each measure 6 centimeters, the other two sides each measure 3 centimeters, and the altitude is 2 centimeters.

Perimeter = 6 + 3 + 6 + 3
= 18 centimeters

Area   = 6 × 2
= 12 square centimeters

Rule 4:   The area of a trapezoid is equal to the product of the altitude and one half the sum of the bases.

Example 5:   Find the perimeter and the area of the following trapezoid:

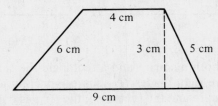

Perimeter = 9 + 4 + 6 + 5
= 24 centimeters

Area   = altitude × $\frac{1}{2}$ (base$_1$ + base$_2$)

= 3 × $\frac{1}{2}$ (9 + 4)

= 3 × $\frac{1}{2}$ × 13

≐ 19.5 square centimeters

Rule 5: The area of a polygon can be found by dividing the polygon into triangles or rectangles, finding the areas of the regions, and summing them.

Example 6: Find the perimeter and the area of the following irregular quadrilateral:

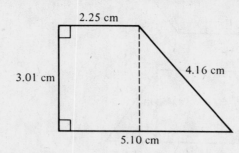

Perimeter = 3.01 + 2.25 + 4.16 + 5.10
= 14.52 cm

Area = area of rectangle + area of triangle

$$= (3.01 \times 2.25) + \left(\frac{1}{2} \times 2.85 \times 3.01\right)$$

= 6.77 cm² + 4.29 cm²
= 11.06 cm²
≐ 11.1 cm²

EXERCISES 2.4

Name the following polygons and find their perimeter and their area.

1.

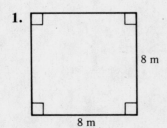

8 m

8 m

2.

2.3 cm

.048 m

3.

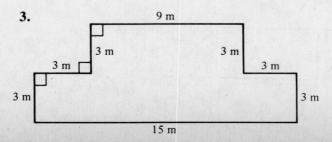

9 m

3 m   3 m

3 m   3 m

3 m   3 m

15 m

4.

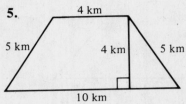

5.

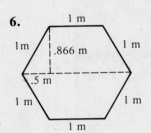

6.

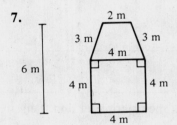

7.

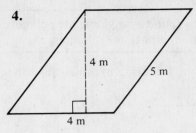

8.

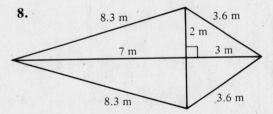

9.

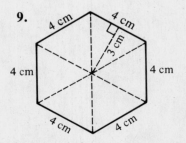

10.

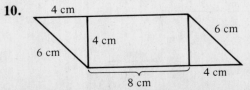

11.

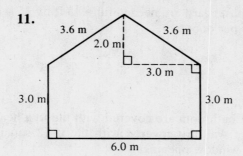

12.

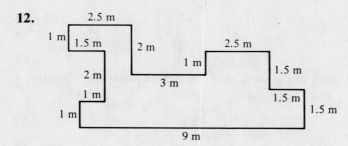

13.

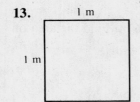

14.

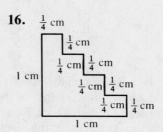

15.

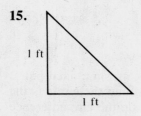

16.

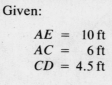

17. Find the perimeter and the area of triangle ACF:

Given:

$AE = 10$ ft
$AC = 6$ ft
$CD = 4.5$ ft

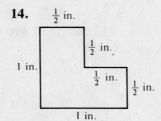

18. A garage footing 18 inches wide has outside dimensions of 25 feet by 30 feet. Find the perimeter of the inside dimensions.

19. How many lineal feet of baseboard are needed for a 13 ft by 15 ft room with three doors, allowing 3 ft per door?

20. The walls of a 3 ft by 4 ft bathroom are covered with tile to a height of 5 ft. How many square feet of wall are covered with tile if 12 square feet are allowed for the door and window openings?

21. Find the area of the surface of a concrete sidewalk measuring 4 ft 8 in. wide and 58 ft long.

22. A gable roof has a rise of 8 ft 4 in. and a span of 14 ft 6 in. Find the area of each gable end.

2.5   Circles

A *circle* is a figure consisting of all points an equal distance from a fixed point. The fixed point is the *center* of the circle. The distance around the circle is the *circumference*. The distance from the center to any point on the circumference is the *radius*. Any part of the circumference is an *arc*.

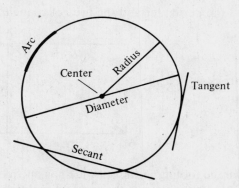

The line segment passing through the center with both end-points on the circumference is the *diameter*. The length of the diameter is twice the *radius*.

A line containing only one point on the circumference is *tangent* to the circle. A line containing two points on the circumference is *secant* to the circle.

A fixed ratio exists between the circumference and the diameter of any circle. That ratio is represented by the Greek letter π and is approximately 3.1416. The exact value of π (pi) is a decimal that does not terminate and does not repeat; π, then, is an irrational number.

Rule 1:   The length of the circumference of a circle is the product of π and the diameter.

$C = \pi \times$ diameter

Since the diameter is twice the radius, the circumference also equals twice the product of π and the radius.

$C = 2 \times \pi \times$ radius

Example 1:   Find the circumference of a circle 5.34 cm in diameter.

$C = \pi \times$ diameter
$\doteq 3.14 \times 5.34$ cm
$\doteq 16.8$ cm

When using the calculator:

Enter	Press	Enter	Press	Display
π	$\boxed{\times}$	5.34	$\boxed{=}$	16.7761045

Round off the circumference to three significant figures, 16.8.

Rule 2:   The area of a circle is the product of π and the square of the radius.

$A = \pi \times$ (radius)^2

Example 2:   Find the area of a circle 6.80 cm in diameter.

Since the diameter equals 6.80 cm, the radius equals 3.40 cm. The square of the radius is 11.56 cm^2. Therefore,

$A = \pi \times$ (radius)^2
$\doteq 3.14 \times (3.40$ cm$)^2$
$\doteq 3.14 \times 11.56$ cm^2
$\doteq 36.3$ cm^2

On the calculator:

Enter	Press	Enter	Press	Press	Display
π	$\boxed{\times}$	3.4	$\boxed{x^2}$	$\boxed{=}$	36.316811

Round off the area to three significant figures, 36.3

Angles may be measured in radians as well as in degrees. A *radian* is the measure of an angle whose vertex is at the center of a circle and that intercepts an arc on the circle equal in length to the radius of the circle.

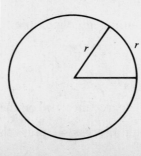

To determine how many times the radius may be laid off along the circumference of a circle, divide the circumference of a circle by its radius. The circumference of any circle is equal to 2π times its radius.

$$C = 2\pi r$$

$$\frac{C}{r} = \frac{2\pi r}{r} = 2\pi$$

Therefore, there are 2π radians in one complete rotation.

Since one degree was defined to be $\frac{1}{360}$ of a complete rotation,

$$1° = \frac{1}{360} \text{ of } 2\pi \text{ radians}$$

$$1° = \frac{\pi}{180} \text{ radian}$$

If $\frac{\pi}{180}$ radian $= 1°$;

$$1 \text{ radian} = \frac{1°}{\frac{\pi}{180}} = \frac{180°}{\pi}$$

Rule 3:   To change from degree measure to radian measure, multiply the number of degrees by $\frac{\pi}{180}$.

Example 3:   Express $120°$ as radian measure.

$$\frac{120}{1} \cdot \frac{\pi}{180} = \frac{\overset{2}{\cancel{120}}}{1} \cdot \frac{\pi}{\underset{3}{\cancel{180}}} = \frac{2}{3}\pi \text{ radians}$$

An approximation of radian measure is found on a scientific calculator by using the degree/radian switch located on the top front surface of the calculator. Set the RD button on R. Enter 30. Press D/R key which instructs the calculator to convert $30°$ to radian measure. The display is 2.094395102 which is approximate to ten significant figures and can then be rounded off to the number of significant figures necessary to work.

Rule 4:   To change from radian measure to degree measure, multiply the radian measure by $\frac{180°}{\pi}$.

Example 4:   Express $\frac{\pi}{4}$ radian in degree measure.

$$\frac{\pi}{4} \cdot \frac{180°}{\pi} = \frac{\cancel{\pi}}{\underset{1}{\cancel{4}}} \cdot \frac{\overset{45°}{\cancel{180°}}}{\underset{1}{\cancel{\pi}}} = 45°$$

To convert $\frac{\pi}{4}$ to degree measure on the calculator, set the RD button to D. Enter π, press $\div$, enter 4, press $=$, press the D/R button. The display is 45.

EXERCISES 2.5

Find the circumference and the area of the circles in Exercises 1 through 4.

1. Radius = 2.20 cm

2. Radius = 5.00 m

3. Diameter = 12.4 m

4. Diameter = 8.68 cm

Find the areas of the figures in Exercises 5 through 8:

5.

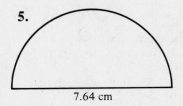

7.64 cm

6. (Find the area of the shaded ring.)

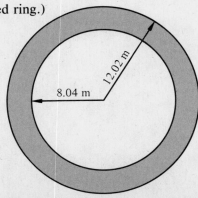

12.02 m

8.04 m

7.

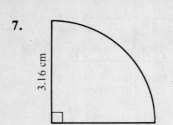

3.16 cm

3.16 cm

8.

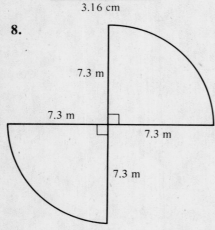

7.3 m

7.3 m

7.3 m

7.3 m

9. A washer 2.0 cm in diameter has a hole 0.5 cm in diameter. Find the area of a face of the washer.

10. Find the area of a circular ring with inside diameter 3.28 cm and outside diameter 5.44 cm.

Convert the angle measure from degree to radian measure. Express the radian measure as (a) exact measure and (b) approximate measure rounded off to four significant figures.

11. 180°   **12.** 60°   **13.** 330°   **14.** 36°

15. 240°   **16.** 225°   **17.** 40°   **18.** 54°

19. 420°   **20.** 72°

Convert the angle measure from radians to degrees:

21. $\dfrac{3\pi}{4}$ radians

22. $\dfrac{\pi}{10}$ radians

23. $\dfrac{2\pi}{3}$ radians

24. $\dfrac{5\pi}{8}$ radians

25. $\dfrac{5\pi}{18}$ radians

26. $\dfrac{7\pi}{4}$ radians

27. $\dfrac{4\pi}{3}$ radians

28. $\dfrac{\pi}{6}$ radians

29. $\dfrac{5\pi}{9}$ radians

30. $\dfrac{\pi}{2}$ radians

31. Find the area of a patio shaped like $\dfrac{1}{4}$ of a circle of radius 8 feet.

32. Find the area of the shaded corner in the figure below:

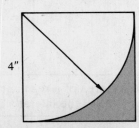

33. Find the area of each 5 inch wide parking stripe in the figure below.

20′

34. A round manhole has an inside diameter of 4 ft 5 in. Find the area in square feet.

2.6 Solids

The solid figures most often encountered in technology are the regular solid figures considered in this section. Technicians are required to know the name, shape, surface area, and volume of these solids.

A regular *prism* is a solid whose ends are equal and parallel polygons and whose sides are rectangles perpendicular to the ends. The ends are called the *bases* of the prism. The perpendicular distance between the bases is the *altitude* of the prism.

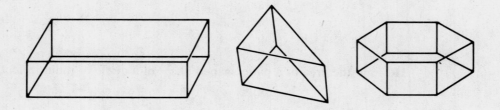

A *cube* is a regular prism whose faces and bases are all equal squares.

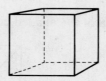

Rule 1: The surface area of a regular prism is the sum of the areas of its faces; that is, the two bases and the lateral sides.

Example 1: Find the surface area of a regular prism whose base is a rectangle 5 cm by 7 cm and whose altitude is 3 cm.

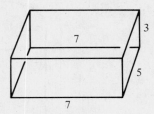

Area of a base = 5 cm × 7 cm = 35 cm²
Area of a side lateral side = 5 cm × 3 cm = 15 cm²
Area of front or back lateral side = 7 cm × 3 cm = 21 cm²
Surface area = (2 × area of base) + (2 × area of side lateral side)
            + (2 × area of front lateral side)
= (2 × 35 cm²) + (2 × 15 cm²) + (2 × 21 cm²)
= 70 cm² + 30 cm² + 42 cm²
= 142 cm²

Example 2: Find the surface area of a cube if one of its edges is 2 meters.

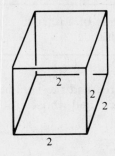

Area of one face = 2 m × 2 m = 4 m²

Since the cube has 6 equal faces,

Surface area = 6 × 4 m² = 24 m²

Rule 2: The volume of a regular prism is the product of the area of the base and the altitude.

Example 3: Find the volume of a regular prism whose base is a triangle of area 20 cm² and whose altitude is 6 cm.

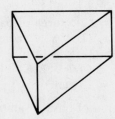

Volume = area of base × altitude
= 20 cm²   × 6 cm
= 120 cm³

A regular circular *cylinder* is a solid, similar to a regular prism, whose bases are circles.

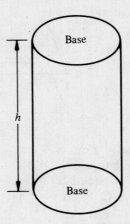

Rule 3:   The surface area of a regular circular cylinder is twice the area of the base added to the product of the circumference of the base and the altitude.

Rule 4:   The volume of a regular circular cylinder is the product of the area of the base and the altitude.

Example 4:   Find the surface area and the volume of a regular circular cylinder whose base has diameter 6.42 cm and whose altitude is 11.2 cm.

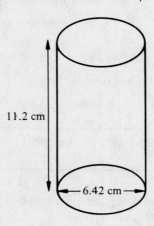

$$
\begin{aligned}
\text{Surface area} &= (2 \times \text{area of base}) + (\text{circumference} \times \text{altitude}) \\
&= (2 \times \pi \times \text{radius}^2) + (\pi \times \text{diameter} \times \text{altitude}) \\
&\doteq (2 \times 3.14 \times 3.21^2) + (3.14 \times 6.42 \times 11.2) \\
&\doteq 64.7 + 226 \\
&\doteq 291 \text{ cm}^2
\end{aligned}
$$

$$
\begin{aligned}
\text{Volume} &= \text{area of base} \times \text{altitude} \\
&= (\pi \times \text{radius}^2) \times \text{altitude} \\
&\doteq (3.14 \times 3.21^2) \times 11.2 \\
&\doteq 3.14 \times 10.3 \times 11.2 \\
&\doteq 362 \text{ cm}^3
\end{aligned}
$$

On the calculator:

Enter	Press	Enter	Press	Enter	Press	Press	Enter	Press	Enter
2	☒	π	☒	3.21	$\boxed{x^2}$	➕	π	☒	6.42

Press	Enter	Press	Display
☒	11.2	➗	290.6356479

Round off the surface area to three significant figures, 291.

On the calculator:

Enter	Press	Enter	Press	Press	Enter	Press	Display
π	☒	3.4	$\boxed{x^2}$	☒	11.2	➗	362.5583905

Round off the volume to three significant figures, 363. Note the difference in approximate answers because a closer approximation of π was used on the calculator. 363 is more accurate than 362.

A regular *pyramid* is a solid with an equilateral polygon as a base and equal isosceles triangles as faces, the latter meeting at a point called the *vertex*.

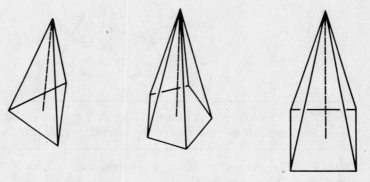

The altitude of the regular pyramid is the perpendicular distance from the vertex to the base.

Rule 5: The surface area of a regular pyramid is the sum of the areas of the lateral sides and the base.

Rule 6: The volume of a regular pyramid is one third the product of the area of the base and the altitude.

Example 5: Find the surface area and the volume of a regular pyramid with a square as the base. An edge of the square is 5.00 cm, the altitude of the pyramid is 7.00 cm, and the altitude of each triangular lateral face is 7.43 cm.

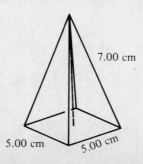

7.00 cm

5.00 cm

5.00 cm

$$\text{Surface area} = \text{area of base} + (4 \times \text{area of lateral side})$$
$$= (5.00)^2 + (4 \times \tfrac{1}{2} \times \text{base} \times \text{altitude of side})$$
$$= 25.0 + (4 \times \tfrac{1}{2} \times 5.00 \times 7.43)$$
$$= 25.0 + 74.3$$
$$= 99.3 \text{ cm}^2$$

$$\text{Volume} = \tfrac{1}{3} \times \text{area of base} \times \text{altitude}$$
$$= \tfrac{1}{3} \times (5.00)^2 \times 7.00$$
$$= 58.3 \text{ cm}^3$$

A regular circular *cone* is a solid, similar to a regular pyramid, whose base is a circle. The altitude of the regular circular cone is the perpendicular distance from the vertex to the center of the circular base.

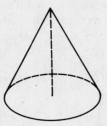

Rule 7: The surface area of a regular circular cone is the sum of the area of the base and one-half the product of the circumference and the lateral height.

Rule 8: The volume of the regular circular cone is one-third the product of the area of the base and the altitude.

Example 6: Find the surface area and the volume of a regular cone whose base is 20.66 m in diameter, altitude is 100.8 m and lateral height is 102.4 m.

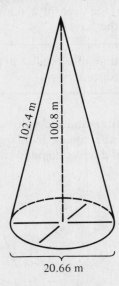

102.4 m

100.8 m

20.66 m

$$\text{Surface area} = \text{area of base} + \left(\frac{1}{2} \times \text{circumference} \times \text{lateral height}\right)$$

$$= (\pi \times \text{radius}^2) + \left(\frac{1}{2} \times \pi \times \text{diameter} \times \text{lateral height}\right)$$

$$= (\pi \times 10.33^2) + (.5 \times \pi \times 20.66 \times 102.4)$$
$$\doteq 3657 \text{ m}^2$$

$$\text{Volume} = \frac{1}{3} \times \text{area of base} \times \text{altitude}$$

$$= \frac{1}{3} \times (\pi \times \text{radius}^2) \times \text{altitude}$$

$$= \frac{(\pi \times 10.33^2) \times 100.8}{3}$$

$$\doteq 11{,}260 \text{ m}^3$$

A *sphere* is a solid on whose surface every point is an equal distance from the *center* of the sphere. The *diameter* of the sphere is the line segment passing through the center of the sphere and having two endpoints on the surface. The *radius* of the sphere is the line segment from the center to any point on the surface.

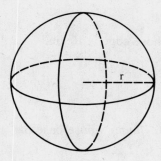

Rule 9: The surface area of the sphere is the product of π and the square of the diameter.

Rule 10: The volume of a sphere is equal to one-third the product of the surface area and the radius.

Example 7: Find the surface area and volume of a sphere with diameter 10.8 cm.

$$\text{Surface area} = \pi \times \text{diameter}^2$$
$$= \pi \times 10.8^2$$
$$\doteq 366 \text{ cm}^2$$

$$\text{Volume} = \frac{1}{3} \times \text{surface area} \times \text{radius}$$

$$= \frac{1}{3} \times 366 \times 5.40$$

$$\doteq 659 \text{ cm}^3$$

Sketch the figure and find the surface area and the volume of the following regular solids:

1. A regular prism with a 5 cm by 6 cm rectangular base and a 7 cm altitude.

2. A regular prism with a 10 m by 12 m rectangular base and 6 m altitude.

3. A cube whose edge is 3 cm.

4. A cube whose edge is 10 cm.

5. A regular cylinder whose base has a diameter of 2.56 m and whose altitude is 6.27 m.

6. A regular circular cylinder whose base has a diameter of 10.3 m and whose altitude is 45.5 m.

7. A regular pyramid whose base is a square with sides of 120 m, and whose altitude is 200 m and lateral height is 300 m.

8. A regular pyramid whose base is a square with 10.4 m edge, and whose altitude is 30.0 m and lateral height is 40.0 m.

9. A regular circular cone whose altitude is 200.0 m, whose base is 50.0 m in diameter and whose lateral height is 201.56 m.

10. A regular circular cone the radius of whose base is 7.94 m, whose altitude is 25.3 m and whose lateral height is 25.96 m.

11. A sphere 41.7 cm in diameter.

12. A sphere 55 m in radius.

13. How many cubic feet of concrete are needed for a concrete column measuring 18 inches in diameter and 9 feet in height?

14. Find the volume of a 2 meter high, 5 m by 6.5 m footing.

15. A wall 10 feet long and 6 feet high is laid with $3\frac{3}{4} \times 2\frac{1}{4} \times 8$ inch standard bricks. Find the number of bricks needed, disregarding mortar joints, and assuming that the largest face of the brick forms the wall.

16. A 12 inch inside diameter pipe is cut into two pieces. What is the capacity (volume) of each piece in cubic inches if one piece is 8 ft 6 in. long?

Unit 2 Self-Evaluation

1. Name the line segment, the rays and the line in the figure below:

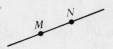

2. Name the parallel lines, the intersecting lines, the coincident lines, the acute angles, the right angles, and the obtuse angles in the figure below, given that $l \parallel m$ and $t \parallel n$:

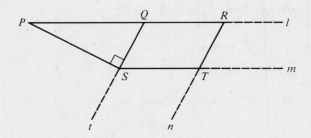

3. Name the vertical angles, the corresponding angles, the alternate interior angles, and the alternate exterior angles in the figure below, given that $l \parallel m$:

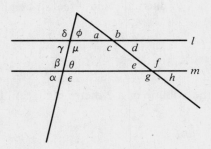

4. Use a protractor to draw 175°.

5. Find the complement of 48°.

6. State whether 137° is an acute or obtuse angle.

7. Find the perimeter and area of the following triangle:

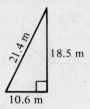

18.5 m

21.4 m

10.6 m

8. Given a right triangle with hypotenuse c, find c given the legs equal to 8 and 15.

9. Find the perimeter and area of a square with edge 14 cm.

10. Find the perimeter and area of a rectangle 6 m by 10 m.

11. Find the perimeter and area of the following parallelogram:

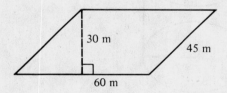

30 m

45 m

60 m

12. Find the circumference and area of a circle with diameter 2.5 cm.

13. Convert $\frac{5\pi}{6}$ radians to degrees.

14. Convert 66° to radians.

In problems 15 through 20, identify the figure and find its surface area and volume:

15.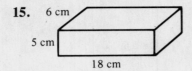
6 cm

5 cm

18 cm

16.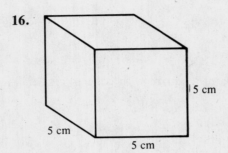

5 cm

5 cm

5 cm

17.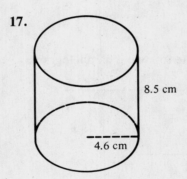

8.5 cm

4.6 cm

18.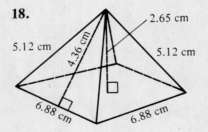

2.65 cm

5.12 cm   4.36 cm   5.12 cm

6.88 cm   6.88 cm

19.

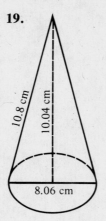

10.8 cm   10.04 cm

8.06 cm

20.

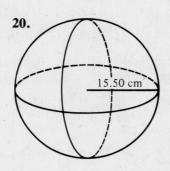

15.50 cm

Basic Algebraic Operations

Unit 3 Objectives
1. To determine the absolute value of a real number and the negative of a real number.
2. To add real numbers.
3. To subtract real numbers.
4. To add and subtract polynomials.
5. To multiply and divide real numbers.
6. To multiply polynomials.
7. To divide polynomials.
8. To remove symbols of grouping.
9. To evaluate algebraic expressions for given values of variables.

3.1   Real Numbers

In preceding sections all subtraction problems have involved subtracting a smaller number from a larger number. $100 - 20 = 80$, $14.31 - 11.22 = 3.09$, and $\frac{7}{5} - \frac{3}{5} = \frac{4}{5}$ are examples of subtraction problems previously encountered. Situations arise in which it becomes necessary to subtract a larger number from a smaller number, resulting in a difference that is less than zero. In order to designate a number less than zero, a minus sign is placed before it.

Examples: If the temperature on a cold day falls 18° below zero, this can be indicated by $-18°$.

A stock drops a point and a half from the closing of yesterday's market. This is indicated in the newspaper by placing $-1\frac{1}{2}$ after the name of the stock.

If the velocity of an object being thrown upward is 96 ft/sec, the velocity of the object traveling downward at the same rate can be indicated by -96 ft/sec. For example, a ball dropped from the roof of a building could travel at -96 ft/sec.

Definition 1.   *Negative numbers* are numbers less than zero. They are preceded by a minus sign.

Definition 2.   *Positive numbers* are numbers greater than zero. They may be preceded by a positive sign. If no sign appears before the number, the number is understood to be positive.

Positive numbers are illustrated by using a diagram called a *number line*.

Draw a horizontal line. Select any point on the line and label it 0. Select any point to the right of 0 and label it 1. Using the measure from 0 to 1, label the natural numbers, 2, 3, 4, 5, and so on, to the right of 0.

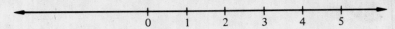

Definition 3.   The positive integers and the negative integers, together with zero, are called *integers*. I = {. . . −3, −2, −1, 0, 1, 2, 3, . . .}

The natural numbers, 1, 2, 3, . . ., are called *positive integers*. They are located to the right of 0 on the number line. For every positive integer there is a number which is the same distance from 0, but in the opposite direction. These numbers are called *negative integers,* and are located to the left of 0 on the number line.

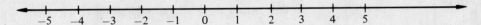

Any number to the right of a given number on the number line is greater than (>) the given number. For example, given the number 2, 4 > 2 because 4 lies to the right of 2 on the number line. Any number to the left of a given number on the number line is less than (<) the given number. For example, given −2, −4 < −2 because −4 lies to the left of −2 on the number line.

The symbols > and < are symbols of *inequality*. Using these symbols, *a* is said to be positive if $a > 0$; *a* is said to be negative if $a < 0$.

Definition 4.   A *rational number* is any number which can be expressed as the quotient of two integers, that is, a/b, where b ≠ 0.

Fractions, such as $\frac{3}{4}$ and $-\frac{3}{4}$ or $\frac{5}{2}$ and $-\frac{5}{2}$, may also be illustrated on the number line.

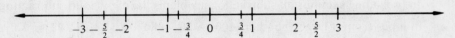

Examples:

$$\frac{3}{7}$$

$$-\frac{11}{15}$$

$$4 = \frac{4}{1}$$

$$0 = \frac{0}{6}$$

$$.25 = \frac{25}{100}$$

Definition 5.   An *irrational number** is any number which cannot be expressed as the quotient of two integers.

Examples: Using a calculator, verify that $\sqrt{5}$, $\sqrt[3]{7}$, π, and −3.121221222. . . are non-terminating.

$\sqrt{5} \doteq 2.236068$ (on the calculator)
$\sqrt[3]{7} \doteq 1.9128071$
$\pi \doteq 3.1415927$
$-3.121221222. . .$

* Irrational numbers will be studied in Unit 10. Examples in this unit will be limited to real numbers that are rational.

Definition 6. Rational numbers, together with irrational numbers, are called *real numbers*.

For every real number there is another number that is the same distance from 0 on the number line, but in the opposite direction. 4 is 4 units to the right of 0, and -4 is also 4 units from 0, but to the left of 0. -4 is called the negative of 4. Similarly, 4 is called the negative of -4. The negative of a number is indicated by putting the symbol, $-$, in front of the number.

Examples:

Number	Negative of the Number
3	-3
-6	$-(-6)$ or 6
9	-9
-5	$-(-5)$ or 5

Note that the negative of a number is not always a negative number.

Definition 7. The *absolute value* of a real number is the value of the number without regard to its sign. It is always positive.

The numbers 4 and -4 have the same absolute value, 4. This is equivalent to saying that the distance from 0 to 4 is equal to the distance from 0 to -4 on the number line. To designate the absolute value of a number symbolically, two vertical bars are used. The absolute value of -4, written symbolically, is $|-4|$.

In general, $|a| = \begin{cases} a \text{ if } a \geq 0. \\ -a \text{ if } a < 0. \end{cases}$

Examples:

$|7| = 7$, since $7 > 0$.
$|3| = 3$, since $3 > 0$.
$|0| = 0$, since $0 = 0$.

$|-7| = -(-7) = 7$, since $-7 < 0$.
$|-3| = -(-3) = 3$, since $-3 < 0$.

EXERCISES 3.1

Determine which of the given numbers is the largest:

1. $4, -3, -7$ **2.** $-3, 0, -1$ **3.** $-4, -5, -6$ **4.** $6, -1, -6$

Insert the proper sign, $>$ or $<$, between the given numbers:

5. $8 \quad 2$ **6.** $-2 \quad 4$ **7.** $0 \quad -2$ **8.** $-4 \quad -5$

Evaluate:

9. $|+8|$ **10.** $|-3|$ **11.** $|0|$ **12.** $\left|\dfrac{2}{3}\right|$

13. $\left|-\dfrac{7}{8}\right|$ **14.** $|-1.5|$

Give the negative of each of the following numbers:

15. -11 **16.** 16 **17.** 10 **18.** $-\dfrac{7}{8}$

19. $\dfrac{3}{4}$ **20.** -50

Use a signed number to represent:

21. A temperature of 12°C below zero.

22. An elevation of 4.5 ft above sea level.

23. A profit of $1530.

24. An overdraft in the bank of $275.68.

3.2 Addition of Real Numbers

Addition is a basic operation in algebra. The following are properties of the real numbers for addition and multiplication:

1. Closure: If a and b are real numbers, then $a + b \in$ the real numbers, and $a \cdot b \in$ the real numbers.

2. Cummutative: If a and b are real numbers, $a + b = b + a$, and $a \cdot b = b \cdot a$.

3. Associative: If a, b, and c are real numbers, then $(a + b) + c = a + (b + c)$, and $(a \cdot b) \cdot c = a \cdot (b \cdot c)$.

4. Additive identity: There is a unique number, 0, of the real numbers such that for every real number a, $a + 0 = a$, and $0 + a = a$.

5. Multiplicative identity: There is a unique number, 1, of the real numbers, such that for every real number a, $a \cdot 1 = a = 1 \cdot a$.

6. Additive inverse: For every member, a, of the real numbers, there is a unique real number, $-a$, such that $a + (-a) = 0$, and $(-a) + a = 0$.

7. Multiplicative inverse: For every real number a, $a \neq 0$, there is a unique real number, $\dfrac{1}{a}$, such that $a \cdot \dfrac{1}{a} = 1$, and $\dfrac{1}{a} \cdot a = 1$.

8. Distributive: If a, b, and c are real numbers, then $a \cdot (b + c) = (a \cdot b) + (a \cdot c)$.

Positive numbers have been defined to be numbers greater than zero and therefore numbers to the right of zero on the number line. In adding two positive numbers, for example, 3 + 2, move three units to the right of zero on the number line. Then move two units to the right of three. Thus, 3 + 2 = 5.

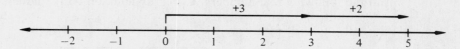

Negative numbers have been defined to be numbers less than zero, and therefore, numbers to the left of zero on the number line. In adding two negative numbers, for example −3 + (−2), move three units to the left of zero on the number line. Then move two more units to the left of a negative three. Thus, −3 + (−2) = −5. Note that in this example parentheses were placed around −2. Parentheses must be used whenever one sign is followed by another sign. The use of parentheses in other cases is optional.

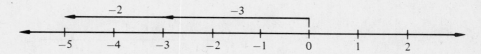

In adding one negative and one positive number, for example −3 + 5, move three units to the left of zero on the number line. Then move five units to the right of a negative three. Thus, −3 + 5 = 2.

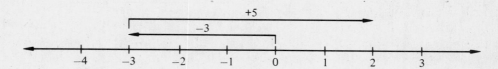

These examples suggest the following rules for addition of real numbers:

Rule 1:   To add two real numbers with like signs, add their absolute values. The sum is a real number prefixed by their common sign.

Rule 2:   To add two real numbers with unlike signs, subtract the smaller absolute value from the larger absolute value. The sum is a real number prefixed by the sign of the number having the larger absolute value.

Example, Rule 1:   6 + 10 = 16

Example, Rule 1:   −4 + (−6) = −10. The absolute value of a −4 is 4. The absolute value of −6 is 6. Since 4 + 6 = 10, the sum, according to Rule 1, is −10.

Example, Rule 2:   6 + (−8) = −2. The absolute value of 6 is 6. The absolute value of −8 is 8. Subtracting 6 from 8 yields 2. Since 8 is larger than 6, the sign of the sum is the sign of 8, which is negative.

Example, Rules 1, 2:

$$2 + 4 + (−8) + (−1) = 6 + (−8) + (−1)$$
$$= −2 + (−1)$$
$$= −3$$

Rule 1: 2 + 4 = 6
Rule 2: 6 + (−8) = −2
Rule 1: −2 + (−1) = −3

EXERCISES 3.2

Add the following real numbers. Check by addition on the number line.

1. $4 + 7$

← ———— *11* ————→

2. $3 + (-7)$

← ———— *-4* ————→

3. $7 + (-3)$

← ———— *4* ————→

4. $(-2) + (-3)$

← ———— *-5* ————→

5. $(-4) + 6$

← ———— *2* ————→

6. $6 + (-5)$

← ———— *1* ————→

7. $(-9) + 3$

← ———— *-6* ————→

8. $(-8) + (-7)$

← ———— *-15* ————→

9. $(-3) + (-5) + (-2)$ ⟵――――――-10――――――⟶

10. $(-6) + 8 + (-6)$ ⟵――――-4――――⟶

11. $9 + (-8) + 4$ ⟵――――5――――⟶

Find the sum of each of the following:

12. $457 + (-300) = 157$ **13.** $-7,005 + 2,006 = 4,991$

14. $\$3.04 + (-\$1.49) = \$1.55$ **15.** $-8.96 + 4.21 = -4.75$

16. $9 + (-6) + 3 = 6$ **17.** $8 + (-1) + 6 + (-5) = 8$

18. $72\% + (-31\%) = 41\%$ **19.** $-10 + (-2) + (-8) + 6 = -14$

20. $\frac{2}{3} + \left(-\frac{1}{3}\right)$ **21.** $12 + 4 + (-8) + (-6) = 2$

22. $8 + (-6) + 6 + (-8)$ 0

23. $-\frac{3}{5} + \frac{7}{10} = \frac{1}{10}$

24. $-\frac{2}{3} + \left(-\frac{3}{8}\right) = \frac{25}{24}$

25. $\frac{3}{7} + \left(-\frac{5}{14}\right)$

26. A mine elevator started at an elevation of 13.62 meters below sea level and was raised 24.3 meters. What was the elevation of the elevator after it was raised?

27. Two forces have y-components of 160 kg and -140 kg. What is the total of the y-components?

28. The 6 P.M. reading on a thermometer showed 8.3° below zero. At 8 P.M. the temperature had dropped 4.7°. What was the temperature reading at 8 P.M.?

29. A contractor made a profit of $16,380 during the first six months of the year, but he had a loss of $2990 during the last six months. What was his profit or loss for the year?

30. A certain stock rose $\frac{1}{2}$ point on Monday, dropped $1\frac{1}{4}$ points on Tuesday, remained unchanged on Wednesday, rose $\frac{7}{8}$ of a point on Thursday, and fell 1 point on Friday. What was the net change in the stock at the end of the week?

3.3 Subtraction of Real Numbers

Eight minus two is written algebraically as $8 - 2$. To subtract 2 from 8, find a number that when added to 2 will equal 8. The number is 6 since $2 + 6 = 8$. On the number line, note that 8 is 6 units to the right of 2. Therefore, $8 - 2 = 6$.

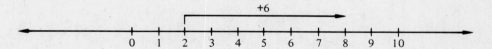

To subtract a negative two from eight, $8 - (-2)$, find a number that when added to a -2 will equal 8. On the number line, 8 is 10 units to the right of -2. Therefore, $8 - (-2) = 10$.

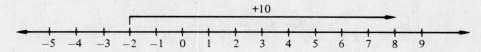

To subtract two from a negative eight, $-8 - 2$, find a number that when added to 2 will equal -8. On the number line -8 is 10 units to the left of 2. Therefore, $-8 - 2 = -10$.

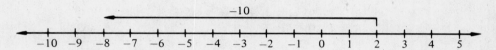

Note that the result of subtracting 2 from -8 yields the same result as adding -2 to -8.

$$-8 - 2 = -10$$
$$-8 + (-2) = -10$$

Definition. For any two real numbers, a and b, $a - b = a + (-b)$.

This definition allows us to rewrite any statement of subtraction as a statement of addition.

When subtracting two positive numbers, the difference is positive if a smaller number is subtracted from a larger number. The difference is negative if a larger number is subtracted from a smaller number. Study the following examples:

$a - b = a + (-b)$	or	Larger number $-$ Smaller number
$8 - 6 = 8 + (-6) = 2$	or	$8 - 6 = 2$
$9 - 8 = 9 + (-8) = 1$	or	$9 - 8 = 1$
$7 - 5 = 7 + (-5) = 2$	or	$7 - 5 = 2$
$9 - 6 = 9 + (-6) = 3$	or	$9 - 6 = 3$
$\frac{3}{5} - \frac{1}{5} = \frac{3}{5} + \left(-\frac{1}{5}\right) = \frac{2}{5}$	or	$\frac{3}{5} - \frac{1}{5} = \frac{2}{5}$

$a - b = a + (-b)$	or	Smaller number $-$ Larger number
$6 - 8 = 6 + (-8) = -2$	or	$6 - 8 = -2$
$8 - 9 = 8 + (-9) = -1$	or	$8 - 9 = -1$
$5 - 7 = 5 + (-7) = -2$	or	$5 - 7 = -2$
$6 - 9 = 6 + (-9) = -3$	or	$6 - 9 = -3$
$\frac{1}{5} - \frac{3}{5} = \frac{1}{5} + \left(-\frac{3}{5}\right) = -\frac{2}{5}$	or	$\frac{1}{5} - \frac{3}{5} = -\frac{2}{5}$

When subtracting a positive number from a negative number the difference is always a negative number. Study the following examples:

$a - b = a + (-b)$	or	A negative number $-$ a positive number
$-8 - 6 = -8 + (-6) = -14$	or	$-8 - 6 = -14$
$-9 - 8 = -9 + (-8) = -17$	or	$-9 - 8 = -17$
$-7 - 5 = -7 + (-5) = -12$	or	$-7 - 5 = -12$
$-9 - 6 = -9 + (-6) = -15$	or	$-9 - 6 = -15$
$-\dfrac{3}{5} - \dfrac{1}{5} = -\dfrac{3}{5} + \left(-\dfrac{1}{5}\right) = -\dfrac{4}{5}$	or	$-\dfrac{3}{5} - \dfrac{1}{5} = -\dfrac{4}{5}$

The negative of a negative number is positive. For example, $-(-6)$ is 6. By definition of subtraction, to subtract a negative number add the negative of a negative number. Study the following examples:

$$a - b = a + (-b)$$
$$8 - (-6) = 8 + 6 = 14$$
$$-8 - (-6) = -8 + 6 = -2$$
$$9 - (-8) = 9 + 8 = 17$$
$$-9 - (-8) = -9 + 8 = -1$$
$$7 - (-5) = 7 + 5 = 12$$
$$-7 - (-5) = -7 + 5 = -2$$
$$6 - (-9) = 6 + 9 = 15$$
$$-6 - (-9) = -6 + 9 = 3$$

EXERCISES 3.3

Subtract the following real numbers:

1. $7 - 5$

2. $3 - 8$

3. $4 - (-2)$

4. $-5 - 6$

5. $-8 - 3$

6. $-4 - (-5)$

7. $-7 - (-3)$

8. $-5 - (-9)$

9. $4 - (-4)$

10. $6 - 6$

11. $8 - (-20)$

12. $-6 - 5$

13. $14 - (-14)$

14. $-8 - (-8)$

15. $-368 - (-340)$

16. $-8004 - 5214$

17. $\$2.41 - (-\$1.04)$

18. $6.28 - 9.13$

19. $9 - 8 - (-1) - (-2)$

20. $13 - 6 - 2 - (-3)$

21. $(-69\%) - (-21\%)$

22. $59\% - (-21\%)$

23. $-11 - (-10) - 8$

24. $-7 - 5 - (-12)$

25. $-\frac{5}{9} - \left(-\frac{2}{9}\right)$

26. $-\frac{7}{10} - \left(-\frac{7}{15}\right)$

27. $\frac{2}{3} - \left(-\frac{5}{8}\right)$

28. $-\frac{5}{21} - \frac{3}{14}$

29. On a cold day in Green Bay, the temperature fell from 13°C to −14°C. How many degrees did the temperature fall?

30. On the first of the month, Mr. Allen deposited $205.64 in his checking account, which initially had a balance of $482. On the 15th of the month he withdrew $135.80, and on the 30th of the month he withdrew $102. What was his balance at the end of the month?

31. Helium boils at $-268.9°C$. Oxygen boils at $-183.0°C$. Subtract the boiling point of oxygen from the boiling point of helium.

32. A 150 kg force has an x-component of -40.8 kg, and a 100 kg force has an x-component of 39.7 kg. Subtract the x-component of the 100 kg force from the x-component of the 150 kg force.

3.4 Addition and Subtraction of Polynomials

In algebra, letters are often used to represent numbers. The following definitions will provide terminology necessary in working with letters.

Definition 1. A letter such as x, y, a, or b, which represents a unknown number, is called a *variable*. If the same letter is used to denote different numbers, different subscripts are used, such as T_0, T_2, or T_a.

Definition 2. An *algebraic expression* is any combination of numbers and variables whose exponents are either zero or a positive integer, and symbols of operation.

Examples: Examples of algebraic expressions are

$$2m + n;\ R_0 - 3R;\ \frac{x + y}{z};\ 7d;\ abc$$

When no numerical coefficient is written before a variable, the numerical coefficient is understood to be one. In the algebraic expression, $2m + n$, the numerical coefficient of n is not expressed, but it is understood to be one.

When two or more letters, such as abc, are written without a sign separating the letters, it means, a times b times c.

Definition 3. A *term* is a single number or the product of a number and one or more variables. The algebraic expression, $3x - 2xy + 4$, has three terms: $3x$; $-2xy$; and $+4$.

Definition 4. A *coefficient* of a number is an indicated factor of a number. In the terms $6cd$, 6 is the numerical coefficient of cd, and cd is the variable coefficient of 6.

Definition 5. A *polynomial* in x is an algebraic expression in which the variable x does not appear in the denominator of any term.

$$x + 5;\ \frac{2}{3}x - 4;\ x^2y - 3xy + x$$

Definition 6. A polynomial consisting of only one term, such as $3x$, is called a *monomial*.

Definition 7. A polynomial consisting of two terms, such as $3T - 4$, is called a *binomial*.

Definition 8. A polynomial consisting of three terms, such as $4W^2 + 2W - 6$, is called a *trinomial*.

Definition 9. Terms that differ only in their numerical coefficients, such as $4a$ and $-2a$, are called *like terms*. The variable and the exponent of the variable must be the same.

By using the distributive property, it is possible to add like terms.

Example 1:
$$3x + 5x = (3 + 5)x$$
$$= 8x$$

Example 2:
$$-7y^2 + 2y^2 = (-7 + 2)y^2$$
$$= -5y^2$$

The addition of like terms may be used to add polynomials.

Example 3:

$(3a + 2b) + (6a - 7b) = 3a + 6a + 2b + (-7b)$	Commutative and associative properties
$= (3 + 6)a + [2 + (-7)]b$	Distributive
$= 9a + (-5)b$	Add
$= 9a - 5b$	Definition of subtraction

This example may be done more easily as follows:

$(3a + 2b) + (6a - 7b) = 3a + 2b + 6a - 7b$	Parentheses preceded by + may be omitted. Combine (add or subtract) like terms: $3a + 6a = 9a$; $2b + (-7b) = -5b$.
$= 3a + 6a + 2b - 7b$	
$= 9a - 5b$	

Example 4:

$(2R - T) + (3R + 7T) + (5R - 3T)$	
$= 2R - T + 3R + 7T + 5R - 3T$	Parentheses preceded by + may be omitted.
$= 2R + 3R + 5R - T + 7T - 3T$	Combine like terms.
$= 10R + 3T$	

Example 5: $(3M + 2N) - (5M - 2N)$

To subtract two polynomials, use the definition for subtraction, $a - b = a + (-b)$.

$$(3M + 2N) - (5M - 2N) = (3M + 2N) + [-(5M - 2N)]$$
$$= (3M + 2N) + (-5M + 2N)$$
$$= 3M + (-5M) + 2N + 2N$$
$$= [3 + (-5)]M + (2 + 2)N$$
$$= -2M + 4N$$

This example can be worked more easily by noting that when parentheses preceded by a minus sign are dropped the sign of each term within the parentheses is changed.

$$(3M + 2N) - (5M - 2N) = 3M + 2N - 5M + 2N$$
$$= -2M + 4N$$

Example 6: $(2a^2 + 5b) + (3a^2 - 7b) - (2a^2 - 4b) = 2a^2 + 5b + 3a^2 - 7b - 2a^2 + 4b$
$$= 3a^2 + 2b$$

EXERCISES 3.4

Add or subtract the following polynomials:

1. $(3R + 4) + (7R - 9)$

2. $(4P - 3Q + 5) + (-2P + 9Q - 2)$

3. $(3m + 5) - (7m + 8)$

4. $(3r + s - 2t) - (r - s + 3t)$

5. $(-5m - 6mn + n) + (3mn - n - m)$

6. $(2.3x^2 + 3.4x^3) + (6.7x^2 - 8.1x^3)$

7. $(5C + C^2) + (2C - 3C^2) + (4C - 6C^2)$

8. $(-2Y - Y^2 + 4Y^3) - (2Y^3 + 4Y^2 + 3Y)$

9. $(2mn + n + 3m) - (-n + 2mn - 4m)$

10. $(a + b + c) - (-a + b - c)$

11. $(R - 4R^2 + 4) - (2R - 3R^2 + 4)$

12. $(2 + a - a^2) + (a + a^2 + 2a^3) + (2 - a - a^2)$

13. $(2T_3 + T_1 - T) - (2T_3 - T_1 + T)$

14. $\left(\frac{2}{3}R + \frac{7}{5}R^2 + R^3\right) + \left(\frac{5}{9}R + \frac{3}{10}R^2 - \frac{1}{4}\right) - \left(-\frac{1}{3} - \frac{7}{10}R^3 - R^2\right)$

15. $(2P_1 - 3P_0 + 2P) - (4P_1 + 5P_0)$

16. $(c - 2d + 7) - (3c + 3d + 1)$

17. $(2xy^2 + 3xy - 5x^2y) + (xy^2 - 2x^2y + 6xy) - (4xy + 2xy^2 - 3x^2y)$

18. $(3T_0 - 2T - 6) - (4T + 8)$

19. $(5Z_0 - 28) - (-4Z_1 - 10) + (16Z_0 - 12Z_1)$

20. $(24f - 13f^2 + 5) + (31f^2 + 17f - 11)$

21. $(.034a - 7.2b) + (-.091a + 3.71ab) + (.413a - 2.66b)$

22. $(16M_1 - 18M_0 + 2M) - (-11M_1 - 3M_0 + 5M) - (7M_1 + M_0 - M)$

23. $(3P + 4P^2 + 7) + (3P - P^2 + 5P^3) - (2 - 8P^2 + 2P^3)$

24. Subtract $5p + 2g - 3$ from the sum of $2p - 6g + 5$ and $3p + 4g - 7$.

25. Subtract $3a^2 + 2a - 4$ from the sum of $4 - 5a^2 + 9a$ and $2 - a - 7a^2$.

26. A body travels a distance of $6s^2 + 3s$. A second body travels a distance $5s^2 - 4s + 3$. How much farther did the first body travel?

27. A freely falling body falls $-\frac{1}{2}gt^2 - 5t$ the first second and $-\frac{1}{2}gt^2 - 14t - 3$ the next second. What distance has the body fallen at the end of 2 seconds?

28. If the width of a rectangle is w and the length is $(w + 7)$, find the perimeter of the rectangle.

29. If a square measures $(4s - 1)$ on each side, find the perimeter of the square.

30. Find the supplement of an angle measuring $(2a - 30)$ degrees.

3.5 Multiplication and Division of Real Numbers

Recall that in Unit 1, multiplication was defined as repeated addition. $3 \cdot 5 = 5 + 5 + 5 = 15$. Using the concept that multiplication is repeated addition, positive and negative numbers can be multiplied.

$$3 \cdot (-5) = (-5) + (-5) + (-5) = -15$$

To multiply $(-5) \cdot 3$, the commutative property for multiplication of real numbers is used to restate the problem as $3 \cdot (-5)$, which is equal to -15.

The product of two negative numbers is a positive number. For example, $(-3)(-5) = +15$. This is verified as follows:

$$(-3)(-5) + (-3)(+5) = -3[(-5) + (+5)] \quad \text{**By the distributive property.**}$$
$$(-3)(-5) + (-15) = -3[0] \quad \text{**As shown above.**}$$
$$(-3)(-5) + (-15) = 0$$

The statement that $(-3)(-5) + (-15) = 0$ implies that $(-3)(-5)$ must equal $+15$.

The above example suggests the following rules for multiplication of real numbers:

Rule 1: To multiply two positive real numbers, or two negative real numbers, multiply their absolute values. The product is a positive real number.

Example 1: $(+3)(+2) = +6$

Example 2: $(-3)(-2) = +6$

Rule 2: To multiply two real numbers with opposite signs, multiply their absolute values. The product is a negative real number.

Example 3: $(+7)(-2) = -14$

Example 4: $(-7)(+2) = -14$

In Unit 1, $6 \div 7$ was defined to be equal to $6 \cdot \frac{1}{7}$. Therefore, the same rules used for multiplication of real numbers are used for division of real numbers.

Rule 3: To divide two positive or two negative real numbers, divide their absolute values. The quotient is a positive real number. To divide two real numbers with opposite signs, divide their absolute values. The quotient is a negative real number.

Example 5: $(+12) \div (+6) = 12 \cdot \dfrac{1}{6} = 2$

Example 6: $(-12) \div (-6) = (-12) \cdot \left(-\dfrac{1}{6}\right) = 2$

Example 7: $(-12) \div (+6) = (-12) \cdot \left(\dfrac{1}{6}\right) = -2$

Every fraction has three signs, the sign of the fraction, the sign of the numerator, and the sign of the denominator. Changing any two of the three signs does not change the value of the fraction.

Example 8: $- \dfrac{+8}{-2} = + \dfrac{-8}{-2} = + \dfrac{+8}{+2}$

$\qquad\qquad\quad 4 = \qquad\; 4 = \qquad 4$

Again, recall that in Unit 1, the rule was followed that the operations of multiplication and division were performed in order from left to right before performing the operations of addition and subtraction. For example, $6 + 8 \div 2 - 4 = 6 + 4 - 4 = 6$. Parentheses are used to indicate that another order is to be followed. For example, $(6 + 8) \div (2 - 4) = 14 \div -2 = -7$.

Rule 4:   In evaluating an algebraic expression, use the following order:

1. Perform all operations inside parentheses.
2. Compute all indicated powers and square roots.
3. Multiply and divide in order from left to right.
4. Add and subtract from left to right.

Example 9: $(-2 + 16) \div (-2) - 3 = 14 \div (-2) - 3$
$$= -7 - 3$$
$$= -10$$

Example 10: $3^2 + 2 - (6 + 4) \div 5 = 3^2 + 2 - 10 \div 5$
$$= 9 + 2 - 10 \div 5$$
$$= 9 + 2 - 2$$
$$= 9$$

Example 11: $(16 \div 2) \times \sqrt{25} + 2^3 \div (-4) = 8 \times \sqrt{25} + 2^3 \div (-4)$
$$= 8 \times 5 \quad + 8 \div (-4)$$
$$= 40 \qquad + \quad (-2)$$
$$= 38$$

Example 12: $(-4964) \div (-73)$

To divide negative numbers using a calculator:

Enter	Press	Press	Enter	Press	Press	Display
4964	+/−	÷	73	+/−	=	68

EXERCISES 3.5

Multiply:

1. $(-3)(6)$ 　　　　　　　　**2.** $(3)(-6)$ 　　　　　　　　**3.** $(-7)(4)$

4. $(7)(-3)$

5. $(-3)(-9)$

6. $(3)(2)$

7. $(-8)(-8)$

8. $(-27)(3)$

9. $(5)(0)$

10. $(0)(7)$

11. $(-82)(47)$

12. $(21)(-32)$

13. $(-406)(-27)$

14. $(-11)(-11)$

15. $(68.2)(-2.03)$

16. $(-80)(20)$

17. $(-95)(-24)$

18. $(2)(-3)(-1)$

19. $(-4)(-1)(-2)$

20. $(-2)(0)(-7)$

21. $(-2)(3)(-1)(3)$

22. $\left(\frac{2}{3}\right)\left(\frac{3}{4}\right)$

23. $\left(-\frac{7}{10}\right)\left(-\frac{1}{4}\right)$

24. $\left(-\frac{5}{8}\right)\left(-\frac{3}{5}\right)$

Divide:

25. $-20 \div 10$

26. $18 \div (-9)$

27. $(-22) \div (-11)$

28. $-24 \div 6$

29. $(-72) \div (-8)$

30. $0 \div 3$

31. $56 \div (-7)$

32. $(-114) \div (-2)$

33. $\left(-\dfrac{1}{3}\right) \div \left(\dfrac{1}{4}\right)$

34. $\dfrac{5}{9} \div \left(-\dfrac{2}{10}\right)$

35. $\dfrac{1}{5} \div .7$

36. $-\dfrac{3}{8} \div \dfrac{9}{10}$

37. $(-.500) \div (-.25)$

38. $(-1.8) \div (.02)$

39. $(-396) \div (-22)$

40. $5070 \div (-65)$

41. $-2(-8) \div (4)(-6)$

42. $12 \div (-2)(6) \div (-2)$

43. $11(2)(-1) \div (-1)$

44. $(3)(5)(-2) \div (-3)(-2)$

45. $(-8 - 4) \div 3$

46. $-6(-6) + 2^2 - 20$

47. $(18 - 2 + 6) \div (6 - 4)$

48. $(-3)(0) + (-2 - 1)$

49. $3^2 - 2^2 \div (-2 + 1)$

50. $(8 - 3 - 2)(6 - 4 - 2)$

51. $-56 \times \sqrt{36} - 27 \div 3^2$

52. $\sqrt{4} - (56 - 3 \times 21) + 4^2$

53. If a flat $4\frac{1}{2}$ in. by $6\frac{1}{4}$ in. plate has a round hole $1\frac{1}{2}$ inches in diameter in it, find the area of the plate rounded to the nearest hundredth.

54. How many cubic yards of earth need to be excavated to install in the ground a septic tank that measures 12 ft long, 10 ft wide, and 6 in. high, if the top of the tank is 4 feet below the ground surface? (Do not allow for form work.)

3.6 Multiplication of Polynomials

Given the monomial x^3, x is referred to as the *base* and 3 is called the *exponent*. By definition,

$$x^3 = x \cdot x \cdot x$$
$$x^2 = x \cdot x$$
$$x^5 = x \cdot x \cdot x \cdot x \cdot x$$

Therefore, $x^3 \cdot x^2 = (x \cdot x \cdot x) \cdot (x \cdot x) = x^5$

Rule 1: When two monomials with like bases are multiplied, add the exponents of the two monomials; the base remains the same.

$$x^m \cdot x^n = x^{m+n}$$

Example 1: $y^4 \cdot y^2 = y^{4+2} = y^6$

Example 2: $a^3 \cdot a = a^{3+1} = a^4$

Note that if the exponent is not written, it is understood to be 1.

When a constant preceeds the variable, such as $6d^2$, it is expanded as $6 \cdot d \cdot d$. Note that only the d is squared. If both the numerical and variable factors are to be squared, parentheses will be used.

Example 3: $2^3 \cdot 2^2 = 2^{3+2} = 2^5 = 32$

Example 4: Write $8f^3$ as a product of factors.

$$8f^3 = 8 \cdot f \cdot f \cdot f$$

Example 5: Write $(5a^2)^2$ as a product of factors.

$$(5a^2)^2 = (5a^2)(5a^2) = (5 \cdot a \cdot a)(5 \cdot a \cdot a)$$

Rule 2: To multiply two monomials with numerical coefficients, multiply the numerical coefficients and then use Rule 1 to multiply the variables.

Example 6: $2m^5 \cdot 3m^2 = (2 \cdot 3)(m^5 \cdot m^2) = 6m^{5+2} = 6m^7$

Example 7: $3bc(2a^2bc^3) = 6a^2b^2c^4$

In multiplying a binomial by a monomial, the distributive property is used.

Example 8: $2r^2(5r^3 + 3r^2) = 2r^2(5r^3) + 2r^2(3r^2)$
$$= 10r^5 + 6r^4$$

Example 9: $-2abc(-8ab + 2bc^2 - 3c^3)$
$$= -2abc(-8ab) - 2abc(2bc^2) - 2abc(-3c^3)$$
$$= 16a^2b^2c - 4ab^2c^3 + 6abc^4$$

Example 10: $5(6t - 11) + t(2t - 8) = 30t - 55 + 2t^2 - 8t$
$$= 2t^2 + 22t - 55$$

EXERCISES 3.6

Multiply:

1. $x^5 \cdot x^6$

2. $a^4 \cdot a^3$

3. $y^7 \cdot y^8$

4. $3 \cdot 3^2$

5. $-3T^2 \cdot 5T^3$

6. $-7R^4 \cdot -4R$

7. $5^2 \cdot 5^3$

8. $-2q(6p^3)$

9. $(-3ab)(4ab^3c)$

10. $(-4d)(2d^3)(3d^4)$

11. $(-2mn)(5m^2)(m^3n)$

12. $(-3F^2)(-2F^4)(-2F)$

13. $\frac{4}{7} D^2 \left(\frac{3}{4} D\right)$

14. $-1.7z^2(4.3z^8)$

15. $4ab^2c(-12ab^2c)$

16. $.03x^2y(.16y^3z)$

17. $\frac{9}{11}R^2S^2\left(-\frac{11}{12}RS\right)$

18. $-3D^3T(-5D^5T^9)$

19. $3(7x + 2)$

20. $-2(5Y + 6)$

21. $6(2R_0 - 4)$

22. $N(8N - 1)$

23. $m(m - 1)$

24. $-y(-2y - 4)$

25. $-2rst(7rs - 3st^2)$

26. $-3b^2c(-2bc + c^2d)$

27. $a^2b(2a - 9b^3c)$

28. $-5x^2(-4x + 7)$

29. $mn(4m - 2n)$

30. $.2R_1(.05R_1 - .3R)$

31. $2(2x^2 - 3x + 6)$

32. $-6(5Y^2 - 7Y + 8)$

33. $2t(3t^2 + t - 1)$

34. $\frac{7}{8}U\left(\frac{8}{9}U^3 + U - 24\right)$

35. $6k(j - 4k)$

36. $-3S_0(-3S_0 + T_1)$

37. $5x^2(5xy + 2y)$

38. $.03WV(.004V - .22W)$

39. $-4de^4f^2(2d^2e^3 - 31e^4f^3 + 9f^4)$

40. $-7Z^3(4.1Z^2 - 6Z - 1.6)$

41. $-6h^2i\left(\frac{2}{3}h^3i + \frac{1}{6}i^2 + 12\right)$

42. $-2(-2R + 5) + 3(4R - 3)$

43. $x(x - 5y + 2) - y(x - 7)$

44. $2(H + 4K - 9) - 6(-2H - 3K + 6)$

45. $a(a - b - 7) - 6(a^2 - 2ab + 7a)$

46. A rectangle measures $2a$ units in length and $(a - 3)$ units in width. Find the area of the rectangle.

47. A triangle has a base of $(2b - 5)$ inches and a height of b inches. Find the area.

48. Electrical power equals voltage times current. If the voltage is E and the current is $(3I - 4)$, find the watts expended in electrical power.

49. Ohm's Law states that the current is equal to the voltage divided by the resistance. Find I, the current, if there is a 110-volt circuit in which the resistance is R.

50. The volume of a sphere is $\dfrac{4\pi r^3}{3}$. Find the volume of a sphere that measures $6r$ in diameter.

3.7   Division of Polynomials

By definition of an exponent, $x^5 = x \cdot x \cdot x \cdot x \cdot x$
$$x^2 = x \cdot x$$
$$x^3 = x \cdot x \cdot x$$

Therefore,

$$x^5 \div x^2 = \frac{x^5}{x^2} = \frac{\overset{1}{\cancel{x}} \cdot \overset{1}{\cancel{x}} \cdot x \cdot x \cdot x}{\underset{1 \quad 1}{\cancel{x} \cdot \cancel{x}}} = x \cdot x \cdot x = x^3$$

$$x^2 \div x^5 = \frac{x^2}{x^5} = \frac{\overset{1}{\cancel{x}} \cdot \overset{1}{\cancel{x}}}{\underset{1 \quad 1}{\cancel{x} \cdot \cancel{x} \cdot x \cdot x \cdot x}} = \frac{1}{x^3}$$

Rule: When two monomials with like bases are divided, subtract the exponents of the two monomials; the base remains the same.

$$x^m \div x^n = x^{m-n} \quad \text{if } m > n$$

$$x^m \div x^n = \frac{x^m}{x^n} = \frac{1}{x^{n-m}} \quad \text{if } n > m$$

Example 1: $x^8 \div x^4 = x^{8-4} = x^4$

Example 2: $\dfrac{x^2}{x^7} = \dfrac{1}{x^{7-2}} = \dfrac{1}{x^5}$

Example 3: $\dfrac{5^3}{5^6} = \dfrac{1}{5^{6-3}} = \dfrac{1}{5^3} = \dfrac{1}{125}$

Example 4: $\dfrac{-14a^3}{2a^2} = -7a^{3-2} = -7a$

Example 5: $\dfrac{p^6 q^7 r^8}{p^7 q^{10} r^2} = \dfrac{r^{8-2}}{p^{7-6} q^{10-7}} = \dfrac{r^6}{pq^3}$

Recall that to divide a by b, a is multiplied by $\dfrac{1}{b}$. For example, $9 \div 3 =$ $9 \cdot \dfrac{1}{3} = 3$. To divide any polynomial by a monomial, multiply the polynomial by one over the monomial.

Example 6: $(12m^4 - 10m^3 + 8m^2) \div 2m^2$

$$= (12m^4 - 10m^3 + 8m^2)\dfrac{1}{2m^2}$$

$$= 12m^4\left(\dfrac{1}{2m^2}\right) - 10m^3\left(\dfrac{1}{2m^2}\right) + 8m^2\left(\dfrac{1}{2m^2}\right)$$

$$= \dfrac{12m^4}{2m^2} - \dfrac{10m^3}{2m^2} + \dfrac{8m^2}{2m^2}$$

$$= 6m^2 - 5m + 4$$

Example 7: When dividing a polynomial by a monomial, work can be shortened by going directly to the third step.

$$(80p^7 - 70p^5 + 40p^4) \div 10p^3 = \dfrac{80p^7}{10p^3} - \dfrac{70p^5}{10p^3} + \dfrac{40p^4}{10p^3}$$
$$= 8p^4 - 7p^2 + 4p$$

EXERCISES 3.7

Divide:

1. $x^9 \div x^3$

2. $a^8 \div a^2$

3. $y^{10} \div y^8$

4. $16b^2 \div 2b^5$

5. $(20z^4) \div (10z)$

6. $48a^7 \div 2a^4$

7. $\dfrac{12D^3}{D^2}$

8. $\dfrac{-8r^2 s^3}{4rs^7}$

9. $\dfrac{2j^4}{3j^2}$

10. $\dfrac{16z^2}{4z}$

11. $\dfrac{-10S_0{}^5}{-2S_0}$

12. $\dfrac{8.1a^7}{-9a}$

13. $\dfrac{-105r^4s^8t^4}{3r^2s^{10}}$

14. $\dfrac{-216c^2}{6}$

15. $\dfrac{4R^2 - 8}{2}$

16. $\dfrac{21k^3 - 15k^2}{-3k}$

17. $\dfrac{12T^4 - 3T^2}{T^3}$

18. $\dfrac{20d^2 + 10d + 5}{5}$

19. $\dfrac{27R^4 - 12R^2 + 9R}{-3R^3}$

20. $\dfrac{-100xy^4 - 20y^{10}}{-10y^3}$

21. $\dfrac{14m^9 + 28m^8 - 56m^6 - 49m^4}{7m^3}$

22. $\dfrac{-6p^4q^2 + 8p^3q^3 + 36p^2q}{-2p^2q}$

23. $\dfrac{10a^4 - 4a^3 - 6a^2 + 4a - 2}{-2a^3}$

24. $\dfrac{C^2D^3 + 4C^3D^2 - CD}{-CD^2}$

25. The number of board feet (*BF*) in a wooden board is found by taking the product of its width (*w*) and thickness (*t*) in inches and its length (*L*) in feet, divided by 12. Write a formula for board feet.

26. Using the formula in problem 25, find *BF* if $w = 2x$, $t = 6x + 4$, and $L = 10$.

27. The tension (*T*) on a wire rope basket hitch is the product of the weight lifted (*W*) and the length of the choker (*L*) divided by the product of the number of chokers (*N*) and the vertical distance from the load to the hook (*V*). Write the formula for tension.

28. Using the formula in problem 27, find *T* if $W = 2.5a$, $L = 6a$, $N = 2$, and $V = a^2$.

3.8   Grouping Symbols

Parentheses ( ), brackets [ ], and braces { } are symbols of grouping used in algebraic expressions.

Rule 1:   When removing more than one symbol of grouping, remove the innermost symbol first and combine like terms.

Example 1: $7 - [7x - (2x + 4)]$ **Remove parentheses, the innermost symbol.**
$= 7 - [7x - 2x - 4]$ **Combine like terms.**
$= 7 - [5x - 4]$ **Remove brackets.**
$= 7 - 5x + 4$ **Combine like terms.**
$= 11 - 5x$

Example 2: $8p - \{2p^2 - [4p(3p - 6)]\} = 8p - \{2p^2 - [12p^2 - 24p]\}$
$= 8p - \{2p^2 - 12p^2 + 24p\}$
$= 8p - \{-10p^2 + 24p\}$
$= 8p + 10p^2 - 24p$
$= 10p^2 - 16p$

EXERCISES 3.8

Simplify:

1. $5m - [6 - (2m - 3)]$

2. $6 + [7a - (2a - 5)]$

3. $3T + 4 - [4T - (2T + 8)]$

4. $-[7D + (3D - 2)]$

5. $-4 - 2[m - (2m + 6)]$

6. $2ab - b[9a - 4 + 2(a - 1)]$

7. $-(2H + 6) + 3[4H - 2(5 + H)]$

8. $-[6T + 4(3T - 2)]$

9. $(2F + 3) - \{[3F + (8 - 2F)] + 5\}$

10. $10 - \{6p - [4 - (3p - 5)]\} - [4 - (p - 2)]$

11. $2r - 3[4s + 2(3s - 7)] + 7$

12. $2a - \{2a - [5a - (3a - 8) - 8] + 4\}$

13. $6p - 5\{2p - 5[p - 2(3p - 1) - 4] + 11\}$

14. $-2K[5K - 7(K + 1) + 2K(K - 1)] + 2(3 - K^2)$

15. $-2\{4x - [6x - (3x + 4) - 7x] + 2\}$

16. In insulating buildings, the heat gain or loss (H) is equal to the product of the U value times the difference in temperature between the inside (T_i) and the outside (T_o) of the building, times the area of the wall in square feet. Write the formula for H for a 12 ft by 9 ft wall, and simplify.

17. The volume (V) of excavation needed is equal to the product of the area of the lot and the difference between the existing elevation (E_o) and the elevation (E_f) wanted. Write the formula for V for a 250 ft by 300 ft lot, and simplify.

18. Write a formula for the area of an L-shaped deck. The L-shape is formed by a square measuring $4x$ inches on the side from which a square corner measuring $(x - 1)$ inches on the side has been removed.

19. Find the volume of a concrete column of diameter $(2a + 3)$ feet and height $12a$ feet.

3.9 Evaluation of Algebraic Expressions

To evaluate an algebraic expression for given values of the unknowns, replace each unknown with its proper value and carry out the indicated operations.

In carrying out the indicated operations, use the following order:

1. Perform all operations within innermost symbol of grouping.
2. Compute all indicated powers.
3. Multiply and divide in order from left to right.
4. Add or subtract from left to right.

Example 1: $5^2 - 5(8 + 2) + 20$

$5^2 - 5(10) + 20$	**Add within parentheses.**
$25 - 5(10) + 20$	**Compute power.**
$25 - 50 + 20$	**Multiply.**
$- 25 + 20$	**Subtract.**
-5	**Add.**

Example 2:   Evaluate the algebraic expression $3x + y$, given $x = -2$ and $y = 4$.

$$3x + y = 3(-2) + 4 = -6 + 4 = -2$$

Example 3:   Evaluate the algebraic expression, $\dfrac{-3m(n + 1)}{5mn}$, where $m = 4$ and $n = -2$.

$$\frac{-3m(n + 1)}{5mn} = \frac{-3(4)(-2 + 1)}{5(4)(-2)}$$

$$= \frac{-3(4)(-1)}{-40}$$

$$= \frac{12}{-40} = -\frac{3}{10}$$

Example 4:   The formula for determining safe pipe pressure, p, in kilograms per square centimeter for a pipe is $p = \dfrac{2st}{D}$, where s is the tensile strength of the pipe in kilograms per square centimeter, t is the thickness of the pipe in centimeters, and D is the diameter in centimeters. Given that the tensile strength is 5000 kg per square centimeter, the diameter is 5 cm, and the thickness is 1.6 cm, find the safe pipe pressure.

$$p = \frac{2st}{D} = \frac{2(5000)(1.6)}{5} = 3200 \text{ kilograms per square centimeter}$$

Example 5:   The formula for determining the horsepower of an electric motor is $h.p. = \dfrac{EI}{750}$, where E is the voltage and I is the current measured in amperes. Find the horsepower when the voltage is 250 and the amperage is 2.

$$h.p. = \frac{EI}{750} = \frac{250(2)}{750} = .67$$

EXERCISES 3.9

If $x = 2$ and $y = -1$, find the value of each expression:

1. $4x + 6$

2. $3y + 2$

3. $3y - 6$

4. $x + y$

5. $6x - 2$

6. $2x + y$

7. $x + 7y$

8. $3x + 2y$

9. $-5x + 4y$

10. $-3x - 3y$

11. $6x(-3y)$

12. $\dfrac{10y}{5x}$

13. $4(2x - 6y)$

14. $\dfrac{8(-3x + 6y)}{15x}$

15. $2x^2$

16. x^2y^2

17. $-7xy^2$

18. $-5x^2y$

If $a = -3$, $b = -1$, and $c = 2$, find the value of each of the following algebraic expressions:

19. $a + b + c$

20. $a - b - c$

21. $-a + b - c$

22. $a^2 + b^2 - c^2$

23. $4a^2 + 3c^2$

24. $-2a(4b - c)$

25. $3ab(2a - 5c + 10)$

26. $4 - a(b + c)$

27. $\dfrac{b - c}{a - b}$

28. $-3 - 2a(b - 3c)$

29. Formulas for the area and perimeter of a rectangle are: $A = lw$ and $p = 2(l + w)$

    a. Find the area of a rectangular piece of sheet metal that is 3.66 meters by 2.44 meters.

    b. A rectangular field is 170 meters by 340 meters. Find the area of the field and how much fence is needed to enclose the field.

30. The mechanical advantage (MA) of the hydraulic press is the force from the large piston (F_l) divided by the force of the small piston (F_s): $MA = \dfrac{F_l}{F_s}$. Find the mechanical advantage, given:

    a. $F_l = 6{,}000$ kg, $F_s = 300$ kg

    b. $F_l = 238{,}500$ kg, $F_s = 2{,}650$ kg

31. The formula for converting from Fahrenheit to Centigrade temperature is: $C = \dfrac{5}{9}(F - 32°)$. Convert the following to Centigrade:

    a. $F = 68°$ **b.** $F = 212°$ **c.** $F = -4°$

32. The length (L) of a drive belt between two pulleys whose radii are R and r and where D represents the distance between centers of the pulley is: $L = 2D + \dfrac{13}{4}(R + r)$. Find the length of a drive belt, given:

    a. $R = 12\frac{1}{4}$ cm, $r = 23\frac{3}{4}$ cm, $D = 136$ cm

    b. $R = 10.16$ cm, $r = 20.32$ cm, $D = 0.6871$ m

33. The diametral pitch of a gear states the size of the tooth and can be likened to the threads per inch determining the size of a screw. The formula for determining the diametral pitch is: $D = \dfrac{n + N}{2C}$, where N and n are the number of teeth in two respective gears, and C is the distance between their centers. Find the diametral pitch, given:
    a. $n = 20$, $N = 30$, $C = 2.5$

    b. $n = 17$, $N = 28$, $C = 2.3$

34. Stress (S) is the force (F) divided by the cross-sectional area (A) over which the force acts. $S = \dfrac{F}{A}$. Find the stress, given:
    a. $F = 1860$ kg, $A = 32$ cm²

    b. $F = 432$ kg, $A = 20.3$ cm²

35. The area of a trapezoidal region is found by using the formula, $A = \dfrac{1}{2} h(b + b')$. Find the area, given:
    a. $h = 16.8$ m, $b = 64.7$ m, $b' = 24.1$ m

    b. $h = 31.1$ cm, $b = 81.2$ cm, $b' = 40.4$ cm

36. $\dfrac{1}{F} = (n - 1)\left(\dfrac{1}{R_1} + \dfrac{1}{R_2}\right)$ is referred to as the Lensmaker's equation. F is the distance between the center of a lens and its focal points, R_1 and R_2. The index of refraction is denoted by n. Find $\dfrac{1}{F}$, given:
    a. $n = 1.656$, $R_1 = 18.2$ cm, $R_2 = 22.4$ cm

b. $n = 1.656$, $R_1 = 22.1$ cm, $R_2 = 16.2$ cm

37. The efficiency (Eff) of a Carnot engine operating between two absolute temperatures, T_1 and T_2, is expressed by the formula, $Eff = 1 - \dfrac{T_2}{T_1}$. Express Eff in percent, given:
   a. $T_1 = 1200°$, $T_2 = 750°$

   b. $T_1 = 1400°$, $T_2 = 580°$

38. The resistance of copper wire in ohms is equal to 10.2 times the length of the wire, divided by the square of the diameter of the wire, $R = 10.2 \dfrac{1}{d^2}$.
   Find R, given:
   a. $l = 420$ m, $d = 6$ mil

   b. $l = 300$ m, $d = 4$ mil

39. The centrifugal force in pounds exerted by a rotating object is
   $$F = \frac{4\pi^2 R N^2 W}{g}.$$
   Find F if the radius of rotation (R) is 4 feet, the speed of revolution (N) is 650 revolutions per second, the weight (W) of the object is 3 pounds, and the force of gravity (g) is 32 ft/sec^2.

40. A window is shaped as a square with a semicircle on top. If one side of the square is 10 ft long, find the area of the window.

Unit 3 Self-Evaluation

Simplify the following:

1. $|-13|$ 　　　　　　**2.** $-11 + (-2)$ 　　　　**3.** $\dfrac{7}{8} + \left(-\dfrac{2}{5}\right)$

4. $21 - (-13)$ 　　　　**5.** $-8 - 4$ 　　　　　　**6.** $\left(-\dfrac{9}{11}\right)(121)$

7. $54 \div (-6)$ 　　　　**8.** $\dfrac{-6a^2b^3}{-2ab^5}$ 　　　　　**9.** $16 - (-8) \div 4 - 6$

10. $[16 - (-8)] \div 4 - 6$

11. $(-5ab)(16a^2b^2c)$

12. $-2mn(4m - 3n)$

13. $(5p + q - 6) + (-11p - q + 4) + (4p - q - 2)$

14. $(6R_1 - 2R_2 + 10) - (2R_1 - 2R_2 - 8)$

15. $2(H - 2K) - 3H(K + 5) - K(-2H - 1)$

16. $\dfrac{60d^5 - 80d^3 + 120d}{-4d}$

17. $20 - \{5e - [2 - 6(2e - 1)] + 1\} - [7 - 2(e - 2)]$

18. If $r = 5$, $s = -2$, $t = 3$, find the value of $-2(5r - s) + 4t$.

19. Find the area of a rectangular piece of plywood which is 72.13 by 45.60 meters.

20. Total resistance (R) of three resistances connected in parallel is found by the formula, $R = \dfrac{R_1R_2R_3}{R_1R_2 + R_2R_3 + R_1R_3}$. Find R, given: $R_1 = 10$, $R_2 = 5$ and $R_3 = 8$.

Linear Equations

Unit 4 Objectives

1. To solve linear equations.
2. To solve linear equations that contain parentheses, fractions, and decimals.
3. To solve linear inequalities.
4. To change the subject of formulas.
5. To write algebraic representations of statements.
6. To solve word problems.
7. To solve problems using ratio and proportion.
8. To solve problems using direct variation, inverse variation, and joint variation.

4.1   Equations

A piece of wire measuring 42 meters in length is to be cut into three pieces. One piece is to be three times the length of the shortest piece and another piece is to be two meters longer than the shortest piece of wire. Into what size pieces will the original piece of wire be cut? In order to solve this problem, and other problems encountered in the work of a technician, sentences are translated into *equations*.

Sentences which are either true or false are called *statements*. The sentence "$4 + 5 = 9$" is a true statement, while the sentence "$4 + 6 = 9$" is a false statement. A sentence which contains at least one variable is called an *open sentence*. $k + 2 = 10$ is an open sentence. The truth or falsity of the statement cannot be determined until k has been replaced. If k is replaced by 4, the sentence is false. If k is replaced by 8, the sentence is true.

Open sentences and statements that express equality are called *equations*. The open sentence "$g - 7 = 11$" is an equation. $g - 7$ is called the *left member* of the equation, and 11 is called the *right member* of the equation.

Equations are said to be either identities or conditional equations. An *identity* is an equation that is true for all values of the unknown. For example, $x + 1 = x + 1$ is an identity. It is true for every replacement of x.

A *conditional equation* is an equation that is false for at least one replacement of the variable. An example of a conditional equation is: $s + 8 = 10$. This sentence is true only under the condition that $s = 2$, since $2 + 8 = 10$. If s were replaced by 4, the sentence would not be true since $4 + 8 \neq 10$.

A *linear equation* is an equation that has the unknown (or unknowns) to the first degree. The equation, "$m + 2 = 5$," is an example of a condition equation that is linear. The variable m is to the first degree.

Any value given to the variable in an equation that will make the sentence true is called a *solution,* or *root* of the equation. Replace the variable in the equation $x + 1 = 4$ with the numbers 1, 2, 3, and 4 and determine the truth or falsity of the sentence. When the variable x is replaced by 3, the sentence is true. Therefore, 3 is said to be a solution of the equation.

Equations that have identical solutions are called *equivalent equations.* The following are equivalent equations since each has the same solution, 5:

$$x = 5$$

$$x + 5 = 10$$

$$5x = 25$$

Finding the solution of conditional equations is usually referred to as "solving the equation." Many conditional linear equations can be solved simply by inspection. It is obvious that $y + 7 = 9$ is a true statement only if $y = 2$. However, all equations cannot be solved simply by inspection. It is not obvious that 4 is the solution to the equation $3x + 2 = 6x - 10$. Therefore, it becomes necessary to establish some rules that will help transform an equation into an equivalent equation whose solution can be determined by inspection.

Rule 1: The same polynomial added to, or subtracted from, both members of an equation produces an equivalent equation.

Rule 2: Multiplying or dividing both members of an equation by the same non-zero polynomial produces an equivalent equation.

The *symmetric property* of equality is also helpful in solving equations. This property states that if the left member of an equation is equal to the right member, the right member is also equal to the left member. For example, if $4 = Y$, then $Y = 4$.

The following are examples of Rule 1, Rule 2, and the symmetric property.

Example 1: Solve the equation "$x - 6 = 8$," for x.

$$x - 6 = 8$$
$$x - 6 + 6 = 8 + 6 \qquad \textbf{Rule 1: Add 6 to both members of the equation.}$$
$$x = 14$$

The result may be checked by substituting the solution, 14, in the original equation.

$$x - 6 = 8$$
$$14 - 6 = 8$$
$$8 = 8$$

Example 2: Solve the equation "$12 = T + 4$," for T.

$$12 = T + 4$$
$$12 - 4 = T + 4 - 4 \qquad \textbf{Rule 1: Subtract 4 from both members of the equation.}$$
$$8 = T$$
$$T = 8 \qquad\qquad\qquad \textbf{Symmetric property}$$

Check:

$$12 = T + 4$$
$$12 = 8 + 4$$
$$12 = 12$$

Example 3:   Solve the equation "$3y = 15$," for y.

$$3y = 15$$
$$\frac{3y}{3} = \frac{15}{3} \qquad \textbf{Rule 2: Divide both members of the equation by 3.}$$
$$y = 5$$

Check:

$$3y = 15$$
$$3(5) = 15$$
$$15 = 15$$

Example 4:   Solve the equation "$\dfrac{m}{7} = 4$," for m.

$$\frac{m}{7} = 4$$
$$7 \cdot \frac{m}{7} = 7 \cdot 4 \quad \textbf{Rule 2}$$
$$m = 28$$

Check:

$$\frac{m}{7} = 4$$
$$\frac{28}{7} = 4$$
$$4 = 4$$

Example 5:   Solve the equation, "$-\dfrac{2}{3}d = -4$," for d.

$$-\frac{2}{3}d = -4$$
$$-3\left(-\frac{2}{3}d\right) = -3(-4) \quad \textbf{Rule 2}$$
$$2d = 12$$
$$\frac{2d}{2} = \frac{12}{2} \qquad \textbf{Rule 2}$$
$$d = 6$$

Check:

$$-\frac{2}{3}d = -4$$
$$-\frac{2}{3}(6) = -4$$
$$-4 = -4$$

This equation could have been solved in one step by multiplying each side of the equation by $-\dfrac{3}{2}$, the reciprocal of the coefficient of d.

$$-\frac{2}{3}d = -4$$
$$-\frac{3}{2}\left(-\frac{2}{3}d\right) = -\frac{3}{2}(-4) \quad \textbf{Rule 2}$$
$$d = 6$$

Example 6:   Solve the equation "$2T + 5 = 9$," for T.

To transform this equation, both Rule 1 and Rule 2 must be used. Using Rule 1 before Rule 2 will make the equation easier to solve and will reduce the possibility of errors.

$$2T + 5 = 9$$
$$2T + 5 - 5 = 9 - 5 \quad \textbf{Rule 1}$$
$$2T = 4$$
$$\frac{2T}{2} = \frac{4}{2} \quad \textbf{Rule 2}$$
$$T = 2$$

Check:

$$2T + 5 = 9$$
$$2(2) + 5 = 9$$
$$4 + 5 = 9$$
$$9 = 9$$

EXERCISES 4.1

Solve each equation for the given variable and check. Use Rule 1.

1. $x + 7 = 11$ **2.** $N - 3 = 10$ **3.** $y - 4 = 10$

4. $8 = a + 2$ **5.** $7 = y + 7$ **6.** $D + 2 = -5$

7. $b - 8 = -2$ **8.** $-2 = T + 4$ **9.** $-10 + p = -1$

10. $Z_1 - 6 = -6$ **11.** $K - .11 = -.7$ **12.** $T_0 + .9 = -1.82$

13. $5 + d_2 = -5$ **14.** $b + 2 = 1$ **15.** $r + 7 = 13$

16. $m - 9 = -7$ **17.** $W - 10 = -5$ **18.** $-3 = 15 + x$

19. $s - 7 = -8$ **20.** $D - .15 = .72$ **21.** $13 + R = -11$

22. $K - 1 = 1$ **23.** $g + .4 = .8$ **24.** $y - 10 = 4$

Solve each equation for the given variable and check. Use Rule 2.

25. $8y = 48$ **26.** $2K = -16$ **27.** $45 = 5W$

28. $54 = -9a$ **29.** $-10y = -40$ **30.** $2b = 7$

31. $15 = \dfrac{a}{3}$ **32.** $\dfrac{R_0}{9} = -2$ **33.** $1 = \dfrac{x}{-4}$

34. $\dfrac{2}{3} Z_1 = 20$ **35.** $\dfrac{3}{5} P = -12$ **36.** $-\dfrac{7}{9} x = -14$

37. $4 = \dfrac{5}{7} D$ **38.** $\dfrac{3}{7} T = \dfrac{9}{14}$ **39.** $\dfrac{4}{9} a = \dfrac{3}{2}$

40. $\dfrac{7}{8} I_0 = -\dfrac{5}{16}$ **41.** $-\dfrac{3}{4} y = -\dfrac{5}{11}$ **42.** $\dfrac{3}{5} = 7b$

43. $\dfrac{e}{5} = -\dfrac{2}{5}$ **44.** $-11 = 11p$

Solve each equation for the given variable and check. Use Rule 1 and Rule 2.

45. $4x + 1 = 9$

46. $3s - 7 = -1$

47. $10 = 3a - 2$

48. $-2 = \frac{2}{3}R + 6$

49. $7y - 1 = -15$

50. $\frac{x}{5} + 4 = -16$

51. $7 = -11Z + 7$

52. $\frac{3}{5}B - 1 = 7$

53. $4Q_0 - 6 = 6$

54. $11 = \frac{m}{9} - 16$

55. $2r - 18 = -22$

56. $9y + 50 = -4$

57. $5T_1 - 3 = 0$

58. $14b + 1 = -13$

59. $-9 = -\frac{3}{7}e + 5$

60. $\frac{D}{3} - 1 = -7$

61. $\frac{3}{11}H - 17 = 2$

62. $\frac{a}{2} + 6 = -3$

63. $\frac{p}{9} + 16 = 11$

64. $\frac{7}{8}V + 2 = 1$

4.2   Solving Linear Equations

Equations to be solved will often be complicated by parentheses, by fractions, and by having a variable on both sides of the equation. The order of the following steps will simplify finding the solution of such equations.

Step 1: Remove any parentheses in the given equation.

Step 2: If there are fractions in the equation, multiply both sides of the equation by a common denominator.

Step 3: Combine similar terms on each side of the equation.

Step 4: If the variable appears on both sides of the equation, eliminate the variable from either side by using Rule 1.

Step 5: Use Rule 1 and Rule 2 to solve.

Example 1: Solve the equation $2p = 6 - p$ for p.

Since the variable p appears on both sides of the equation, it may be eliminated from the right member by adding p to both members.

$$2p = 6 - p$$
$$2p + p = 6 - p + p \quad \textbf{Rule 1}$$
$$3p = 6$$
$$p = 2 \qquad\qquad \textbf{Rule 2}$$

Check:

$$2p = 6 - p$$
$$2(2) = 6 - 2$$
$$4 = 4$$

Example 2: Solve the equation $2(y - 3) = 4y$ for y.

$$2(y - 3) = 4y$$
$$2y - 6 = 4y \qquad\qquad \textbf{Remove parentheses}$$
$$-2y + 2y - 6 = -2y + 4y \quad \textbf{Rule 1}$$
$$-6 = 2y$$
$$-3 = y \qquad\qquad \textbf{Rule 2}$$
$$y = -3 \qquad\qquad \textbf{Symmetric Property}$$

Check:

$$2y - 6 = 4y$$
$$2(-3) - 6 = 4(-3)$$
$$-6 - 6 = -12$$
$$-12 = -12$$

Example 3: Solve $\frac{x}{3} + \frac{2}{7} = -\frac{1}{21}$ for x.

The least common denominator for 3, 7, and 21 is 21. Therefore, each member of the equation is multiplied by 21.

$$\frac{x}{3} + \frac{2}{7} = -\frac{1}{21}$$

$$21 \cdot \left(\frac{x}{3} + \frac{2}{7}\right) = 21\left(-\frac{1}{21}\right) \quad \textbf{Rule 1}$$

$$21 \cdot \frac{x}{3} + 21 \cdot \frac{2}{7} = -1$$

$$7x + 6 = -1$$

$$7x = -7 \qquad\qquad \textbf{Rule 1}$$

$$x = -1 \qquad\qquad \textbf{Rule 2}$$

Check:

$$\frac{x}{3} + \frac{2}{7} = -\frac{1}{21}$$

$$-\frac{1}{3} + \frac{2}{7} = -\frac{1}{21}$$

$$-\frac{7}{21} + \frac{6}{21} = -\frac{1}{21}$$

$$-\frac{1}{21} = -\frac{1}{21}$$

Example 4: Solve $.3t + .35 = 2.45$ for t.

Decimals may be eliminated from a problem in a manner similar to the above method. Since the equation involves tenths and hundredths, the decimals may be eliminated from the equation by multiplying both members by one hundred.

$$\begin{aligned}
.3t + .35 &= 2.45 \\
100(.3t + .35) &= 100(2.45) \qquad \textbf{Rule 1} \\
30t + 35 &= 245 \\
30t &= 210 \qquad \textbf{Rule 1} \\
t &= 7 \qquad \textbf{Rule 2}
\end{aligned}$$

Check:

$$\begin{aligned}
.3t + .35 &= 2.45 \\
.3(7) + .35 &= 2.45 \\
2.1 + .35 &= 2.45 \\
2.45 &= 2.45
\end{aligned}$$

EXERCISES 4.2

Solve each equation for the given variable and check:

1. $6r = 3r + 9$

2. $-3d + 20 = d$

3. $6b - 5 = 2b + 7$

4. $5y - 3 = 4y + 2$

5. $4a - 3 = 12 - a$

6. $-2M + 5 = 13 - 2M$

7. $6d - 2 = 3d + 1$

8. $4x + 3 = 12 + x$

9. $6 - 2Z = 3Z - 4$

10. $8 + D = 8 - 5D$

11. $6 + 3Z_0 = Z_0 - 18$

12. $5 - b = 8b - 13$

13. $2(W + 3) = -2$

14. $-3(2f - 4) = -6(f - 1)$

15. $6(2x + 1) = 3(x + 8)$

16. $2(M - 5) = M - 3$

17. $6 - p = 6 + 2p$

18. $6R_1 - 4 + 2 = 3R_1 + 1$

19. $3 + 6C = 24 - C$

20. $3s + 5 - 7s = -3$

21. $\dfrac{3a}{2} - \dfrac{1}{6} = \dfrac{4}{3}$

22. $\dfrac{3}{5}I + \dfrac{3}{10} = -\dfrac{9}{20}$

23. $\dfrac{5Q}{2} - \dfrac{1}{3} = \dfrac{Q}{3} - \dfrac{1}{6}$

24. $\dfrac{y}{5} - \dfrac{y}{2} = 9$

25. $\dfrac{3a}{2} + \dfrac{a}{3} = \dfrac{7}{6}$

26. $\dfrac{s}{2} + \dfrac{s - 1}{5} = \dfrac{7s - 2}{10}$

27. $\dfrac{D}{4} - \dfrac{3D + 1}{5} = \dfrac{D + 3}{10}$

28. $\dfrac{x}{5} - \dfrac{1}{15} = \dfrac{4}{3}$

29. $\dfrac{10p}{3} = \dfrac{8p + 2}{2}$

30. $\dfrac{8K + 3}{6} - \dfrac{7K - 1}{4} = -\dfrac{1}{2}$

31. $\dfrac{3(t - 3t)}{9} = -\dfrac{16}{4}$

32. $\dfrac{5b - 3b}{4} = \dfrac{8 + 2}{5} - 2$

33. $.2r - .32 = -.6r$

34. $7.5 + .25x = 2$

35. $.05m + .4 = .05 - .3m$

36. $.5b - .05 = .3b + .03$

37. $-.0001 - .6x = .0099 + .4x$

38. $.001x + .24 = 6.1$

39. $-2(4 - x) = x + 3(x - 2)$

40. $\dfrac{3(W - 5)}{4} - \dfrac{2(W - 2)}{6} = W + 1$

41. $-3(f + 1) + 2 = 2(f + 7)$

42. $3 + 2(3T - 1) = 7(2T + 1) + T$

43. $-\dfrac{2}{5}(L_0 - 1) = \dfrac{1}{2}(2L_0 + 1)$

44. $-2(6y + 5) = -3(8y - 1)$

4.3 Inequalities

$x + 6 = 10$ is an open sentence that expresses equality. An open sentence that does not express equality is called an *inequality*. The open sentence, $x + 6 < 10$, is an example of an inequality since it states that x plus 6 is less than 10. To solve the equation, $x + 6 = 10$, 6 is subtracted from both members of the equation to obtain an equivalent equation, $x = 4$. Similarly, to solve the inequality, $x + 6 < 10$, 6 is subtracted from both members to yield the equivalent inequality:

$$x + 6 - 6 < 10 - 6$$
$$x < 4.$$

Since there are an infinite number of real numbers less than 4, the inequality, $x + 6 < 10$, has an infinite number of solutions. This may be illustrated on the number line as follows:

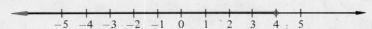

The heavy line and arrow on the number line indicate the real numbers less than 4. The number 4 is circled, but not filled in, to indicate that the number 4 is not a solution to the given inequality.

In solving equations, both members of the equation are often multiplied or divided by the same non-zero number to obtain an equivalent equation. To examine the possibility for inequalities, use the inequality, $4 < 8$. Multiply both members by 2:

$$4 < 8$$
$$2 \cdot 4 < 2 \cdot 8$$
$$8 < 16$$

Divide both members by 2:

$$4 < 8$$
$$\frac{4}{2} < \frac{8}{2}$$
$$2 < 4$$

Multiply both members by -2:

$$4 < 8$$
$$-2 \cdot 4 > -2 \cdot 8$$
$$-8 > -16$$

When multiplying by a -2, the symbol of inequality must be changed from less than to greater than in order for the equivalent inequality to be true. This is referred to as changing the *sense* of the inequality.

Dividing by -2 also changes the sense of the inequality.

$$4 < 8$$
$$\frac{4}{-2} > \frac{8}{-2}$$
$$-2 > -4$$

Rule 1: The same polynomial added to, or subtracted from, both members of an inequality produces an equivalent inequality.

Rule 2: Multiplying or dividing both members of an inequality by the same positive polynomial produces an equivalent inequality.

Rule 3: Multiplying or dividing both members of an inequality by the same negative polynomial produces an equivalent inequality if the sense of the inequality is changed.

Example 1: Solve the inequality, $x - 7 > -5$, and graph the solution on the real number line.

$$x - 7 > -5$$
$$x - 7 + 7 > -5 + 7 \quad \textbf{Rule 1}$$
$$x > 2$$

Example 2: Solve the inequality, $\frac{1}{6}x \geq -1$, and graph the solution on the real number line.

$$\frac{1}{6}x \geq -1$$

(The symbol $\geq$ is read "greater than or is equal to.")

$$6 \cdot \frac{1}{6}x \geq 6 \cdot (-1) \quad \textbf{Rule 2}$$

$$x \geq -6$$

(The circle at -6 is filled in to indicate that -6 is a solution of the given equation.)

Example 3: Solve the inequality, $-2x > 9$, and graph the solution on the real number line.

$$-2x > 9$$
$$\frac{-2x}{-2} < \frac{9}{-2} \quad \textbf{Rule 3}$$
$$x < -\frac{9}{2}$$

Example 4: Solve the inequality, $2x + 4 \leq 3x + 6$, and graph the solution on the real number line.

$$2x + 4 \leq 3x + 6$$
$$-3x + 2x + 4 \leq -3x + 3x + 6 \quad \textbf{Rule 1}$$
$$-x + 4 \leq 6$$
$$-x + 4 - 4 \leq 6 - 4 \quad \textbf{Rule 1}$$
$$-x \leq 2$$
$$-1 \cdot -x \geq -1 \cdot 2 \quad \textbf{Rule 3}$$
$$x \geq -2$$

Example 5: One object has velocity equal to $(25 + 20t)$. A second object has velocity equal to $(97 - 16t)$. At what time t will the velocity of the first object be smaller than the velocity of the second object?

$$v_1 = 25 + 20t$$
$$v_2 = 97 - 16t$$
$$v_1 < v_2$$
$$25 + 20t < 97 - 16t$$
$$20t + 16t < 97 - 25$$
$$36t < 72$$
$$t < 2$$

When $t < 2$, the velocity of the first object will be smaller than the velocity of the second object.

EXERCISES 4.3

Solve the following inequalities and graph the solutions on the real number line:

1. $x + 5 > -1$ $\longleftrightarrow$

2. $x - 6 < -2$ $\longleftrightarrow$

3. $\frac{2}{3} x \leq -6$ $\longleftrightarrow$

4. $11x > 33$ $\longleftrightarrow$

5. $-4x \leq -16$ $\longleftrightarrow$

6. $-x \leq -5$ $\longleftrightarrow$

7. $2x > 0$ $\longleftrightarrow$

8. $-3x < 0$ $\longleftrightarrow$

9. $5x + 2 < 4x - 1$ ←——————————————→

10. $7x - 2 > 4x - 1$ ←——————————————→

11. $9x + 1 > 11x - 5$ ←——————————————→

12. $2(x + 3) \geq 3x + 1$ ←——————————————→

13. $\frac{x}{4} - \frac{1}{2} < \frac{5}{8} + x$ ←——————————————→

14. $3(2x - 7) \leq 5(x - 3)$ ←——————————————→

15. If the velocity of one object is $(14 - 15t)$ feet per second and the velocity of a second object is $(27 - 2t)$ ft/sec, at what time t will the velocity of the second object be greater than the velocity of the first object?

16. The velocity of a moving object is equal to the initial velocity plus the acceleration times the time; that is, $v = v_o + at$. If one object has initial velocity of 45 meters per second with an acceleration of 9.8 meters per second squared and a second object has an initial velocity of 85 meters per second with an acceleration of (-9.8) meters per second squared, for what value of t will the velocity of the second object be smaller than the velocity of the first object?

17. The amount of pressure is equal to the force applied divided by the area. The force applied to an area of 7 square inches is 1 unit less than the force applied to an area of 10 square inches; that is, $a_1 = 10$, $f_1 = x$, $a_2 = 7$, $f_2 = x - 1$. Find the value of f_1 if the pressure of the first one is smaller than the pressure of the second one.

18. Rate equals the distance divided by the time. It takes one vehicle x hours to travel 500 km and a second vehicle 2 more hours than the first vehicle to travel 750 km. If the rate of the first vehicle is greater than the rate of the second vehicle, what is the maximum amount of time required for the first vehicle to travel 500 km?

19. The price of a job minus the estimated job expenses minus the overhead is equal to the profit. What must the price of a job be set at if the job expenses are $32,500 and the overhead is $2900, to ensure a minimum profit of $4225?

20. If the distance from job site A to job site B is 23.5 miles and the distance from job site B to job site C is 13.7 miles, what is the minimum distance to get from job site A to job site C?

4.4 Changing the Subject of Formulas

When basic facts in technology are represented by equations, they are often referred to as *formulas*. For example, it is stated that Fahrenheit temperature is equal to $\frac{9}{5}$ centigrade temperature plus 32; $F = \frac{9}{5}C + 32$.

This equation is referred to as a formula for temperature conversion. The formula has been solved for F in order to convert from centigrade to Fahrenheit measure. It is possible, by using Rule 1 and Rule 2, to solve the formula for C in order to convert from Fahrenheit to centigrade.

Example 1: $F = \frac{9}{5}C + 32$, solve for C.

$$\frac{9}{5}C + 32 = F \qquad \textbf{Symmetric property}$$

$$\frac{9}{5}C + 32 - 32 = F - 32 \qquad \textbf{Rule 1}$$

$$\frac{9}{5}C = F - 32$$

$$\frac{5}{9} \cdot \frac{9}{5}C = \frac{5}{9}(F - 32) \quad \textbf{Rule 2}$$

$$C = \frac{5}{9}(F - 32)$$

Example 2: The gas law used in chemistry states that $PV = RT$; solve for V.

$$PV = RT$$
$$\frac{PV}{P} = \frac{RT}{P} \quad \textbf{Rule 2}$$
$$V = \frac{RT}{P}$$

Example 3: The circumference of a circle is the product of π and the diameter of the circle, $C = \pi d$. If the circumference of a circle is given to be 18.84 cm, determine the diameter.

$$C = \pi d$$
$$18.84 \text{ cm} \doteq 3.14\,d$$
$$\frac{18.84}{3.14} \text{ cm} \doteq d$$
$$6 \text{ cm} \doteq d$$

If several problems are to be worked with the circumference given, it would simplify work to change the subject of the formula before substituting the given value.

Example 4: Find the diameter of each of the following circles, given:

    a. $C = 18.84$ cm
    b. $C = 0.628$ m
    c. $C = 84.78$ cm
    d. $C = 3.2028$ m

In order to simplify the computation, first change the subject of the formula from C to d.

$$C = \pi d$$
$$\frac{C}{\pi} = d \quad \textbf{Rule 2}$$
$$d = \frac{C}{\pi} \quad \textbf{Symmetric property}$$

Substituting the values of the given circumferences:

    a. $d \doteq \dfrac{18.84 \text{ cm}}{3.14} \doteq 6 \text{ cm}$

    a. $d \doteq \dfrac{0.628 \text{ m}}{3.14} \doteq .2 \text{ m}$

    c. $d \doteq \dfrac{84.78 \text{ cm}}{3.14} \doteq 27 \text{ cm}$

    d. $d \doteq \dfrac{3.2028 \text{ m}}{3.14} \doteq 1.02 \text{ m}$

EXERCISES 4.4

Solve each of the following formulas for the indicated variable.

1. $E = IR$, for R

2. $D = rt$, for r

3. $PD = WD$, for P

4. $I = prt$, for t

5. $I = \dfrac{En}{7}$, for E

6. $L = L_0(1 + at)$, for t

7. $Q = P(Q_2 - Q_1)$, for Q_1

8. $V_2 = V_1 + at$, for a

9. $S = \frac{1}{2}gt^2$, for g

10. $p_1v_2 = p_2v_1$, for v_1

11. $1 = \dfrac{yd}{mR}$, for m

12. $A = \frac{1}{2}bh$, for h

13. $V = lwh$, for h

14. $P = 2l + 2w$, for w

15. $M = \dfrac{Lt + g}{t}$, for g

16. $W = \dfrac{2P + H}{R}$, for P

17. $V = \dfrac{V_t + V_0}{2}$, for V_0 **18.** $C = 2\pi r$, for r

Formula	Given:	Solve for, and find:
19. $A = \frac{1}{2} h(a + b)$	$a = 6$ $b = 11$ $A = 68$	h
20. $I = \dfrac{E}{R}$	a. $I = 1.6, R = 12$	E
	b. $I = 1.5, R = 4.6$	E
	c. $I = 2.3, R = 12.4$	E
21. $V = V_0 + at$	$V = 39$ $V_0 = 75$ $a = -9$	t
22. $V = \frac{1}{2} lw(D + d)$	$l = 8.3$ $w = 7.2$ $D = 2.1$ $V = 152.4$	d

Formula	Given:	Solve for, and find:
23. $P_2 = \dfrac{P_1 T_2}{T_1}$	$P_2 = 27$ $T_1 = 250$ $T_2 = 225$	P_1
24. $S = \dfrac{n}{2}(a + l)$	a. $S = 195$ $n = 20$ $a = -8$	l
	b. $S = 80$ $n = 8$ $a = 3$	l
	c. $S = 12$ $n = 90$ $a = \dfrac{1}{5}$	l
25. $k = \dfrac{1}{2} mv^2$	a. $v = 3$ $k = 180$	m
	b. $v = 5$ $k = 100$	m

4.5   Algebraic Representations of Statements

In the preceding sections, statements in algebraic language were given in the form of an equation. The equation "$x + 6 = 8$" told us that if 6 were added to the unknown, the result would be equal to 8.

Many of the problems encountered by the technician will be in stated form. Before they can be solved, it will be necessary to translate the statements into equations. For example, the sum of an unknown number and 29 is 37. This statement is expressed algebraically as "$x + 29 = 37$."

Example 1: A number 5 more than an unknown number is equal to 9. With n as the unknown number, the statement can be expressed in algebraic language as "$n + 5 = 9$."

Example 2: "Three times an unknown number is 8 greater than the unknown" is translated into algebraic language as "$3n = n + 8$."

EXERCISES 4.5

Translate the following statements into algebraic language and solve the equation for the unknown.

1. A number 3 more than twice the unknown is 7.

2. A number one fourth the size of the unknown is 6.

3. Two thirds of the unknown, increased by 4, is 12.

4. Twelve divided by the unknown equals two.

5. The sum of an unknown number and 28 is 54.

6. Eight more than an unknown number is twenty-two.

7. If three times an unknown number is increased by nine, the result is fifteen.

8. If an unknown number is added to 32, the sum is 58.

9. 8 subtracted from an unknown is 10.

10. Three fifths of an unknown number increased by 7 is 13.

11. A number 5 more than 6 times the unknown is 41.

12. An unknown number divided by 10 is 68.

13. The sum of 24 and an unknown number is 67.

14. If an unknown number is added to 13, the sum is 63.

15. If the sum of 8 and an unknown number is divided by 2, the result is 9.

16. One eighth of an unknown decreased by 5 is 2.

17. Four times an unknown number is 4.

18. One-half an unknown number is 6 more than the unknown.

19. An unknown number divided by 8 is 10.

20. A number $\frac{3}{4}$ the size of the unknown is 15.

21. Steam radiation (S) is equal to the sum of one half the area of the glass (G), one twentieth of the wall (W) area, and one two-hundredth of the cubical content (C) of the room. Write the formula for S.

22. The length (L) of a bell-and-spigot cast iron water pipe is equal to the difference of the center-to-center distance (D) and the sum of the length of the elbow (A) and the length of the fitting (B). Write the formula for L.

23. The volume of a hemispherical dome is half of the volume of a sphere of the same radius (r). Write formula for the volume of a dome.

24. The length (L) of the rafters of a roof is equal to the square root of the sum of the squares of the rise (r_v) and the run (r_h). Write the formula for L.

4.6 Word Problems

Common types of word problems encountered by the technical student are stated below. The following steps should be taken in handling such problems.

1. Read carefully.
2. Put into algebraic language.
3. Solve and check.

Example 1: A board 48 cm long is cut into two pieces. One piece of the board is twice as long as the other. What is the length of each piece?

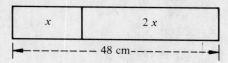

Let x = the length of one piece of the board.
Let $2x$ = the length of the other piece.
$$x + 2x = 48$$
$$3x = 48$$
$$x = 16$$

Therefore, 16 cm is the length of one piece of the board and 2 times 16, or 32 cm, is the length of the other piece.

Check: One piece, 16 cm, is twice as long as the other piece, which is 32 cm in length, and together they equal 48 cm.

Example 2: Find the dimensions of a piece of sheet metal if its length is 14 cm longer than its width, and its perimeter is 100 cm.

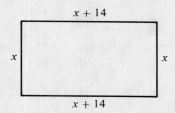

Let x = the width of the piece of sheet metal.
Then $x + 14$ = the length.
$2w + 2l$ = the perimeter of a rectangle.
$$2x + 2(x + 14) = 100$$
$$2x + 2x + 28 = 100$$
$$4x + 28 = 100$$
$$4x = 72$$
$$x = 18 \text{ cm}$$
Therefore, the width = $x = 18$ cm
and the length = $x + 14 = 18 + 14 = 32$

Check:

$$2(18 \text{ cm}) + 2(32 \text{ cm}) = 100 \text{ cm}$$
$$36 \text{ cm} + 64 \text{ cm} = 100 \text{ cm}$$
$$100 \text{ cm} = 100 \text{ cm}$$

Example 3: A weight of 300 g is placed 10 cm from the fulcrum of a lever. How far from the fulcrum must a 150 g weight be placed for the lever to be balanced?

Let x = the distance from the 150 g weight to the fulcrum.

Since the length of one lever arm times the force applied to that side must equal the length of the other lever arm times the force applied to the other side of the fulcrum to maintain balance, we have the statement, $(F_1 \cdot L_1) = (F_2 \cdot L_2)$

$$150 \cdot x = 300 \cdot (10)$$
$$150x = 3000$$
$$x = 20 \text{ cm}$$

Check: $(F_1 \cdot L_1) = (F_2 \cdot L_2)$

$$(150)(20) = (300)(10)$$
$$3000 = 3000$$

Example 4: How many liters of a 20% solution of acid should be added to 40 liters of an 80% solution of acid to obtain a 60% solution?

Let $\quad n =$ number of liters of the 20% solution.
$\quad 40 =$ number of liters of the 80% solution.
$n + 40 =$ number of liters of the 60% solution.

$$20\% \text{ of } n = .20n \qquad = \text{amount of acid in 20\% solution.}$$
$$80\% \text{ of } 40 = .80(40) \qquad = \text{amount of acid in 80\% solution.}$$
$$60\% \text{ of } (n + 40) = .60(n + 40) = \text{amount of acid in 60\% solution.}$$

pure acid in 20% solution	+	pure acid in 80% solution	=	pure acid in 60% solution	
.20n	+	.80(40)	=	.60(n + 40)	
20n	+	80(40)	=	60(n + 40)	**Rule, 2 multiply by 100**
20n	+	3200	=	60n + 2400	**Remove parentheses**
		3200	=	40n + 2400	**Rule 1, subtract 20n**
		800	=	40n	**Rule 1, subtract 2400**
		20	=	n	**Rule 2, divide by 40**

Check:

$$20 \text{ liters of } 20\% \text{ solution of acid} = 4 \text{ liters of acid}$$
$$40 \text{ liters of } 80\% \text{ solution of acid} = 32 \text{ liters of acid}$$
$$40 + 20, \text{ or } 60 \text{ liters of } 60\% \text{ solution of acid} = 36 \text{ liters of acid}$$
$$4 \text{ liters} + 32 \text{ liters} = 36 \text{ liters}$$

EXERCISES 4.6

Solve and check each of the following problems:

1. A piece of wire 36 meters long is cut so that one piece is 14 meters longer than the other. Find the lengths of the two pieces.

2. The perimeter of a rectangle whose width is $\frac{3}{8}$ its length is 176 cm. Find its dimensions.

3. A board 38 cm long is cut into three pieces. If the longest piece is 4 times the length of the shortest piece and the third piece is 2 cm longer than the shortest piece, find the length of each piece of the board.

4. A metal pipe 8 ft long is to be divided into two pieces. Find the length of each piece if one is three-eighths as long as the other.

5. A triangle has a perimeter of 41 cm. One side is 4 cm longer than the shortest side and 3 cm shorter than the longest side. Find the dimensions of the triangle.

6. A board 0.6 m long is cut so that one piece is 4 cm longer than the other. Find the length of each piece.

7. If a weight of 375 mg is 25 cm from the fulcrum of a lever, how far from the fulcrum must a weight of 625 mg be placed to balance the lever?

8. A 60-mg weight is on a lever 10 cm from one side of the fulcrum. Where should a weight of 75 mg be placed to give equilibrium?

9. How long is the side of a square whose perimeter is 28 cm?

10. What is the length of the rafters of a roof that has a rise of 8 feet, a run of 10 feet, and an overhang of 1 foot? (Use the Pythagorean Theorem.)

11. When the length of a side of a square is increased by 4 cm and its width is decreased by 2 cm, the perimeter is $2\frac{1}{2}$ m. Find the original dimensions of the square.

12. Five lamps connected in series have a resistance of 1250 ohms. The first has a resistance of 240 ohms; the second, 260 ohms; and the other three are alike. Find the resistance of each of the like lamps. The total resistance of two or more series-connected resistors is equal to the sum of the individual resistances.

13. How many liters of a 10% salt solution must be added to 60 liters of a 30% salt solution to obtain a 20% solution?

14. How many ounces of nuts at 50 cents per ounce should be mixed with 12 ounces of nuts at 40 cents per ounce to give a mixture worth 42 cents per ounce?

15. How much coffee worth $4.20 a kg should be added to 140 kg of coffee worth $5.70 a kg to obtain a mixture worth $4.80 a kg?

16. How much pure alcohol should be added to 24 ml of a 35% solution to obtain a 50% solution?

17. How many pounds of candy worth $4.55 a lb should be mixed with 40 lb of candy worth $7.70 a lb to obtain a mixture worth $6.30 a lb?

18. How many liters of low-lead gasoline worth 35 cents a liter should be blended with 600 liters of high-test gasoline worth 40 cents a liter to produce a mixture worth 38 cents a liter?

19. How many rolls of wallpaper 18 inches wide and 8 yards long are needed for a room 9 feet high, 12 feet wide, and 14 feet long? (Do not deduct for window and door openings.)

20. How much paint is required for a silo 10 feet in diameter and 25 feet high, if it has a conical roof with a rise of 3 feet and an overhang of 1 foot?

21. The area of a triangle with sides a, b, and c is given by the formula

    $A = \sqrt{S(S - a)(S - b)(S - c)}$

    where $S = \frac{1}{2}(a + b + c)$.

    Using this formula, find the area of a triangle measuring 20 inches, 25 inches, and 7 inches.

22. The safe bearing power (s) in pounds when a drop hammer is used is

    $s = \dfrac{2wh}{s + 1}$

    What is the safe bearing power of a pile that penetrates (s) $\frac{3}{4}$ inch under the fall (w) of a 2000-lb hammer dropping (h) 15 feet?

4.7 Ratio and Proportion

In comparing the area of a piece of sheet metal that is 12 square meters to a piece 15 square meters in area, it may be stated that the ratio of the area of the first piece to the area of the second is 12 to 15. This ratio may be written as $12 \div 15$, $\frac{12}{15}$, or 12:15. When we say that the ratio of the two pieces of sheet metal is $\frac{12}{15}$ or $\frac{4}{5}$, this is to say that the area of one piece of sheet metal is $\frac{4}{5}$ the area of the other. Quantities being compared in ratios should be like quantities, such as 12 square meters to 15 square meters. For example, the area of 12 square meters could not be compared to an area of 15 square centimeters.

The *ratio* of one quantity to another like quantity is the quotient of one quantity divided by the other.

Example 1:   Express each of the following as a ratio in lowest terms:

   a. 3 tons to 7 tons

    $\frac{3}{7}$ or 3:7

   b. 6%

    $\frac{6}{100} = \frac{3}{50}$ or 3:50

Example 2:   The diameters of two pulleys are $4\frac{1}{8}$ centimeters and $6\frac{1}{2}$ centimeters, respectively. What is the ratio of the diameter of the smaller pulley to that of the larger pulley?

$$\frac{4\frac{1}{8}}{6\frac{1}{2}} = \frac{\frac{33}{8}}{\frac{13}{2}} = \frac{33}{52} \text{ or } 33:52$$

Equal ratios are called *proportions*. The ratio $\frac{12}{15}$ is equal to the ratio $\frac{4}{5}$. Therefore, $\frac{12}{15} = \frac{4}{5}$ is a proportion.

The proportion $\frac{a}{b} = \frac{c}{d}$ is read "*a* is to *b* as *c* is to *d*." *a* and *d* are called the *extremes,* and *b* and *c* are called the *means.* In any proportion, *the product of the extremes is equal to the product of the means.* That is, $ad = bc$.

$\frac{a}{b} = \frac{c}{d}$ may be written as $\frac{a}{c} = \frac{b}{d}$, or $\frac{d}{b} = \frac{c}{a}$. Note that in each case, the means or extremes were interchanged but that $ad = bc$.

The proportion $\frac{a}{b} = \frac{c}{d}$ may also be rewritten as $\frac{b}{a} = \frac{d}{c}$. Note that the ratios were inverted, and $ad = bc$.

Example 3:   Find *x* in the proportion, $\frac{x}{4} = \frac{5}{2}$

   $\frac{x}{4} = \frac{5}{2}$

   $2x = 20$ **The product of the extremes is equal to the product of the means.**

   $x = 10$

Example 4:   The relationship between two gears may be expressed by the following proportion:

$$\frac{T_A}{T_B} = \frac{S_B}{S_A}$$

T_A and T_B are the number of teeth in gears *A* and *B*, respectively. S_A and S_B represent the revolutions per minute (*rpm*) of the respective gears. If a gear with 20 teeth rotates at 160 rpm, at how many rpm would a gear with 16 teeth revolve under the same conditions?

$$\frac{T_A}{T_B} = \frac{S_B}{S_A}$$

$$\frac{20}{16} = \frac{S_B}{160}$$

$16 \, S_B = 20(160)$ **The product of the means is equal to the product of the extremes.**

$$S_B = 200$$

Therefore, the gear with 16 teeth would revolve at 200 rpm.

Example 5: 21 is what percent of 35?

21 is x% of 35

The ratio of 21 to 35 is equal to the ratio of x to 100 since percent means "as compared to 100."

$$\frac{21}{35} = \frac{x}{100}$$

$$35x = 2100$$

$$x = \frac{2100}{35}$$

$$x = 60$$

Therefore, 21 is 60% of 35.

In a right triangle, if one acute angle is equal to an acute angle of another right triangle, the triangles are *similar*. The corresponding sides of similar triangles are proportional.

Example 6: Given the similar triangles:

Find: x

$$\frac{3}{6} = \frac{5}{x}$$

$$3x = 30$$

$$x = 10$$

EXERCISES 4.7

Express each of the following as a ratio in lowest terms:

1. 6 cm to 7 cm

2. 12 g to 16 g

3. 3 m to 10 m

4. 2 l to 4l

5. 7%

6. 13%

7. 3 m to 450 cm

8. 2 m to 4000 mm

9. $\frac{1}{2}$ g to 48 kg

10. $3\frac{1}{2}$ m to $4\frac{1}{5}$ m

11. $15\frac{1}{2}$ km to 6 km

12. 5 kg to 40 g

13. The pitch of a gable roof is the ratio of the rise of the rafter to the span of the rafter. The span of a rafter is twice the run of the rafter. Find the pitch of the roof if the rise is 12 feet and the run is 13 feet.

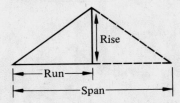

14. Find the pitch of a roof if the rise is 200 cm and the span is 7 m.

15. A large gear has 90 teeth, while a smaller gear has only 35 teeth. What is the ratio of the number of teeth on the smaller gear to the number on the larger gear?

16. What is the ratio of the speed of one motor turning at 1500 revolutions per minute (rpm) to the speed of another motor turning at 4000 revolutions per minute (rpm)?

Given the following proportions, solve for the given unknowns:

17. $\dfrac{E}{2} = \dfrac{15}{3}$

18. $\dfrac{4}{3} = \dfrac{T_1}{6}$

19. $\dfrac{11}{5} = \dfrac{f}{10}$

20. $\dfrac{p}{7} = \dfrac{15}{4}$

21. $\dfrac{7}{m} = \dfrac{1}{11}$

22. $\dfrac{14}{9} = \dfrac{7}{R}$

23. $\dfrac{3.2}{.2} = \dfrac{t}{1.1}$

24. $\dfrac{.84}{z} = \dfrac{.6}{.5}$

Translate the following problems into proportions and solve:

25. x is to 4 as 6 is to 11.

26. 9 is to 11 as e is to 6.

27. If a 4-inch wall 50 square feet in area requires 300 bricks, how many bricks are needed to build a 4-inch wall 500 square feet in area?

28. If a compressor with a pulley of diameter $4\frac{1}{4}$ inches runs at 500 rpm, at how many rpm does the motor run if the diameter of the motor is 3 inches?

29. A square bar of steel 2.5 m long weighs 18 kg. Find the weight of the same type of bar measuring 3.8 m long.

30. If a wire 3000 feet long has a resistance of 0.41 ohm, find the resistance of the same type of wire if it measures 400 feet.

31. Find 2% of 28.

32. What percent of 240 is 21.6?

33. If 15.26 is 7% of a number, what is the number?

34. A pump discharging 5 gallons of water per minute fills a tank in 25 hours. How long will it take a pump discharging 25 gallons per minute to fill the tank?

35. A full tank holds 10,000 liters. How many liters are in the tank if it is two-thirds full?

36. Find the rise of a rafter of pitch $\frac{3}{8}$ if the run is 14 feet.

37. When gas is heated, it expands in volume; it contracts in volume when cooled. This is expressed as the proportion, $\frac{V_1}{V_2} = \frac{T_1}{T_2}$. The volume is V, and the temperature of the gas on the absolute scale of temperature is T. If the volume of a gas is 300 cm^3 when its absolute temperature is 270°, what is its new volume when it is heated to 370°?

38. Inspectors at a factory check 3.5% of the daily production. If the factory produced 2500 articles, how many of them were inspected?

39. A piece of wood is said to taper when there is a gradual decrease or increase in its diameter. A piece tapers 0.75 inch in 3 inches of length. What length must the piece be to taper 1 inch?

40. Inspectors at a factory rejected 64 items, which represented $1\frac{1}{2}$% of the daily production. How many items were produced that day?

41. If 5 lines of a manuscript consist of 48 words, how many words are there on a page of 28 lines?

42. A general contractor estimated a building to cost $85,700. If the electrical work costs $6500, what percent of the total cost is the electrical work?

43. If a gear with 30 teeth rotates at 150 rpm, at how many rpm would a gear with 15 teeth revolve under the same conditions?

44. Given:

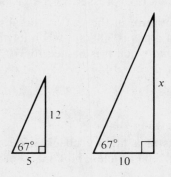

Find: x

45. On a residential building, 8% of the total cost is carpentry. If the total cost of the building is $47,000, what is the cost for carpentry labor?

46. If the scale of a drawing is $\frac{3}{8}''$:$1'$, what is the ratio of the actual object to the drawing? Express your answer as a fraction in lowest terms with whole numbers in the numerator and the denominator.

4.8   Variation

Direct variation means that as one quantity increases, the other quantity increases, or as one quantity decreases, the other quantity decreases. The statement "$y = kx$" is read "y varies directly as x." k is called the *proportionality constant* or the *constant of variation*.

The formula for finding the circumference of a circle, $C = \pi d$, is an example of direct variation. π is the constant of variation. As the diameter of the circle increases, the circumference of the circle increases.

Example 1: If y varies directly as x, and $y = 10$ when $x = 2$, find the proportionality constant.

$$y = kx$$

$k = \dfrac{y}{x}$ **Divide both members of the equation by x to solve for k.**

$$k = \dfrac{10}{2}$$

$$k = 5$$

Example 2: If y varies directly as x, and $y = 20$ when $x = 5$, find y when $x = 6$.

$k = \dfrac{y}{x}$ $\qquad y = kx$

$k = \dfrac{20}{5}$ $\qquad y = 4(6)$

$k = 4$ $\qquad y = 24$

Inverse variation means that as one quantity increases, the other quantity decreases, or as one quantity decreases, the other quantity increases. The statement $y = \dfrac{k}{x}$ is read "y varies inversely as x," or "y is inversely proportional to x.

Boyle's law states that the volume of a gas sample is inversely proportional to the pressure exerted upon it at constant temperature, $p = \dfrac{k}{V}$, or $V = \dfrac{k}{p}$. Increasing the pressure on a gas reduces its volume, and decreasing the pressure increases its volume.

Example 3: If y varies inversely as x, and $y = 8$ when $x = 2$, find the constant of variation.

$$y = \dfrac{k}{x}$$

$k = yx$ **Multiply both members of the equation by x to solve for k.**

$k = 8 \cdot 2$

$k = 16$

Example 4: If y varies inversely as x, and $y = 16$ when $x = 4$, find the value of y when $x = 2$.

$k = yx$ $\qquad y = \dfrac{k}{x}$

$k = 16 \cdot 4$ $\qquad y = \dfrac{64}{2}$

$k = 64$ $\qquad y = 32$

Joint variation means that one quantity varies directly as the product of two or more quantities. $y = kxz$ is read "y varies jointly as x and z."

The formula for finding the area of a triangle, $A = \dfrac{1}{2} bh$, is an example of joint variation, where $\dfrac{1}{2}$ is the proportionality constant. The area of the triangle depends on the product of the altitude and the base of the triangle.

Example 5: If y varies as x and z, and $y = 48$ when $x = 6$ and $z = 2$, find the proportionality constant.

$$y = kxz$$

$k = \dfrac{y}{xz}$ $\qquad$ **Divide both members of the equation by xz to solve for k.**

$k = \dfrac{48}{6 \cdot 2} = \dfrac{48}{12}$

$k = 4$

Example 6: If y varies as x and z, and $y = 30$ when $x = 5$ and $z = 3$, find y when $x = 7$ and $z = 4$.

$$k = \frac{y}{xz} \qquad\qquad y = kxz$$

$$k = \frac{30}{5 \cdot 3} = \frac{30}{15} \qquad y = 2 \cdot 7 \cdot 4$$

$$k = 2 \qquad\qquad y = 56$$

EXERCISES 4.8

Write the given statements as equations containing a constant of proportionality:

1. V varies directly as P.

2. T varies inversely as P.

3. I varies directly as the square of R.

4. y varies inversely as the cube of x.

5. A varies directly as l and w.

6. A varies directly as the product of b and c and inversely as the square of d.

7. The volume of a circular cone varies as the product of its height and the square of the radius of its base.

8. Newton's law of gravitation states that the force, F, of attraction between two masses, m_1 and m_2, varies directly as the product of the masses, and inversely as the square of the distance, r, between the masses.

9. Force varies as the product of mass and acceleration.

10. Power varies directly as work and inversely as time.

Find the proportionality constant, given:

11. y varies directly as x, and $y = 25$ when $x = 5$.

12. y varies directly as x, and $y = 3$ when $x = 9$.

13. y varies directly as x, and $y = 4$ when $x = 23$.

14. y varies inversely as x, and $y = 4$ when $x = 11$.

15. a varies inversely as b, and $a = 12$ when $b = 3$.

16. R varies inversely as S, and $R = 2.5$ when $S = 4$.

17. y varies as x and z, and $y = 40$ when $x = 2$ and $z = 10$.

18. m varies as n and p, and $m = 8$ when $n = 4$ and $p = 3$.

19. W varies as X and Y, and $W = 6$ when $x = 4.8$ and $Y = 10$.

Find the value of the indicated variable in each of the following:

20. If y varies directly as x, and $y = 60$ when $x = 15$, find y when $x = \frac{1}{2}$.

21. If W varies directly as L_0, and $W = 44$ when $L_0 = 22$, find W when $L_0 = 60$.

22. If y varies inversely as x, and $y = 13$ when $x = 3$, find the value of y when $x = 117$.

23. If R_1 varies inversely as R_2, and $R_1 = 64$ when $R_2 = 4$, find R_1 when R_2 has a value of 12.

24. If y varies as x and z, and $y = 90$ when $x = 4$ and $z = 15$, find y when $x = 6$ and $z = 10$.

25. If F varies as H and K, and $F = 200$ when $H = 50$ and $K = \frac{1}{2}$, find F when $H = 40$ and $K = \frac{1}{10}$.

26. The number of machines built varies directly with the labor hours. If 2400 labor hours are required to build 12 machines, how many labor hours are necessary to build 200 machines?

27. The revolutions per minute vary inversely with the diameter of a pulley. If the driver pulley is 100 mm in diameter, spinning at 1300 rpm, what must the diameter of the driven pulley be to spin at 3000 rpm?

28. In gable roofs having the same rise, the pitch varies inversely with the span. What is the pitch of a roof with span 30 ft if a roof of equal rise and a span of 40 ft has a pitch of $\frac{1}{3}$?

29. If the elevation and the run of the stairway are the same, the number of steps required varies inversely with the size of the riser. If 25 steps are required with a rise of 6.8 inches, how many steps are required with a rise of 7.3 inches?

30. A business found that the number of sales varies inversely with the price of the item. If a price of $10.95 each produces yearly sales of 13,500 items, lowering the price to $9.19 would produce what volume of sales?

31. The heat loss varies inversely with the amount of insulation in the walls. If $\frac{3}{4}$ in. of insulation on a brick wall with gypsum board produces a U-value of 0.19, which is an indicator of heat loss, what will the U-value be when $1\frac{5}{8}$ in. of insulation is used on the same wall?

32. The circumference of a circle varies directly with the diameter of a circle. If a circle 8 feet in diameter has a circumference of 25 feet, what will the circumference be for a circle 3.5 feet in diameter?

33. For printers, the margin area varies inversely with the amount of printed area. On a standard sheet, if the margin area is 25.2 sq in. and the printed area is 68.3 sq. in., what will be the allowable printed area if the margin area is 32.7 sq. in.?

34. The diameter of a pulley varies inversely with the number of revolutions per minute. If a crankshaft $4\frac{3}{8}$ in. in diameter turns at 1125 rpm, find the rpm of the generator pulley if its diameter is $4\frac{1}{8}$ in.

35. The size of a drawing varies directly with the actual measurements. If a scale of 1:3 is used, to what length must a line be drawn if the actual measurement is 96 in.?

36. The number of feet of roadway lighted by an automobile varies directly with the height of the headlights. If headlights placed at 20 inches above the ground cause 22 feet of lighting, at what height must the headlights be placed to have 34 feet of highway lighted?

37. The time required for a crane to lift an object is directly related to the weight of the object and the distance it must be lifted. If it takes the crane 2.5 minutes to lift 4000 pounds through 60 feet, how long will it take the crane to lift 6000 pounds through 55 feet?

38. Force varies directly as the mass and indirectly as the square of the distance. If a force of 25 pounds is required to move a mass of 3.8 slugs over 4 feet, find the force required to move a mass of 14.1 slugs over 3.5 feet.

Express as equations and solve:

39. If a varies jointly as b and the square of c, and inversely as d; and $a = 48$ when $b = 4$, $c = 2$, and $d = 6$, find the value of a when $b = 3$, $c = \frac{1}{3}$, and $d = 2$.

40. The distance a freely falling body travels varies directly as the square of time. If a body falls 257.6 meters in 4 seconds, how far will it fall in 6 seconds?

41. Distance varies directly as rate and time. If a train travels 300 km in 5 hours, how far will it travel in 7 hours?

42. The horsepower of a gasoline engine varies as the diameter squared of each cylinder and the number of cylinders. If 6 cylinders, each having a diameter of 6 cm, provide 60 hp., what horsepower is provided by 8 cylinders, each with a diameter of 5 cm?

43. Resistance of copper wire (in ohms) varies directly as the length of the wire, in meters, and inversely as the square of the diameter of the wire in mils. If 200 meters of copper wire with a diameter of 5 mils has a resistance of 856 ohms, what is the resistance of 150 meters of copper wire with a diameter of 6 mils?

Unit 4 Self-Evaluation

Solve Equations 1 through 5 for the given variable:

1. $R_2 - 9 = -3$

2. $\frac{3}{7} m = -9$

3. $\frac{x}{3} - 4 = 2$

4. $-2(4p + 1) = 3(4 - 2p)$

5. $\frac{t}{3} - \frac{2t + 1}{7} = \frac{t + 4}{42}$

6. Solve the formula for K_{ts}:
$$K = q(K_{ts} - 1) + 1$$

7. Translate the following statement into algebraic language and solve the equation for the unknown: "If two thirds of an unknown is decreased by twelve, the result is twelve."

8. The perimeter of a triangle is 71 cm. The longest side is 5 times the shortest side. The third side is 8 cm longer than twice the shortest side. Find the length of each side.

9. Express as a ratio: 11 meters to 15 meters.

10. Solve the proportion for R: $\frac{3}{8} = \frac{5}{R}$.

11. Two boys weigh a total of 110 pounds. They balance on a seesaw when one is 5 feet from the fulcrum and the other one is 4 feet from the fulcrum. How much does each boy weight?

12. A scale on a map states that $\frac{3}{8}$ of a km represents 3 cm. If the distance measured between two towns is $2\frac{1}{2}$ cm, how far apart are the two towns?

13. If y varies directly as x, and $y = 27$ when $x = 9$, find y when $x = \frac{1}{2}$.

14. The weight a horizontal beam can bear varies inversely as the length between the supports. If a 20-m beam can bear 2400 kg, how many kg can a 16-m beam bear?

15. Solve the given formula for a: $A = \frac{1}{2}h(a + b)$

16. Change the subject of the given formula to a: $V = V_0 + at$

Solve the inequalities 17 through 20 and graph the solutions on the real number line:

17. $x - 4 < -5$

18. $-16x > -32$

19. $14x - 7 < 2(8x + 1)$

20. The velocity of a certain object is given by $v = 50 + 32t$. For what values of t is the object descending; that is, when is the velocity less than zero?

Graphs

Unit 5 Objectives
1. To express given data using circle graphs, bar graphs, and line graphs.
2. To plot points in the rectangular coordinate system.
3. To find the distance between two points.
4. To graph first degree equations in two variables.
5. To find the equation of a straight line.

unit 5

Graphs are an indispensable aid in the interpretation of technical relationships and principles. Relationships existing between two or more quantities are more evident when presented in graphic form than when stated in a table.

Consider the table presenting the sound absorption coefficient of different materials:

Open window	1.00	Acoustic plaster	.25
Acousti-celotex	.82	Floor	.08
Felt	.70	Chalkboard	.06
Draperies	.50	Linoleum	.03
Carpet	.40	Marble	.01

When this table is presented in graphic form, the materials having similar sound absorption coefficients are easier to identify.

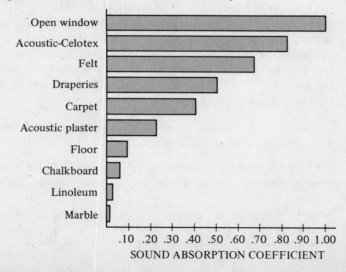

Figure 5.1

SOUND ABSORPTION COEFFICIENT

The formula $s = \frac{1}{2} gt^2$ gives the distance a freely falling body has traveled in t seconds, when the fall starts from rest and g is 9.75 meters per second squared. This physics principle can be better understood when presented in graphical form, as seen in Figure 5.2.

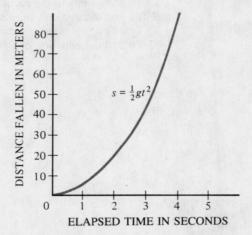

Figure 5.2   ELAPSED TIME IN SECONDS

Several conclusions can be drawn from the graph. The longer the time the body has been falling, the longer the distance it has traveled. In any given second, the body has traveled a greater distance than in the preceding second. These observations might not have been so obvious from the formula alone.

Economists use graphs to represent data collected in surveys. Scientists use graphs to represent data collected in experiments. Technicians and engineers use graphs to represent collected data as well as to find solutions to certain problems.

5.1   Graphical Methods for Collected Data

A *graph* is a pictorial representation of the relationship between collected data. It summarizes the data in such a way that, by a quick glance at the graph, a meaningful understanding of the relationships among the data can be achieved.

Three types of graphs will be studied in this section: the *circle graph,* the *bar graph,* and the *line graph.*

The *circle graph* is used to show the relationship between a whole and its parts. Consider a survey to find out how a factory worker spends his annual salary. The factory worker spends his salary for food, shelter, clothing, transportation, medical bills, recreation, and miscellaneous. This survey is best represented by a circle graph because the whole circle can represent the annual salary, which can then be divided into different parts according to what proportion is spent in each category.

Rule 1:   To draw a circle graph:

Step 1:   Compute the percentage that each part is of the whole.

Step 2:   Draw a circle and divide it into sectors proportionally according to the percentages obtained in Step 1.

Step 3:   Label the graph by writing in each sector the category and the corresponding percentage.

Example 1: A factory worker has an annual salary of $20,000. He spends $5000 on shelter, $4000 on food, $2000 on clothing, $2000 on transportation, $1000 on medical bills, $2000 on recreation and $4000 on miscellaneous. Draw a circle graph.

Step 1:

Category	Amount	Percentage
Shelter	$ 5000	25%
Food	4000	20%
Clothing	2000	10%
Transportation	2000	10%
Medical Bills	1000	5%
Recreation	2000	10%
Miscellaneous	4000	20%
Total	$20,000	100%

Steps 2 and 3: Use a protractor to subdivide the circle into the following angles:

$$25\% \text{ of } 360° = 90°$$
$$20\% \text{ of } 360° = 78°$$
$$10\% \text{ of } 360° = 36°$$
$$5\% \text{ of } 360° = 18°$$

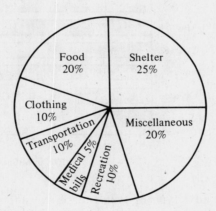

Figure 5.3

Another common graphing technique is the *bar graph*. A bar graph consists of parallel bars, either horizontal or vertical, each representing a fixed category. The length of each bar is drawn proportionally to the amount in each category.

Consider the problem of representing graphically the areas of the seven regions of the world. A bar graph would lend itself to this problem since each bar can represent a different continent and, by comparing the lengths of the bars, the relation among their areas can be perceived.

Rule 2: To draw a bar graph:

Step 1: Determine a suitable scale so that all the different amounts in the various categories will be included.

Step 2: Determine the length of each bar in the scale so that the length of the bar is proportional to the given amount.

Step 3: Construct the bars.

Step 4: Label the graph.

Example 2:   Draw a bar graph for the following data representing the areas in hectares (represented in millions) of seven regions of the world.

Region	Area (in millions of sq. hectares)
Africa	3.3
Asia	2.7
Europe	0.5
North America	2.4
Oceania	0.9
South America	1.8
U.S.S.R.	2.0

Step 1:   To determine a suitable scale it is necessary to notice the smallest (.5) and the largest (3.3) values. An appropriate scale would be in millions of square hectares ranging from .3 million to 3.3 millions.

Step 2:   To determine the length of each bar in terms of the scale, divide 3.3 by .3. On the horizontal line mark off equal segments.

Steps 3 and 4:

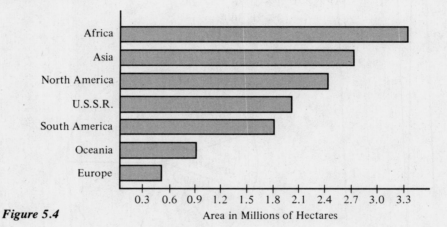

Figure 5.4
Area in Millions of Hectares

The next type of graph to be considered is the *line graph*. It is used to compare the changes in one quantity with the changes in another quantity. Often, one of the quantities is continuous, such as a time interval. A classic example of a problem where a line graph is used is in representing the population of the United States from 1900 to 1970. In this case the years are the continuous variable.

Rule 3:   To draw a line graph:

Step 1:   Determine a suitable scale for the two variables, one represented on the horizontal axis and the other on the vertical axis.

Step 2:   Draw and label both axes to scale.

Step 3:   Plot the points corresponding to the given data.

Step 4:   Connect the points of Step 3 with straight lines.

Example 3:   Draw a line graph for the following data:

Year	Population (in millions)	Year	Population (in millions)
1900	76	1940	131
1910	92	1950	151
1920	105	1960	179
1930	122	1970	203

Step 1: On the horizontal scale, place the years ranging from 1900 to 1970 at intervals of 10. On the vertical scale, place the population in millions ranging from 60 to 220 in intervals of 20.

Steps 2, 3, and 4:

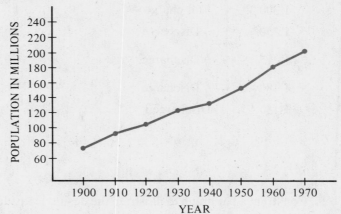

Figure 5.5

EXERCISES 5.1

1. Construct a circle graph to illustrate land use in a certain town as given by the following table:

Land use	Percent
Residential	53%
Industrial	13%
Commercial	8%
Parks	9%
Streets	10%
Vacant	7%

2. Construct a circle graph to illustrate the field of study of students within the technical division of a college as given by the following table:

Field of study	Number
Drafting	195
Air-conditioning	286
Electronics	175
Building construction	100
Automotive	85
Aeronautics	52

3. Construct a circle graph to illustrate the composition of a metal alloy given to be:

Metal	Percent
Copper	45%
Aluminum	27%
Zinc	1%
Tin	3%
Lead	24%

4. Construct a bar graph to illustrate the readings of the hydrometer showing the condition of the battery in accordance with the following table:

Reading	Condition
1.300	Full charge
1.250	$\frac{1}{4}$ Discharge
1.215	$\frac{1}{2}$ Discharge
1.180	$\frac{3}{4}$ Discharge
1.150	Discharged

5. Construct a bar graph to illustrate the desirable inside temperature in different media given by the following table:

Type of room or building	Temperature (°C)
Steam bath	43
Hospital operating room	29
Bathroom	27
Paint shop	27
Hospital	22
Public building	21
Residence	21
School	21
Store	18
Factory	18
Machine shop	17
Gymnasium	14
Boiler shop	12

6. Construct a bar graph to illustrate the stopping distances for automobiles at various speeds given by the following table:

Speed (kph)	Meters traveled before stopping
18	5
44	13
84	25
135	41
197	60

7. Construct a line graph to illustrate the relationship between the thread diameter and the number of threads to 1 centimeter in a tap drill as given by the following table:

Thread diameter (centimeters)	Threads per centimeters
$\frac{1}{4}$	10
$\frac{1}{2}$	6.5
$\frac{3}{4}$	5
1	4
$1\frac{1}{4}$	3.5
$1\frac{1}{2}$	3
$1\frac{3}{4}$	2.5
2	2.2

8. Construct a line graph to illustrate temperature readings at different times of the day as given by the following table:

Time	Temperature
7 A.M.	21°C
9 A.M.	25°C
11 A.M.	26°C
1 P.M.	27°C
3 P.M.	28°C
5 P.M.	26°C
7 P.M.	22°C

9. Construct a line graph to illustrate the relationship between the power and the current when the resistance is 3 ohms, as given by the following table:

Current (amperes)	Power (watts)
−5	75
−4	48
−3	27
−2	12
−1	3
0	0
1	3
2	12
3	27
4	48
5	75

10. Construct a line graph to illustrate the relationship between the number of BTU's required to produce a certain temperature.

Temperature in Degrees Farenheit	BTU's
50	1000
80	1500
105	2000
120	2500
125	3000
130	4000

5.2   Rectangular Coordinate System

In all technical areas, formulas relating two variables occur frequently. For example:

$E = 2W$, Hooke's Law, shows the relationship between the weight applied to a spring and the elongation of the spring.

$C = \pi d$ shows the relationship of the diameter to the circumference of a circle.

$p = \dfrac{F}{A}$ shows the relationship of force to the pressure, given a constant area.

$i = \dfrac{V}{R}$, Ohm's Law, shows the relationship between the potential difference across the ends of a conductor and the following current, given the resistance of the conductor.

These formulas can be more fully understood when represented by means of a graph.

Consider a special case of Hooke's Law, $E = 2W$, relating weight applied to a spring to the elongation of the spring. When no weight is applied, there is no elongation of the spring, because $2 \cdot 0 = 0$. A weight of 1 kg causes an elongation of 2 cm because $2 \cdot 1 = 2$. A weight of 2 kg causes an elongation of 4 cm because $2 \cdot 2 = 4$. A weight of 3 kg causes an elongation of 6 cm because $2 \cdot 3 = 6$. These relationships can be expressed as *ordered pairs* of numbers as follows:

$$(0,0), (1,2), (2,4), (3,6)$$

The first number in the ordered pair is the weight in kg applied to the spring and the second number is the elongation in cm caused by weight expressed by the first number.

A set of ordered pairs is called a *relation*. The set of left members of the ordered pairs is called the *domain* of the relation, and the set of right members is called the *range* of the relation.

For the relation, {(0,0), (1,2), (2,4), (3,6)},
the domain is {0,1,2,3} and
the range is {0,2,4,6}.

A *function* is a relation such that for every element of the domain there is one and only one element of the range.

{(0,0), (1,2), (2,4), (3,6)} is a function.

The relation, {(0,1), (3,2), (0,4)}, is not a function since 0 is associated with two elements in the range, 1 and 4.

In order to graph the formula, $E = 2W$, it is necessary to associate each ordered pair with a point in the plane. A *rectangular coordinate system,* or *Cartesian coordinate system,* is obtained by taking two perpendicular lines, called the *axes*. The point of intersection of the two lines is called the *origin*. One number line is horizontal and is referred to as the *horizontal axis*. On the horizontal axis, points to the right of the origin are positive and points to the left of the origin are negative. The other number line is vertical and is referred to as the *vertical axis*. On the vertical axis, points above the origin are positive and points below the origin are negative (see Figure 5.6).

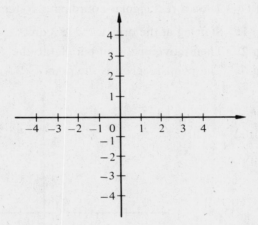

Figure 5.6

Any point in the plane can be designated by an ordered pair. The left member of the ordered pair is called the *abscissa,* It represents the distance and the direction along the horizontal axis. The right member of the ordered pair is called the *ordinate*. It represents the distance and the direction along the vertical axis. The abscissa and ordinate together are called the rectangular coordinates of a point, P, in the plane. For example, (2,4) is the point 2 units away from the vertical axis in the positive direction, and 4 units away from the horizontal axis in the positive direction. The point is represented by a dot (·) on a rectangular coordinate system in the appropriate place (see Figure 5.7).

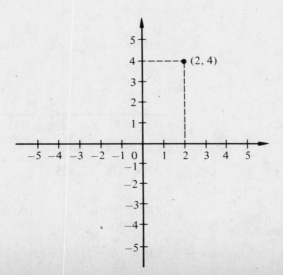

Figure 5.7

Rule 1:   To plot a point on a rectangular coordinate system when the point is given as an ordered pair:

Step 1:   Starting at the origin, move along the horizontal axis the number of units to the right or to the left indicated by the value and sign of the first coordinate of the ordered pair.

Step 2:   From the point reached in Step 1, move up or down, parallel to the vertical axis, the number of units indicated by the sign and value of the second coordinate of the ordered pair.

Step 3:   Mark the point reached in Step 2 by a dot and label it by writing the ordered pair beside the dot.

Example 1:   Plot $(-3, 1)$ on a rectangular coordinate system.

Step 1:   Starting at the origin, move 3 units to the left along the horizontal axis.

Step 2:   Then move up 1 unit parallel to the vertical axis.

Step 3:   The point reached is the point $(-3, 1)$ (see Figure 5.8).

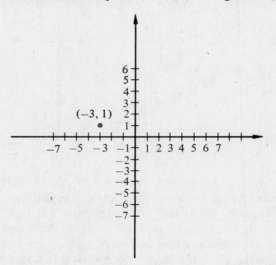

Figure 5.8

Example 2:   Plot the following points on a rectangular coordinate system: $\{(-5, 2), (0, 1), (4, -4), (-6, -6), (-2, 0)\}$. State the domain and range of the relation. Is the relation a function?

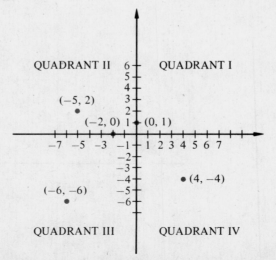

Figure 5.9

The domain is $\{-5, 0, 4, -6, -2\}$. The range is $\{1, -4, -6, 0, 2\}$. The relation is a function.

The horizontal and vertical axes divide the plane into four regions, or *quadrants*. In Quadrant I, both the first and second coordinates are positive; in Quadrant II, the first coordinate is negative and the second coordinate is positive; in Quadrant III, both the first and second coordinates are negative; and in Quadrant IV, the first coordinate is positive and the second coordinate is negative.

Example 3: In which quadrants are the following points: $(-2,3)$ $(8,-1)$, $(-5,-7)$?

$(-2,3)$ is in Quadrant II, because the first coordinate is negative and the second coordinate is positive.
$(8,-1)$ is in Quadrant IV because the first coordinate is positive and the second coordinate is negative.
$(-5,-7)$ is in Quadrant III because both coordinates are negative.

Example 4: Plot points $(-4,-3)$, $(5,-2)$, and $(-2,2)$. Connect them in order using line segments and describe the resulting geometric figure.

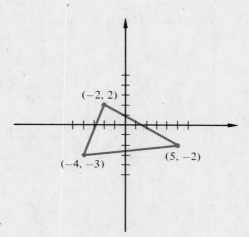

Figure 5.10

The resulting geometric figure is a triangle.

Example 5: A formula for changing degrees Centigrade to degrees Farenheit is given by

$$F = \frac{9}{5}C + 32$$

Determine a set of ordered pairs where the first coordinate represents degrees Centigrade and the second coordinate represents degrees Farenheit, for the following degrees Centigrade: $C = 0, 25, 40, 60, 100$. Graph the ordered pairs on a rectangular coordinate system.

When $C = \quad 0, F = \frac{9}{5}(0) \quad + 32 = \quad 0 + 32 = \quad 32 \rightarrow (0,32)$

When $C = \quad 25, F = \frac{9}{5}(25) \quad + 32 = \quad 45 + 32 = \quad 77 \rightarrow (25,77)$

When $C = \quad 40, F = \frac{9}{5}(40) \quad + 32 = \quad 72 + 32 = 104 \rightarrow (40,104)$

When $C = \quad 60, F = \frac{9}{5}(60) \quad + 32 = 108 + 32 = 140 \rightarrow (60,140)$

When $C = 100, F = \frac{9}{5}(100) + 32 = 180 + 32 = 212 \rightarrow (100,212)$

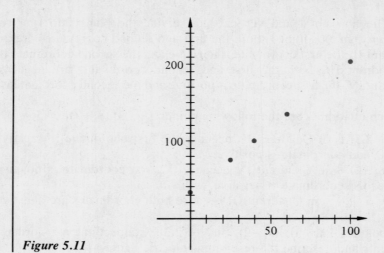

Figure 5.11

The Pythagorean Theorem is used to find the distance between two points in the plane.

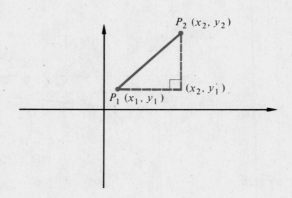

Figure 5.12

The distance, $\overline{P_1P_2}$, is the hypotenuse of a right triangle.
The legs are $x_2 - x_1$, and $y_2 - y_1$.
Therefore, $\overline{P_1P_2}^2 = d^2 = (x_2 - x_1)^2 + (y_2 - y_1)^2$.

$$d = \sqrt{(x_2 - x_1)^2 + (y_2 - y_1)^2}$$

Rule 2:   To find the distance between two points in the number plane:

Step 1:   Find the difference of the first coordinates and square it.

Step 2:   Find the difference of the second coordinates and square it.

Step 3:   Add the results of Steps 1 and 2.

Step 4:   Take the square root of the result of Step 3.

Example 6:   Find the distance between $(-3,2)$ and $(4,-1)$.

Step 1: $[4 - (-3)]^2 = (4 + 3)^2 = 7^2 = 49$

Step 2: $(-1 - 2)^2 = (-3)^2 = 9$

Step 3: $49 + 9 = 58$

Step 4: $\sqrt{58} \doteq 7.62$

EXERCISES 5.2

In Exercises 1 through 9, plot the following points on a rectangular coordinate system:

1. $(-1,3)$
2. $(0,5)$
3. $(-4,-4)$
4. $(6,2)$
5. $(3,-3)$
6. $(-1,0)$
7. $(-5,-2)$
8. $(0.5,-2.5)$
9. $\left(-\frac{1}{4}, -\frac{1}{4}\right)$

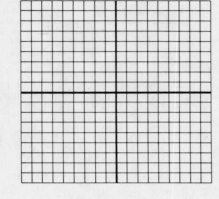

In Exercises 10 through 17, draw each of the geometric figures, given the coordinates of its vertices, and tell which figure it is:

10. $(-4,0), (0,4), (4,0)$

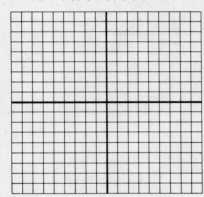

11. 11. $(-4,-1), (0,-4), (3,-1), (0,2)$

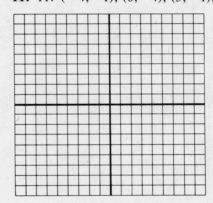

12. $(5,3), (5,1), (-2,-4), (-2,-1)$

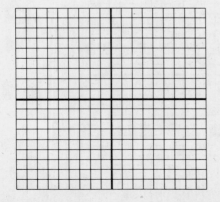

13. $(-4,1), (-5,-1), (2,-5), (3,-3)$

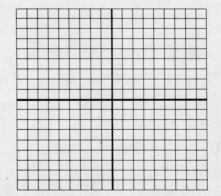

14. $(-3, -2)$, $(-2, 2)$, $(4, 2)$

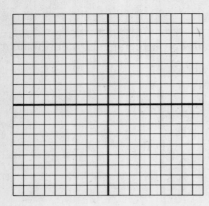

15. $(0, 4)$, $(-2, 0)$, $(0, -6)$, $(2, 0)$

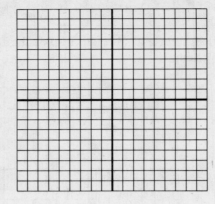

16. $(-6, -3)$, $(6, -3)$, $(-1, 2)$, $(-4, 2)$

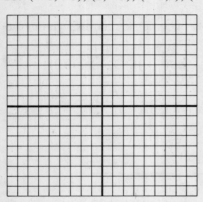

17. $(-5, 2)$, $(3, -1)$, $(3, 5)$

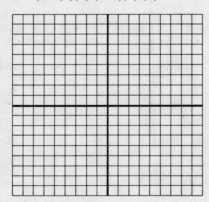

18. Plot five points with first coordinate equal to (-3).

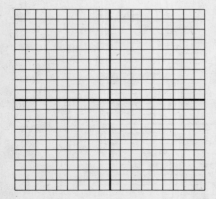

19. Plot five points with first coordinate equal to (2).

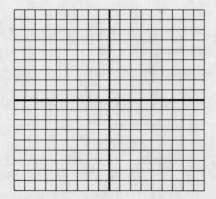

20. Plot five points with second coordinate equal to (4).

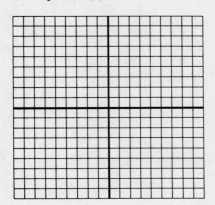

21. Plot five points with second coordinate equal to (−1).

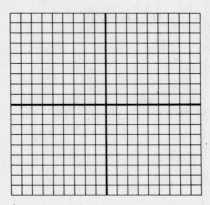

22. In what quadrants is the first coordinate always positive?

23. In what quadrants is the first coordiante always negative?

24. In what quadrants is the second coordinate always positive?

25. In what quadrants is the second coordinate always negative?

26. Give the coordinates of the points specified in the figures below:

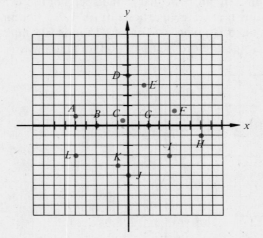

Figure 5.13

State the domain and the range of the following relations, and state whether the relation is a function:

27. {(2,7), (3,9), (4,11)}

28. {(0,1), (1,1), (2,1), (3,1), (4,1)}

29. {(5,10), (7,14), (9,18), (11,22)}

30. {(−1,−1), (3,3), (−5,−5)}

31. {(2,1), (4,3), (6,5), (4,7)}

32. {(1,2), (1,3), (1,4)}

In Exercises 33 through 37, find the distance between the given points:

33. (2,5), (6,6)

34. (−2,3), (−5,1)

35. (−4,−3), (−7,−2)

36. (3,−1), (7,0)

37. (−5,−3), (3,−4)

38. The distance in feet that a vehicle will travel after the brakes are applied is directly proportional to the speed of the vehicle in miles per hour, where the constant of proportionality is 1.2. Determine the ordered pairs when the speed is given as: 5 mph, 15 mph, 25 mph, 35 mph, 45 mph, and 55 mph. Graph the ordered pairs on a rectangular coordinate system.

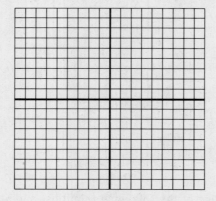

39. The layout for a plate shaped like a triangle has vertices (−5,−2), (−1,3), and (4,−4). Find the length of each side.

40. A drawing for a sheet of zinc, rectangular in shape, has end points (0,0), (13,0), (13,20), and (0,20). Find the mid-point of the diagonal form (0,0) to (13,20).

5.3 Graphing For First Degree Equations in Two Variables

To graph the solutions of the equation, $y = \frac{2}{3} x + 1$, arbitrarily choose values of x, and find the corresponding values of y. For example, let $x = -3, 0, 3,$ and 6.

$$y = \frac{2}{3} x + 1.$$

If $x = -3$, $y = \frac{2}{3} \cdot -3 + 1 = -1.$

If $x = 0$, $y = \frac{2}{3} \cdot 0 + 1 = 1.$

If $x = 3$, $y = \frac{2}{3} \cdot 3 + 1 = 3.$

If $x = 6$, $y = \frac{2}{3} \cdot 6 + 1 = 5.$

These values of x and y are expressed as ordered pairs, (x,y). The relation, $\{(-3,-1), (0,1), (3,3), (6,5), \ldots\}$, has an infinite number of ordered pairs which satisfy the given equation. The domain is the set of real numbers, and the range is the set of real numbers. Since for every element of the domain there is one and only one element of the range, the relation is a function. Plot these points using the horizontal axis as the x-axis and the vertical axis as the y-axis. Draw a line through the points to picture the solution of the equation. The arrows at each end of the line indicate that the line continues infinitely in both directions.

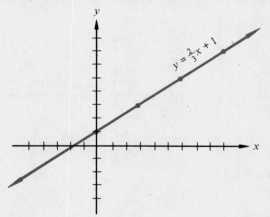

Figure 5.14

The line crosses the y-axis at the point (0,1). 1 is called the *y-intercept* of the line. Examining the ordered pairs, note that as x increases 3, y increases 2. The ratio of the change of y to the change of x is called the *slope* of the line. The slope may be determined by using any two of the points on the line. Choosing (3,3) and (9,7), the slope $= \frac{7-3}{9-3} = \frac{4}{6} = \frac{2}{3}$. The slope of the line, $y = \frac{2}{3} x + 1$, is $\frac{2}{3}$.

In general, an equation of the form $y = mx + b$ (where m and b are real numbers) is the equation of a straight line. This equation is called the *slope-intercept* form of the equation of a straight line. Such first degree equations are called *linear equations*. The number b is the y-intercept. The straight line will cross the y-axis at $(0,b)$. The slope of the line is $m = \dfrac{y_2 - y_1}{x_2 - x_1}$, where (x_1, y_1) and (x_2, y_2) are any two points on the line.

Rule 1:   To graph a linear equation in two variables using the slope-intercept form:

Step 1:   Assign a variable to each axis and solve the equation for the variable assigned to the vertical axis. (When using x and y, assign y to the vertical axis.)

Step 2:   Plot the y-intercept.

Step 3:   Plot several points using the slope of the given line.

Step 4:   Draw a straight line through these points.

Example 1:   Graph the equation, $x + 3y = 12$.

Step 1: $x + 3y = 12$
$$3y = -x + 12$$
$$y = -\frac{1}{3}x + 4$$

Step 2:   The y-intercept is 4. Plot the point $(0,4)$.

Step 3:   The slope of the line is $-\frac{1}{3}$. Starting at $(0,4)$, plot several points. As x increases 3, y decreases 1.

Step 4:   Draw a line through these points.

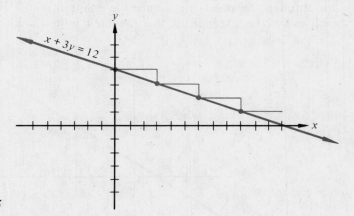

Figure 5.15

Example 2:   Graph the equation, $y = 6$.

This equation is equivalent to the equation, $y = 0x + 6$. The y-intercept is 6 and the slope is 0. As x varies, y is fixed at 6. Therefore, the line is parallel to the x-axis.

Notice that the relation, $\{(-1,6), (0,6), (1,6), (2,6), \ldots\}$ is a function. The domain is the set of real numbers. The range is 6.

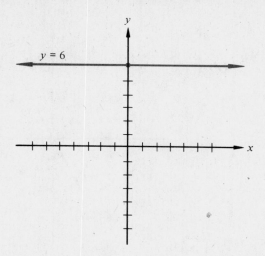

Figure 5.16

Example 3:   Graph the equation, $x = -3$.

x does not vary. Therefore, the slope of this line,

$$m = \frac{y_2 - y_1}{x_2 - x_1} = \frac{y_2 - y_1}{0},$$

is not defined, and the line is parallel to the y-axis.

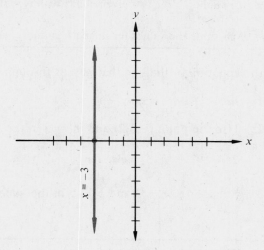

Figure 5.17

Notice that the relation, $\{(-3, -1), (-3, 0), (-3, 1), (-3, 2), \ldots\}$ is not a function since for the member of the domain, -3, there is more than one element of the range.

Example 4:   Graph the formula, $E = 2W$, which shows the relationship of a weight applied to the elongation of a spring.

Step 1:   Assign W to the horizontal axis and E, the vertical axis.

Step 2: $E = 2W + 0$
The E-intercept is 0. Plot the point $(0,0)$.

Step 3:   The slope of the line is $\frac{2}{1}$. Starting at $(0,0)$ plot several points. As W increases 1, E increases 2.

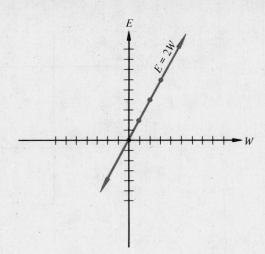

Figure 5.18

Rule 2:   To find the equation of a straight line when given any two points on the line:

Step 1:   Find the slope, $m = \dfrac{y_2 - y_1}{x_2 - x_1}$.

Step 2:   Use either of the two given points with m in the equation $y = mx + b$ and solve the equation for b.

Step 3:   Substitute the values for m and b in the equation, $y = mx + b$.

Example 5:   Find the equation of the line that passes through the points $(2, 11)$ and $(-1, 2)$.

Step 1: $m = \dfrac{y_2 - y_1}{x_2 - x_1} = \dfrac{11 - 2}{2 - (-1)} = \dfrac{9}{3} = 3$

Step 2:   Use the point $(-1, 2)$ and $m = 3$ in the equation $y = mx + b$.
$2 = 3(-1) + b$
$2 = -3 + b$ **Solve for b.**
$5 = b$

Step 3:   Substitute $m = 3$ and $b = 5$ in the equation $y = mx + b$.
$y = 3x + 5$.

Rule 3:   To find the equation of a line given a point and a slope, use the point-slope form of the equation of a straight line, $y - y_1 = m(x - x_1)$.

Example 6:   Find the equation of a line that passes through the point $(2, -4)$ and has a slope of 3.

$y - y_1 = m(x - x_1)$
$y - (-4) = 3(x - 2)$
$y + 4 = 3x - 6$
$y = 3x - 10 \text{ or } 3x - y - 10 = 0.$

EXERCISES 5.3

Graph Exercises 1 through 14:

1. $y = x$

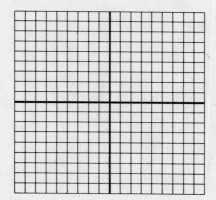

2. $y = 5x$

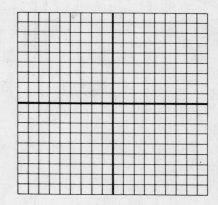

3. $3x + y = 0$

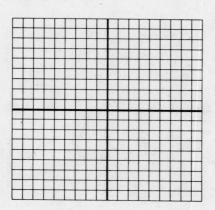

4. $x + 2y = 3$

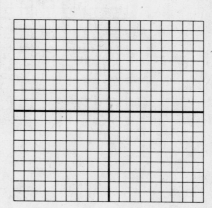

5. $2x + y = 5$

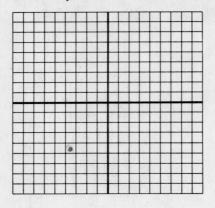

6. $x + y = 0$

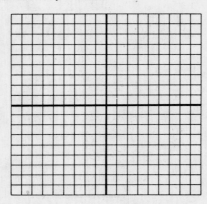

7. $2x + 3y = 2$

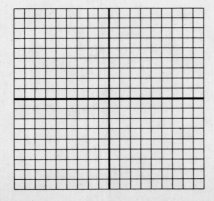

8. $x = 5$

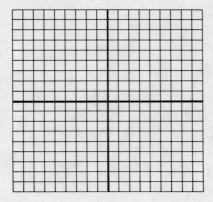

9. $y - 3 = 0$

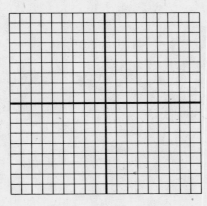

10. Graph $E = RI$(Ohm's Law), where $I = 6$.

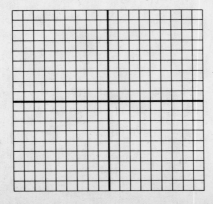

11. Graph $t = \dfrac{h}{4}$, the time of fall of a bomb dropped from an altitude h.

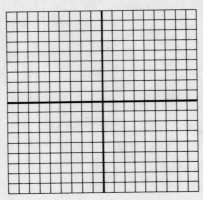

12. The resistance in ohms of a copper wire l foot in length and 40 mils in diameter is given by

$$R = \frac{10.4l}{d^2}$$

Graph the relationship of R and l.

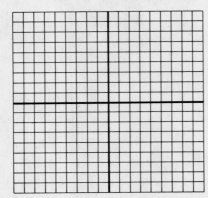

Find the equation of a line that passes through the given points:

13. $(2,5)$ and $(3,7)$

14. $(1,-1)$ and $(5,-9)$

15. $(-6,-2)$ and $(-2,1)$

16. (0,4) and (3,0)

17. (2,4) and (2, −3)

18. (6,5) and (−2,5)

Find the equation of a line that passes through the given point and has slope m:

19. (6, −3), $m = 4$

20. (−2, −1), $m = -3$

21. $\left(\frac{1}{3}, \frac{2}{3}\right)$, $m = 2$

22. (4,5), $m = 0$

Unit 5 Self-Evaluation

1. Draw a bar graph to represent the purchasing power of the U.S. dollar given by the following table:

1970	$0.86
1965	1.06
1960	1.13
1955	1.25
1950	1.39
1945	1.92
1940	2.38

2. Draw a circle graph to represent the proposed expenditure of a federal dollar given by the following table:

National defense	$0.40
Social security	.25
Education	.15
Veterans	.05
Other	.15

3. Draw a line graph to represent the automobile sales for a motor company given by the following table:

1970	2,000,000
1965	1,400,000
1960	1,000,000
1955	750,000
1950	200,000
1945	500,000
1940	180,000

In Exercises 4 through 6, graph the given equations:

4. $x - y = 2$

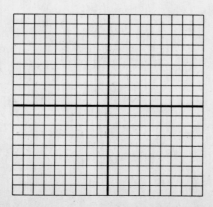

5. $3x + 2y = 1$

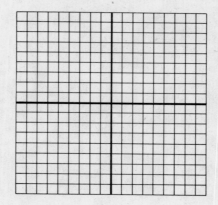

6. $y - 2 = 0$

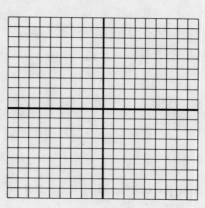

7. Graph $z = pz_1 + c$, given $p = 3$ and $c = 1$.

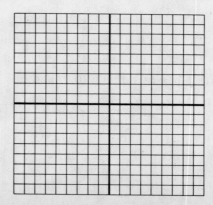

8. Graph $R = \dfrac{P}{D^2}$, given $D = -3$

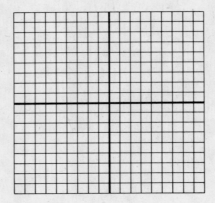

Tell whether or not the following relations are functions. If the relation is a function, state the domain and range:

9. $\{(-8,-2),\ (-6,-8),\ (-2,-6)\}$

10. $\{(-4,-1),\ (-1,-3),\ (-4,-2)\}$

11. $x - 1 = 0$

12. $x + 3y = 4$

13. Find the equation of the line that passes through the points $(-6,-2)$ and $(-8,4)$.

14. Find the equation of the line that passes through the point $(4,-4)$ and crosses the x-axis at $x = 2$.

15. Find the equation of the line that passes through $(0,0)$ and has a slope of -5.

Factoring

Unit 6 Objectives
1. To factor monomials from polynomials.
2. To find the product of the sum and difference of two monomials.
3. To factor the difference of two squares.
4. To find the product of two binomials.
5. To factor trinomials.
6. To square a binomial.
7. To factor perfect square trinomials.
8. To factor the sum or difference of two cubes.
9. To factor by grouping terms.
10. To factor polynomials completely.

unit 6

6.1 Factoring Monomials from Polynomials

Solving equations containing variables to the second degree, or higher, requires knowledge of *factoring*. For example, the equation $\frac{4}{3}\pi r^3 = \pi r^2$ can be solved for r by factoring.

Factoring is the reverse operation of finding a product. To factor an expression means to find the factors whose product is the given expression. In the given expression $2d(3d - 4) = 6d^2 - 8d$, $2d$ and $3d - 4$ are factors of $6d^2 - 8d$.

To factor a polynomial, it is necessary to find the greatest common monomial factor. In most cases, the greatest common factor can be found by inspection. For example, the greatest common factor of 8, 24, and 36 is 4, since 4 is the largest number that divides evenly into all three numbers.

The greatest common factor of R^3, R^5, and R^7 is R^3 because R^3 is the highest power of R that is a factor of R^3, R^5, and R^7. Note that the highest power of R that will divide evenly into R^3, R^5, and R^7 is 3, the smallest exponent of the three R's.

Example 1: To find the greatest common factor of a^2b^4, a^3b^5, and a^5b^3, first note that a and b are common factors. The smallest exponent of a that appears in any term is 2, and the smallest exponent of b is 3. Therefore, the greatest common factor is a^2b^3.

Example 2: To find the greatest common factor of the terms $15a^8$ and $20a^4$, first note that 5 is the greatest common factor of the coefficients 15 and 20. a^4 is the greatest common factor of a^8 and a^4. Therefore, the greatest common factor of $15a^8$ and $20a^4$ is $5a^4$.

Example 3:   Find the greatest common factor of $2(p + q)$ and $r(p + q)$.

The greatest common factor is $(p + q)$ because it appears in both terms.

Rule: To factor a polynomial whose terms contain a common factor:

Step 1:   Inspect each term of the polynomial to determine the greatest common monomial factor.

Step 2:   Divide each term of the polynomial by the greatest common monomial factor.

Step 3:   Write the polynomial as the product of the common factor and the quotient obtained.

Example 4:   Factor $12m^3 - 4m^2 + 2m$

Step 1:   By inspection, the greatest common monomial factor is $2m$.

Step 2: $\dfrac{12m^3 - 4m^2 + 2m}{2m} = \dfrac{12m^3}{2m} - \dfrac{4m^2}{2m} - \dfrac{2m}{2m} = 6m^2 - 2m + 1$

Step 3: $12m^3 - 4m^2 + 2m = 2m(6m^2 - 2m + 1)$

Check: Multiply the two factors to verify that their product is equal to the original polynomial.

$$2m(6m^2 - 2m + 1) = 12m^3 - 4m^2 + 2m$$

Example 5:   Factor $5a^3b - 15a^2b$

Step 1:   By inspection, $5a^2b$ is the greatest common monomial factor.

Step 2: $\dfrac{5a^3b - 15a^2b}{5a^2b} = \dfrac{5a^3b}{5a^2b} - \dfrac{15a^2b}{5a^2b} = a - 3$

Step 3: $5a^3b - 15a^2b = 5a^2b(a - 3)$

Check: $5a^2b(a - 3) = 5a^3b - 15a^2b$

Example 6:   Factor $a(b - 2) + 4(b - 2)$

Step 1:   By inspection, $(b - 2)$ is the greatest common monomial factor.

Step 2: $\dfrac{a(b - 2) + 4(b - 2)}{(b - 2)} = a + 4$

Step 3: $a(b - 2) + 4(b - 2) = (a + 4)(b - 2)$

Check: $(a + 4)(b - 2) = a(b - 2) + 4(b - 2)$

EXERCISES 6.1

Factor and check:

1. $12x - 6y$

2. $12e - 8e^2$

3. $8y^2 + 2y$

4. $4ar^2 - 16a^2$

5. $6a - 42a^2$

6. $3T_r - 6$

7. $4g^2 - 5g^3 + 6g^4$

8. $5ay - 5by + 5cy$

9. $4mn + 6m - 2n$

10. $-p^5 - p^4 + 3p^3$

11. $at - av + aw$

12. $-12ga^2D + 9ga^2d$

13. $3\pi r^2 - 9\pi r^2 h - 3r^2$

14. $17a + 34b - 51c$

15. $6Q^5 - 4Q^4 + 10Q^3 - 8Q^2$

16. $.2T_0 - .4T_1 + 6T_2$

17. $.32R^2 - 1.6R$

18. $\frac{1}{2}f^2 - \frac{3}{4}F - \frac{5}{8}$

19. $\frac{1}{3}a^3b^3 + \frac{5}{9}a^2b^2$

20. $-6bt^2 - 15bt + 12b$

21. $2m^2n + 4m^2n^2 - 10mn^3$

22. $100rs^3 - 30r^2s^2$

23. $51abc + 34bc^2d$

24. $.04kd^3 + .008k^2 - 2kd$

25. $8.1t^6 - 27t^4 + 5.4t^3$

26. $-44j^7 + 77j^5 - 55j^3$

27. $56V_uW^2 + 16V_uW - 16V_s$

28. $16D_0 + 52D + 12D_t$

29. $-r^3s - r^2s^3 - rs^5$

30. $PQ_1 + PQ_2 - PQ_3$

31. $EI - EI^2$

32. $PD - PW$

33. $7V_1 + 14V_2 - 28V$

34. $9fp - pf - 18p$

35. $P_1V_1 + P_1V_2 - 2P_1$

36. $3.14r_1 + 15.7r_2$

37. $Lat + L_0at - Lat^2$

38. $PRT + RTU - PTU$

39. $-6f - 4f_sk$

40. $38RST - 51RST_c$

41. $e(c + d) + 3(c + d)$

42. $R_1(R - 3) + R_2(R - 3)$

43. $m(2n - 5) - 3(2n - 5)$

44. $2T_0(T_1 + T_2) - 3(T_1 + T_2)$

45. $f(g + h) - (g + h)$

46. $(p + q) - r(p + q)$

47. Express the sum of 26 and 39 as a product of 13 and a sum of two numbers. (Hint: Express 26 and 39, each as a product of 13 and another number; factor out 13).

48. The perimeter of a rectangle is equal to the sum of twice the length and twice the width; that is, $P = 2l + 2w$. Use factoring to show that the perimeter equals twice the sum of the length and the width.

49. The formula to find the taper (T) of a machine part in inches per foot is given by:

$$T = \frac{12D}{l} - \frac{12d}{l}$$

Express the formula as a product instead of a difference.

50. Express V as a product, given that

$$V = 3a^2\pi h + 3b^2\pi h + 6\pi h^3$$

6.2   The Product of the Sum and Difference of Two Monomials

The product of the sum of two terms, $x + 5$, and the difference of the same two terms, $x - 5$, may be obtained by using the distributive property:

$$(x + 5)(x - 5) = x(x - 5) + 5(x - 5)$$
$$= x^2 - 5x + 5x - 25$$
$$= x^2 - 25$$

The product of $(x + 5)(x - 5) = x^2 - 25$, the square of x minus the square of 5.

Rule: The product of the sum and difference of the same two terms is the square of the first term minus the square of the second term.

Example 1: $(a + b)(a - b) = a^2 - b^2$

Example 2: $(3F - 2G)(3F + 2G) = 9F^2 - 4G^2$

EXERCISES 6.2

Find the product:

1. $(s + 3)(s - 3)$

2. $(Y + 7)(Y - 7)$

3. $(a - 6)(a + 6)$

4. $(b - 10)(b + 10)$

5. $(x + 1)(x - 1)$

6. $(r - 12)(r + 12)$

7. $(2a + 5)(2a - 5)$

8. $(6y - 1)(6y + 1)$

9. $(5b + 6)(5b - 6)$

10. $(3m + 7)(3m - 7)$

11. $(2R + 5S)(2R - 5S)$ **12.** $(m - n)(m + n)$

13. $(9d + 7e)(9d - 7e)$ **14.** $(2U + 7)(2U - 7)$

15. $(.3p + q)(.3p - q)$ **16.** $(.2a + .5b)(.2a - .5b)$

17. $\left(\frac{3}{5}D + E\right)\left(\frac{3}{5}D - E\right)$ **18.** $\left(\frac{2}{7}T_c - T\right)\left(\frac{2}{7}T_c + T\right)$

19. $\left(\frac{1}{3}r - \frac{1}{5}t\right)\left(\frac{1}{3}r + \frac{1}{5}t\right)$ **20.** $\left(w + \frac{5}{7}\right)\left(w - \frac{5}{7}\right)$

21. Find the product of 21 and 19 using the formula for the product of a sum and a difference of two monomials. (Hint: Express 21 as $(20 + 1)$ and 19 as $(20 - 1)$.)

22. The area (A) of the shaded ring is:

$$A = \pi(r_o + r_i)(r_o - r_i)$$

Express this formula without parentheses.

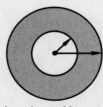

$A = \pi(r_o + r_i)(r_o - r_i)$

6.3 Factoring the Difference of Two Squares

Study the following examples from the previous section:

$$(x + 5)(x - 5) = x^2 - 25$$
$$(a + b)(a - b) = a^2 - b^2$$
$$(3F - 2)(3F + 2) = 9F^2 - 4$$

The product in each case is the difference of two squares. The factors of the difference of two squares are the sum and difference of the same two quantities. For example, to factor $x^2 - 25$, find a number that when squared is x^2. The number is x. x is called the square root of x^2. 5 is the square root of 25 since 5^2 is 25.

Rule: To factor the differences of two squares, take the square root of each term. The factors are the sum and difference of the square roots.

Example 1: Factor $y^2 - 16$.

The square root of y^2 is y.
The square root of 16 is 4.

Therefore, $y^2 - 16 = (y + 4)(y - 4)$.

Example 2: Factor $9p^2 - 4$.

The square root of $9p^2$ is $3p$.
The square root of 4 is 2.

Therefore, $9p^2 - 4 = (3p + 2)(3p - 2)$.

EXERCISES 6.3

Factor:

1. $B^2 - 36$

2. $d^2 - 64$

3. $a^2 - 1$

4. $m^2 - 100$

5. $64 - z^2$

6. $49 - r^2$

7. $25p^2 - 4$

8. $4t^2 - 25$

9. $121E^2 - 49$

10. $25x^2 - 144$

11. $4a^2 - 9y^2$

12. $49m^2 - 36n^2$

13. $\frac{1}{4} R^2 - T^2$

14. $\frac{1}{100} a^2 - b^2$

15. $\frac{1}{9} x^2 - y^2$

16. $\frac{9}{25} S^2 - \frac{4}{49}$

17. $.04 - t^2$

18. $.16 - z^2$

19. $64f^4g^2 - 9h^2$

20. $1 - a^{10}$

21. Write 400 minus 25 as the difference of two perfect squares and factor it.

22. Find the area of the shaded region by expressing it as the difference of the area of the larger square minus the area of the smaller square; then factor it.

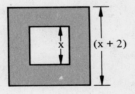

6.4 The Product of Two Binomials

Two binomials, $2h + 3$ and $h - 4$, were previously multiplied by using the distributive property:

$$(2h + 3)(h - 4) = 2h(h - 4) + 3(h - 4)$$
$$= 2h^2 - 8h + 3h - 12$$
$$= 2h^2 - 5h - 12$$

$2h \cdot (-4)$, or $-8h$, and $3 \cdot h$ or $3h$ are referred to as the *cross-products*.

Rule: To multiply two binomials:

Step 1: Multiply the first term of each binomial.

Step 2: Find the sum of the two cross-products.

Step 3: Multiply the last term of each binomial.

The product of the two binomials will be the sum of these results.

Example 1: Multiply: $(2h + 3)(h - 4)$

  Step 1: $2h \cdot h = 2h^2$

  Step 2: $-8h + 3h = -5h$

  Step 3: $3 \cdot -4 = -12$

$$\overbrace{\underbrace{(2h + 3)\ (h}_{3h} - 4)}^{-8h}$$

  Therefore, $(2h + 3)(h - 4) = 2h^2 - 5h - 12$.

Example 2: Multiply: $(7x - 2)(3x - 5)$

$$\overbrace{\underbrace{(7x - 2)\ (3x}_{-6x} - 5)}^{-35x} = (7x \cdot 3x) + (-6x - 35x) + (-2 \cdot -5)$$
$$= 21x^2 - 41x + 10$$

EXERCISES 6.4

Find the product:

1. $(x + 4)(x + 3)$ **2.** $(y + 7)(y + 2)$

3. $(2a + 5)(3a + 1)$ **4.** $(5b + 1)(4b + 3)$

5. $(d - 4)(d - 5)$ **6.** $(m - 1)(m - 7)$

7. $(6r - 5)(3r - 2)$ **8.** $(10 - 3t)(2 - t)$

9. $(R + 7)(R - 3)$ **10.** $(z - 11)(z + 5)$

11. $(b - 11)(b + 4)$ **12.** $(5 - n)(2 + n)$

13. $(8P - 3)(5P + 2)$

14. $(5x + 2y)(5x + 4y)$

15. $(3R_0 + 2R)(3R_0 + 2R)$

16. $(2r - t)(2r + 11t)$

17. $\left(\frac{1}{3}f + \frac{1}{2}\right)\left(\frac{2}{3}f + \frac{1}{2}\right)$

18. $\left(\frac{3}{4}x - \frac{1}{5}\right)\left(\frac{3}{4}x - \frac{1}{5}\right)$

19. $(8p - 3q)(7p + q)$

20. $(12Q + 5)(12Q - 7)$

21. $(1.6s + 2.3)(2.3s - .4)$

22. $(16.78c - .2d)(.3c + 5.3d)$

23. $(2r - 3s)(3r + 5t)$

24. $(5dc - 4e)(8dc - 3e)$

25. The motion of an object is given by $(16t - 48)(16t + 64)$. Express this as a sum.

26. Find the shaded area and express it without parentheses.

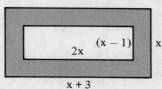

6.5   Factoring Trinomials

In the preceding section, two binomials were multiplied. Examine the signs in the following examples:

$$(p + 7)(p + 3) = p^2 + 10p + 21$$
$$(p - 7)(p - 3) = p^2 - 10p + 21$$

Note that when the last term of the trinomial is positive, the sign of the second term in both binomial factors is the same as the sign of the middle term of the trinomial.

$$(y + 4)(y - 5) = y^2 - y - 20$$
$$(y - 4)(y + 5) = y^2 + y - 20$$

Note that when the sign of the last term of the trinomial is negative, the second term of one binomial factor is positive and the second term of the other binomial factor is negative.

Rule: To factor a trinomial of the form $ax^2 + bx + c$, when $a = 1$:

Step 1:   Factor x^2, the first term of the trinomial, $(x \quad)(x \quad)$.

Step 2:   If c is positive, find factors of c whose sum is equal to b; signs of the factors are alike and are the same as the sign of b.

Step 3:   If c is negative, find factors of c whose difference is equal to b; signs of the factors are opposite and the larger factor takes the sign of b.

Example 1:   Factor the trinomial $x^2 + 5x + 6$.

Step 1:   Consider factors of x^2, the first term of the trinomial.

$$(x \quad)(x \quad)$$

Step 2:   When the last term of the trinomial is positive, consider factors of the last term, 6, whose sum is 5, the coefficient of the middle term.
    The factors of 6 are: 6 and 1,
                                         2 and 3.

The factors of 6 selected are 2 and 3, since their sum is 5. Both signs of the factors, 2 and 3, are positive since 5 is positive.

$$(x + 2)(x + 3)$$

To check, multiply the two factors, $(x + 2)(x + 3)$, to see that their product is the original trinomial, $x^2 + 5x + 6$.

Example 2:   Factor the trinomial $m^2 + 4m - 12$.

Step 1:   Consider factors of m^2, the first term of the trinomial.

$$(m \quad)(m \quad)$$

Step 2:   When the last term of the trinomial is negative, consider factors of the last term, 12, whose difference is 4, the coefficient in the middle term of the trinomial.
    Factors of 12 are: 12 and 1,
                                         3 and 4,
                                         2 and 6.

The factors of 12 selected are 2 and 6, since their difference is 4. The factors 2 and 6 have opposite signs. 6, the larger factor, is positive since 4 is positive:

$$(m + 6)(m - 2)$$

Check: $(m + 6)(m - 2) = m^2 + 4m - 12$

The preceding examples all involved trinomials in which the coefficient of the squared term was 1. Factoring the trinomial $6d^2 - 17d - 14$ is more difficult because factors of the first and last terms of the trinomial have to be considered. A method of trial and error can be used.

Example 3:   Factor the trinomial $6d^2 - 17d - 14$.

Consider factors of $6d^2$, the first term of the trinomial:

$(6d \quad)(d \quad)$ or $(2d \quad)(3d \quad)$

Now consider all combinations of 14, the last term of the trinomial:

$(6d \quad 14)(d \quad 1)$	$(2d \quad 14)(3d \quad 1)$
$(6d \quad 1)(d \quad 14)$	$(2d \quad 1)(3d \quad 14)$
$(6d \quad 2)(d \quad 7)$	$(2d \quad 2)(3d \quad 7)$
$(6d \quad 7)(d \quad 2)$	$(2d \quad 7)(3d \quad 2)$

Since the last term of the trinomial is negative, select the combination whose difference will yield 17, the coefficient in the middle term.

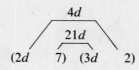

$$
\begin{array}{c}
4d \\
21d \\
(2d \quad 7) \quad (3d \quad 2)
\end{array}
$$

The method of trial and error in the above example may be simplified in the following way. To factor $6d^2 - 17d - 14$:

Step 1:   Multiply the coefficients of the first and last terms:

$6(-14) = -84$.

Step 2:   Find factors of 84 which, when subtracted, will equal -17, the coefficient in the middle term. The factors are -21 and 4 since $-21 \cdot 4 = -84$ and $-21 + 4 = -17$.

Step 3:   The factors, -21 and 4, are the cross-products.

$$
\begin{array}{c}
4d \\
-21d \\
(\quad) \ (\quad)
\end{array}
$$

Step 4:   The only factors of 4 which can be used are 2 and 2. Note that 4 is not a factor of the coefficient of the first term of the trinomial.

$$
\begin{array}{c}
4d \\
-21d \\
(2d \quad) \ (\quad 2)
\end{array}
$$

Step 5:   If $2d$ is a factor of $6d^2$, the other factor must be $3d$. If 2 is a factor of 14, the other factor must be 7.

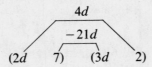

Step 6:   The last term of the trinomial is negative. Therefore, the sign of the second term of one factor is positive and the sign of the second term of the other factor is negative. Because the coefficient in the middle term is negative, the larger product, $21d$, is negative.

$$(2d - 7)(3d + 2)$$

Check: $(2d - 7)(3d + 2) = 6d^2 - 17d - 15.$

Example 4:   Factor $6q^2 - 19q + 15$.

Step 1:   Multiply the coefficient in the first and last terms of the trinomial: $6 \cdot 15 = 90$.

Step 2:   Find factors of 90 which, when added, will equal 19, the coefficient in the middle term. 10 and 9 are the factors, since $10 \cdot 9 = 90$ and $10 + 9 = 19$.

Step 3:   The factors, 10 and 9, are the cross-products.

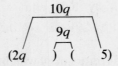

Step 4:   The only factors of 10 which can be used are 2 and 5 since 10 is not a factor of 6. The 2 is placed in the first factor since 2 is a factor of 6, and 5 is not.

$$
\begin{array}{c}
10q \\
9q \\
(2q \qquad) \; (\qquad 5)
\end{array}
$$

Step 5:   If $2q$ is a factor of $6q^2$, the other factor is $3q$. If 5 is a factor of 15, the other factor is 3.

$$
\begin{array}{c}
10q \\
9q \\
(2q \;\; 3) \;\; (3q \;\; 5)
\end{array}
$$

Step 6:   The last term of the trinomial is positive. Therefore, the sign of the second term in both factors is negative, the same as the sign of the middle term of the trinomial.

$$(2q - 3)(3q - 5)$$

Check: $(2q - 3)(3q - 5) = 6q^2 - 19q + 15$

EXERCISES 6.5

Factor the following trinomials. Check your work.

1. $x^2 + 5x + 6$

2. $y^2 + 6y + 5$

3. $p^2 + 18p + 77$

4. $a^2 + 9a + 20$

5. $m^2 - 6m + 8$

6. $z^2 - 10z + 16$

7. $b^2 - 7b + 12$

8. $R^2 - 7R + 10$

9. $c^2 + c - 6$

10. $W^2 - 5W - 14$

11. $n^2 - 2n - 8$

12. $B^2 + 3B - 10$

13. $t^2 + 7t - 18$

14. $y^2 - 10y - 11$

15. $f^2 - 10f + 25$

16. $q^2 - 4q + 4$

17. $T^2 + 7T + 12$

18. $F^2 + 6F + 8$

19. $n^2 + n - 20$

20. $c^2d^2 + cd - 12$

21. $a^2 + ab - 2b^2$

22. $x^2y^2 - 6xy - 7$

23. $5m^2 + 11m + 2$

24. $5f^2 - 16f + 3$

25. $3c^2 + 2c - 5$

26. $2d^2 - d - 3$

27. $6r^2 - r - 5$

28. $6t^2 + 7t - 5$

29. $16V^2 - 11V - 5$

30. $4x^2 - 11x + 6$

31. $16h^2 - 38h - 5$

32. $9K^2 - 34K - 8$

33. $2r^2 - 5r - 3$

34. $4p^2 + 16p + 15$

35. $9X^2 + 3XY - 2Y^2$

36. $3a^2 - 7ab + 2b^2$

37. $3b^2 + 14bc - 5c^2$ **38.** $18t^2 + 33tu + 14u^2$

39. Write the mechanics expression $A^2 - 3AL + 2L^2$ in factored form.

40. The length of a plate changes when the plate is heated; the new length is given by the formula $(t^2 + 300t + 20,000)$. Express the length in factored form.

6.6   The Square of a Binomial

To square a factor means to multiply it by itself. For example, $(a + 3)^2$ means that the factor $(a + 3)$ is multiplied by the factor $(a + 3)$. Study the following examples:

$$(a + 3)^2 = (a + 3)(a + 3) = a^2 + 6a + 9$$
$$(y - 5)^2 = (y - 5)(y - 5) = y^2 - 10y + 25$$
$$(f + 7)^2 = (f + 7)(f + 7) = f^2 + 14f + 49$$
$$(a + b)^2 = (a + b)(a + b) = a^2 + 2ab + b^2$$

Note the products of each example. The first term is the square of the first term of the binomial. The second term is twice the product of the two terms of the binomial. The last term is the square of the second term of the binomial.

Rule: To square a binomial:

Step 1:   Square the first term of the binomial.

Step 2:   Take twice the product of the two terms of the binomial.

Step 3:   Square the last term of the binomial.

Example 1: $(W - 4)^2$

Step 1: W squared is W^2

Step 2:   Twice the product of the two terms of the binomial is $2 \cdot (-4)W = -8W$

Step 3:   The square of the last term of the binomial is 4^2 or 16. Therefore,

$$(W - 4)^2 = W^2 - 8W + 16$$

Example 2: $(3T + 7)^2$

The square of the first term is $9T^2$. Twice the product of the two terms of the binomial is $42T$. The square of the last term is 49.

$$(3T + 7)^2 = (9T)^2 + 2(3T)(7) + (7)^2$$
$$= 9T^2 + 42T + 49$$

EXERCISES 6.6

Perform the indicated operation:

1. $(x + 5)^2$ **2.** $(m - 2)^2$ **3.** $(z - 11)^2$ **4.** $(e + 6)^2$

5. $(2a + 3)^2$ **6.** $(4N - 1)^2$ **7.** $(5g - 2)^2$ **8.** $(3h + 1)^2$

9. $(3D + 5E)^2$ **10.** $(7r - 2s)^2$ **11.** $(6 - T)^2$ **12.** $(8 + d)^2$

13. $(x + y)^2$ **14.** $(a - b)^2$

15. Express the square of 15 as the square of the sum of 10 and 5, and expand.

16. The formula for the shaded area is:

$A = 0.215r^2$

Find the shaded area if the radius is equal to $(x + 3.2)$

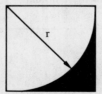

6.7   Factoring Perfect Square Trinomials

In the last section, the factor $a + b$ was squared. The square of the binomial $a + b$ resulted in a trinomial, $a^2 + 2ab + b^2$. This trinomial is referred to as a *perfect square trinomial*. The first and last terms of the trinomial are squares, and the middle term is twice the product of the square roots of the first and last term of the trinomial. The sign of the middle term of the trinomial is always the same as the second term of the binomial being squared. A perfect square trinomial may be factored simply if it is recognized. Study the following trinomials. Which are perfect square trinomials?

1. $x^2 + 6x + 9$
2. $y^2 - 10y + 25$
3. $a^2 + a + 4$
4. $16 + 40c + 25c^2$
5. $f^2 + 6f - 9$

Examples 1, 2, and 4 are perfect trinomials. Example 3 is not since twice the product of the square roots of the first and last terms should be $4a$. Example 5 is not a perfect square trinomial because the last term, -9, is not a square.

Rule: To factor a perfect square trinomial:

Step 1:   Take the square root of the first and last terms of the trinomial.

Step 2:   Write a binomial factor whose terms are the square roots obtained in the first step. The sign separating the two terms is the sign of the middle term of the trinomial.

Step 3:   Square the binomial factor obtained.

Example 1:   Factor $x^2 + 10x + 25$.

Step 1:   The square root of x^2 is x, and the square root of 25 is 5.

Step 2: $(x + 5)$ is the binomial factor.

Step 3:   The square of the binomial factor $(x + 5)$ is $(x + 5)^2$. Therefore,

$$x^2 + 10x + 25 = (x + 5)^2$$

Example 2:   Factor $9e^2 - 24e + 16$.

Step 1:   The square root of $9e^2$ is $3e$, and the square of 16 is 4.

Step 2: $(3e - 4)$ is the binomial factor.

Step 3:   The square of the binomial factor $(3e - 4)$ is $(3e - 4)^2$. Therefore,

$$9e^2 - 24e + 16 = (3e - 4)^2$$

EXERCISES 6.7

Factor the following trinomials after first checking to see that they are perfect square trinomials.

1. $d^2 + 18d + 81$

2. $m^2 - 16m + 64$

3. $a^2 - 2a + 1$

4. $b^2 + 2b + 1$

5. $25 - 10c + c^2$

6. $36 + 12n + n^2$

7. $4a^2 + 28a + 49$

8. $9a^2 - 6a + 1$

9. $25x^2 - 20x + 4$

10. $49y^2 + 42y + 9$

11. $144p^2 - 24p + 1$

12. $81x^2 - 126xy + 49y^2$

13. $\dfrac{4}{9}y^2 - \dfrac{4}{5}y + \dfrac{9}{25}$

14. $121F^2 + 22FG + G^2$

15. $m^2 + 2mn + n^2$

16. $r^2 - 2rs + s^2$

17. $4T^2 + 12TR + 9R^2$

18. $16W^2 - 40WU + 25U^2$

19. $9x^2 + 30xy + 25y^2$

20. $4b^2 + 36bc + 81c^2$

21. A square has area $(25x^2 + 10x + 1)$. Use factoring to find the length of each side.

22. A square steel plate has area $(4x^2 + 52x + 169)$. Find the length of each side.

6.8   Factoring the Sum or Difference of Two Cubes

The factors of $a^3 - b^3$ are $(a - b)$ and $(a^2 + ab + b^2)$. This is verified by multiplying the factors $(a - b)$ and $(a^2 + ab + b^2)$ to see that their product is $a^3 - b^3$.

$$\begin{aligned}(a - b)(a^2 + ab + b^2) &= a(a^2 + ab + b^2) - b(a^2 + ab + b^2) \\ &= a^3 + a^2b + ab^2 - a^2b - ab^2 - b^3 \\ &= a^3 - b^3\end{aligned}$$

In the same manner it can be verified that the product of $(a + b)$ and $(a^2 - ab + b^2)$ is $a^3 + b^3$.

Observe the patterns in the factors of $a^3 + b^3$ and $a^3 - b^3$.

$$a^3 + b^3 = (a + b)(a^2 - ab + b^2)$$
$$a^3 - b^3 = (a - b)(a^2 + ab + b^2)$$

a is the cube root of a^3 since a raised to the third power is a^3. 2 is the cube root of 8 since 2 raised to the third power is 8. $2^3 = 2 \cdot 2 \cdot 2 = 8$. One factor is a binomial and the other factor is a trinomial.

Rule: To factor the sum or difference of two cubes:

Step 1:   Take the cube root of the first and last terms to obtain the binomial factor.

Step 2:   Use the binomial obtained in (1) to obtain the second factor, a trinomial; square the first term of the binomial, take the product of the two terms with the opposite sign, and square the last term of the binomial.

Example 1:   Factor $y^3 + 8$.

The cube root of y^3 is y, and the cube root of 8 is 2. Therefore, the binomial factor is $(y + 2)$. Use $(y + 2)$ to obtain the trinomial factor. Square y. Take the product of y and 2 with the opposite sign, and square 2. The trinomial factor is $(y^2 - 2y + 4)$.

Therefore, $y^3 + 8 = (y + 2)(y^2 - 2y + 4)$

Example 2:   Factor $125 - f^3$.

The cube root of 125 is 5, and the cube root of $-f^3$ is $-f$. Therefore, the binomial factor is $(5 - f)$. To obtain the trinomial factor, square 5, take the product of 5 and $-f$ with the opposite sign, and square $-f$. The factor is $(25 + 5f + f^2)$.

Therefore, $125 - f^3 = (5 - f)(25 + 5f + f^2)$

EXERCISES 6.8

Factor:

1. $x^3 - 27$

2. $p^3 + 64$

3. $125 + m^3$

4. $8 - d^3$

5. $R^3 - 1$

6. $r^3 + 8$

7. $64 - s^3$ **8.** $1 + c^3$ **9.** $8p^3 - q^6$

10. $27a^3 - 1$ **11.** $b^9 + 8$ **12.** $125g^3 + 8h^3$

13. $27 + 8F^3$ **14.** $D^6 - 27$

6.9   Factoring by Grouping Terms

Upon examination of $g^2 + 5g + gh + 5h$, no common monomial factor can be observed. Group the first two terms and the last two terms as follows: $(g^2 + 5g) + (gh + 5h)$. Now, g is a common factor of g^2 and $5g$, and h is a common factor of gh and $5h$. Therefore,

$$(g^2 + 5g) + (gh + 5h) = g(g + 5) + h(g + 5).$$

Observe that $(g + 5)$ is a common factor of the two terms and can be factored as in Section 4.1.

$$g(g + 5) + h(g + 5) = (g + 5)(g + h).$$

The purpose of grouping terms is to find a common factor.

Example 1:   Factor $a^2 - 3a - 4ab + 12b$.

$(a^2 - 3a) + (-4ab + 12)$

$a^2 - 3a - 4ab + 12b = (a^2 - 3a) + (-4ab + 12b)$ **Group the first two and the last two terms.**

$\quad\quad\quad\quad = a(a - 3) - 4b(a - 3)$ **Factor each term.**

$\quad\quad\quad\quad = (a - 3)(a - 4b)$ **Factor out $(a - 3)$.**

Example 2:   Factor $3x - xy - 3y + x^2$.

If we group the first two and the last two terms: $(3x - xy) + (-3y + x^2)$, x is a common factor of $3x - xy$. However, there is no common monomial factor of $-3y + x^2$. So, instead group the first and last terms, and the second and third terms. This will lead to a common factor.

$3x - xy - 3y + x^2 = (3x + x^2) + (-xy - 3y)$

$\quad\quad\quad\quad = x(3 + x) - y(x + 3)$

$\quad\quad\quad\quad = (x + 3)(x - y)$

EXERCISES 6.9

Factor the following polynomials by grouping:

1. $x^2 + xy + 7x + 7y$

2. $ab + a - 3b - 3$

3. $3D + 15 + DE + 5E$

4. $4c - 4 + cd - d$

5. $3x^2 + 3 + x^2y + y$

6. $R + T + RS + ST$

7. $51m - 85 - 6mn + 10n$

8. $3st - 7s + 6rt - 14r$

9. $ac + 9b + bc + 9a$

10. $21x - 7y + 15x^2 - 5xy$

11. $5r^3 - 3qs^2 + 5s^2 - 3qr^3$

12. $a^2 + bc + ab + ac$

6.10 Factoring Completely

All polynomials should be completely factored; that is, factors should be inspected to see that they cannot be factored again by one of the given methods. For example, $2x^2 + 10x + 12$ can be factored to be $(2x + 6)$ and $(x + 2)$. The factor $(2x + 6)$ has a common factor of 2 and can be factored again to be $2(x + 3)$. The trinomial $2x^2 + 10x + 12$ could have been factored more simply if the common factor 2 had been removed first:

$$2x^2 + 10x + 12 = 2(x^2 + 5x + 6)$$
$$= 2(x + 3)(x + 2)$$

Example 1: Factor $4x^2 - 16$.

$$4x^2 - 16 = 4(x^2 - 4)$$
$$= 4(x - 2)(x + 2)$$

Example 2: Factor $6a^3 + 39a^2b - 21ab^2$

$$6a^3 + 39a^2b - 21ab^2 = 3a(2a^2 + 13ab - 7b^2)$$
$$= 3a(2a - b)(a + 7b)$$

Example 3: Factor $D^4 - 16$

$$D^4 - 16 = (D^2 - 4)(D^2 + 4)$$
$$= (D - 2)(D + 2)(D^2 + 4)$$

EXERCISES 6.10

Factor completely:

1. $12ax - 18a^2$

2. $4v^2 - 1$

3. $2ay^3 - 4ay^2 + 2ay$

4. $u^4 - 16$

5. $4h^2 - 12hk + 9k^2$

6. $3T^2 + 5T + 2$

7. $6m^2 + 3m - 900$

8. $R^4 - 5R^2 + 4$

9. $9r^2 - 21r - 18$

10. $b^2 - 2b - 35$

11. $4t^2 + 4t - 32$

12. $15c^2 - c - 28$

13. $16x^2 - 7x - 9$

14. $8b^2x^2 - 12bx + 6abx$

15. $-16dy^2 + 4dy - 4dy^2$

16. $20p^2 - 5$

17. $2f^2 - 162$

18. $6a^2 - 13a - 15$

19. $4cz^2 - 19cz + 21c$

20. $2m^3n - 2m^2n + 10mn$

21. $V^2 - 6V - 135$

22. $15s^2 - 26st + 8t^2$

23. $16y^2 - 49z^2$

24. $25e^2 + 30ef + 9f^2$

25. $11m^2 + 29m - 12$

26. $1 - 8x^3$

27. $9a^2 - 48ab + 64b^2$

28. $b^3 + 27$

29. The area of a circle is given as $(\pi x^2 + 14\pi x + 49\pi)$. Find the radius of the circle.

30. Find the area of the shaded region and express it in factored form. (Hint: Multiply the area of the square times four and subtract it from the area of the circle.)

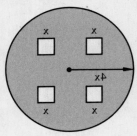

Unit 6 Self-Evaluation

Find the products:

1. $(2m + n)(2m - n)$

2. $(7d - 9)(5d + 7)$

3. $(13r + 2s)(5r - 7s)$

4. $(5 - 9Z)(3 - 11Z)$

5. $(6T + 5)^2$

Factor the following polynomials completely:

6. $64 - x^2$

7. $z^2 + 5z - 66$

8. $15r^3t - 20r^2t^3$

9. $4B^2 + 20B + 25$

10. $27x^3 - y^3$

11. $7f^2 - 63$

12. $T(P + R) - 3(P + R)$

13. $3a^2b^2 - 6a^3b^2 + 9a^3b$

14. $30s^2 - 8s - 8$

15. $W^2 - 18W + 81$

16. $3x^2 + 15xy + 9y^2$

17. $16A^2 - 4$

18. $10g^2 + 11g + 3$

19. $42y^2 - 53y + 15$

20. $6 + 41M + 44M^2$

Algebraic Fractions

Unit 7 Objectives
1. To reduce algebraic fractions to lowest terms.
2. To multiply and divide algebraic fractions.
3. To find the lowest common denominator of two or more algebraic fractions, and to convert them to equivalent fractions with common denominators.
4. To add and subtract algebraic fractions.
5. To solve linear equations involving fractions.

unit 7

Algebraic fractions play an important role in mathematics. They frequently appear in formulas pertaining to physics, chemistry, and every branch of engineering. For example,

$$F = K\frac{W_1 W_2}{D}$$ is the force of attraction between two bodies;

$$\frac{1}{R_t} = \frac{1}{R_1} + \frac{1}{R_2}$$ is the joint conductance of resistors in parallel;

$$\frac{1}{f} = \frac{1}{p} + \frac{1}{q}$$ is a lens formula;

$$P = \frac{W(R - r)}{2R}$$ is a pulley formula;

$$\frac{W}{Q_1} = \frac{T_2}{T_1} - 1$$ is a refrigeration formula.

It is extremely important for the technician to be able to manipulate formulas such as the ones mentioned above. Hence, the need for any person in a technical field to be able to add, subtract, multiply, and divide algebraic fractions, as well as solve algebraic equations.

7.1   Algebraic Fractions in Lowest Terms

An *algebraic fraction* is a fraction in which the numerator and/or the denominator are algebraic expressions.

An algebraic fraction is said to be in *lowest terms* when the numerator and the denominator are polynomials and have no common factors.

When adding, subtracting, multiplying, or dividing algebraic fractions, the result should be given as an algebraic fraction in lowest terms.

Rule: To reduce an algebraic fraction to *lowest terms:*

Step 1: Factor the numerator and the denominator completely.

Step 2: Divide the numerator and the denominator by the factor or factors that they have in common.

The resulting fraction in Step 2 is the equivalent fraction in lowest terms.

Example 1: Reduce to lowest terms $\dfrac{8e}{16e^2 - 32e}$

Step 1: $\dfrac{8e}{16e^2 - 32e} = \dfrac{8e}{16e(e - 2)}$ Factor.

Step 2: $\dfrac{\overset{1\ 1}{\cancel{8e}}}{\underset{2\ 1}{\cancel{16e}}(e - 2)} = \dfrac{1}{2(e - 2)}$ **Divide by common factors 8 and e.**

Therefore, $\dfrac{8e}{16e^2 - 32e} = \dfrac{1}{2(e - 2)}$

Example 2: Reduce to lowest terms $\dfrac{a^2 - 5a + 6}{a^2 + 3a - 10}$.

Step 1: $\dfrac{a^2 - 5a + 6}{a^2 + 3a - 10} = \dfrac{(a - 2)(a - 3)}{(a + 5)(a - 2)}$ Factor.

Step 2: $\dfrac{\overset{1}{\cancel{(a - 2)}}(a - 3)}{(a + 5)\underset{1}{\cancel{(a - 2)}}} = \dfrac{a - 3}{a + 5}$ **Divide by common factor $(a - 2)$.**

Therefore, $\dfrac{a^2 - 5a + 6}{a^2 + 3a - 10} = \dfrac{a - 3}{a + 5}$

EXERCISES 7.1

Reduce to lowest terms:

1. $\dfrac{6x}{2x - 10}$

2. $\dfrac{a^2 - 2a}{5a}$

3. $\dfrac{P - 2R}{P^2 - 2PR}$

4. $\dfrac{2E - 4}{3E - 6}$

5. $\dfrac{r^2 - s^2}{r - s}$

6. $\dfrac{uv - uw}{v^2 - w^2}$

7. $\dfrac{i^2}{i^2 - ir}$

8. $\dfrac{(A + R)^2}{A^2 - R^2}$

9. $\dfrac{x^2 - 4}{3x - 6}$

10. $\dfrac{5a - 15}{a^2 - 9}$

11. $\dfrac{m^2 + 7m + 10}{m + 2}$

12. $\dfrac{D - 3}{D^2 - 6D + 9}$

13. $\dfrac{aA + 4a}{A^2 + 2A - 8}$

14. $\dfrac{x^2 - 2xy - 3y^2}{2x + 2y}$

15. $\dfrac{d^2 + 15df + 54f^2}{d^2 - 36f^2}$

16. $\dfrac{N^2 - 25}{N^2 - 2N - 15}$

17. $\dfrac{Q^2 - Q - 6}{Q^2 - 5Q + 6}$

18. $\dfrac{p^2 - 2p - 8}{p^2 - p - 12}$

19. $\dfrac{2x^2 + 3x - 2}{2x^2 - 3x + 1}$

20. $\dfrac{3R^2 - 5Rr - 2r^2}{R^2 - 7Rr + 10r^2}$

21. $\dfrac{2w^2 - 8}{2w^2 - w - 6}$

22. $\dfrac{10a - 50b}{2a^2 - 12ab + 10b^2}$

23. $\dfrac{5X - 10Y}{X^2 - 4Y^2}$

24. $\dfrac{z^2 - 3z + 2}{z^2 - 4}$

25. $\dfrac{\theta^2 - 5\theta}{\theta^2 - 4\theta - 5}$

26. $\dfrac{w^2\pi^2 - 16w^2}{w\pi^2 + 9w\pi + 20w}$

27. The ideal mechanical advantage is given by the ratio of the distance moved by the force operating the machine to the distance moved by the load. If the distance moved by the force is $2\pi r$ and the distance moved by the load is $\frac{1}{2}(2\pi r - 2\pi R)$, find the ideal mechanical advantage in lowest terms.

7.2 Multiplication and Division of Algebraic Fractions

The operations of multiplication and division of algebraic fractions are performed the same way as in arithmetic. The numerators are multiplied and the denominators are multiplied to give the product fraction. By dividing out all common factors from either numerator or either denominator before multiplying, the resulting product will be a fraction in lowest terms.

Rule 1: To multiply two algebraic fractions:

Step 1: Factor the numerators and the denominators of both fractions completely.

Step 2: Divide either numerator and either denominator by all the factors that they have in common.

Step 3: Multiply the remaining factors in the numerators.

Step 4: Multiply the remaining factors in the denominators.

The resulting product is the algebraic fraction with the result of Step 3 as a numerator and the result of Step 4 as a denominator.

Example 1: Multiply: $\dfrac{4}{5p} \cdot \dfrac{7p^2}{8q}$.

Step 1:   **Not necessary.**

Step 2: $\dfrac{\overset{1}{\cancel{4}}}{\underset{1}{\cancel{5p}}} \cdot \dfrac{7\overset{p}{\cancel{p^2}}}{\underset{2}{\cancel{8q}}} = \dfrac{1}{5} \cdot \dfrac{7p}{2q}$ **Divide by common factors 4 and p.**

Steps 3 & 4: $\dfrac{1}{5} \cdot \dfrac{7p}{2q} = \dfrac{7p}{10q}$ **Multiply numerators. Multiply denominators.**

Therefore, $\dfrac{4}{5p} \cdot \dfrac{7p^2}{8q} = \dfrac{7p}{10q}$.

Example 2: Multiply: $\dfrac{10a^3b^2}{27xyz} \cdot \dfrac{15ay^2}{4b^3x^2}$.

Step 1:   **Not necessary.**

Step 2: $\dfrac{10a^3b^2}{27xyz} \cdot \dfrac{15ay^2}{4b^3x^2} = \dfrac{\overset{5}{\cancel{10}}a^3\overset{1}{\cancel{b^2}}}{\underset{9}{\cancel{27}}x\underset{1}{\cancel{y}}z} \cdot \dfrac{\overset{5}{\cancel{15}}a\overset{y}{\cancel{y^2}}}{\underset{2b}{\cancel{4b^3}}x^2}$ **Divide by common factors.**

Steps 3 & 4: $= \dfrac{25a^4y}{18bx^3z}$ **Multiply.**

Example 3: Multiply: $\dfrac{a^2 + ab - 2b^2}{3x - 6y} \cdot \dfrac{x - 2y}{a^2 - b^2}$.

Step 1: $\dfrac{a^2 + ab - 2b^2}{3x - 6y} \cdot \dfrac{x - 2y}{a^2 - b^2}$

           $= \dfrac{(a - b)(a + 2b)}{3(x - 2y)} \cdot \dfrac{(x - 2y)}{(a + b)(a - b)}$ **Factor.**

Step 2: $= \dfrac{\overset{1}{\cancel{(a - b)}}(a + 2b)}{3\underset{1}{\cancel{(x - 2y)}}} \cdot \dfrac{\overset{1}{\cancel{(x - 2y)}}}{(a + b)\underset{1}{\cancel{(a - b)}}}$ **Divide by common factors.**

Steps 3 & 4: $= \dfrac{a + 2b}{3(a + b)}$ **Multiply.**

Rule 2: To divide two algebraic fractions, multiply the dividend by the reciprocal of the divisor.

Example 4: Divide: $\dfrac{3ab}{4x} \div \dfrac{6abc}{7x}$.

$$\dfrac{3ab}{4x} \div \dfrac{6abc}{7x} = \dfrac{3ab}{4x} \cdot \dfrac{7x}{6abc}$$ Multiply the dividend by the reciprocal of the divisor.

$$= \dfrac{\overset{1\,1\,1}{\cancel{3ab}}}{\underset{1}{4x}} \cdot \dfrac{\overset{1}{\cancel{7x}}}{\underset{2\,1\,1}{\cancel{6abc}}}$$ Factor and divide by common factors.

$$= \dfrac{7}{8c}$$

Example 5: Divide: $\dfrac{r^2 - 4r + 4}{r^2 - 4} \div \dfrac{3r - 6}{5r + 10}$.

$$\dfrac{r^2 - 4r + 4}{r^2 - 4} \div \dfrac{3r - 6}{5r + 10} = \dfrac{r^2 - 4r + 4}{r^2 - 4} \cdot \dfrac{5r + 10}{3r - 6}$$

$$= \dfrac{\overset{1}{\cancel{(r-2)}}\overset{1}{\cancel{(r-2)}}}{\underset{1}{\cancel{(r-2)}}\underset{1}{\cancel{(r+2)}}} \cdot \dfrac{5\overset{1}{\cancel{(r+2)}}}{3\underset{1}{\cancel{(r-2)}}}$$

$$= \dfrac{5}{3}$$

Multiply the dividend by the reciprocal of the divisor.
Factor and divide by common factors.

EXERCISES 7.2

Perform the indicated operations:

1. $\dfrac{2a}{5} \cdot \dfrac{10}{3b}$

2. $\dfrac{3x}{4y} \cdot \dfrac{2y}{9}$

3. $\dfrac{6m}{n} \cdot \dfrac{2s}{3t}$

4. $\dfrac{5r_1 r_2}{2r_3} \cdot \dfrac{10r_1}{r_2 r_3}$

5. $\dfrac{X}{2Y} \cdot \dfrac{3V}{4W}$

6. $\dfrac{12x}{5yz} \cdot \dfrac{xy}{6z}$

7. $\dfrac{9pq}{4} \cdot \dfrac{8p}{3q^2}$

8. $\dfrac{2IR^2}{ir} \cdot \dfrac{i^3}{6Rr}$

9. $\dfrac{r + s}{2r} \cdot \dfrac{6r^3}{(r + s)^2}$

10. $\dfrac{(x + y)^3}{3} \cdot \dfrac{9}{x + y}$

11. $\dfrac{P^2 - 4}{3P} \cdot \dfrac{12PR}{P^2 + 5P + 6}$

12. $\dfrac{a^2 - 1}{ab^3} \cdot \dfrac{a^2b}{a^2 + 3a + 2}$

13. $\dfrac{u^2 - v^2}{u^2 - 9} \cdot \dfrac{u + 3}{u^2 + 2uv + v^2}$

14. $\dfrac{d^2 - 3d - 4}{2d - 6} \cdot \dfrac{4}{d^2 - 1}$

15. $\dfrac{2e^2 - 3e + 1}{2e^2 + 7e - 4} \cdot \dfrac{5e^2 + 17e - 12}{e^2 - 1}$

16. $\dfrac{3g^2 + 8gh - 3h^2}{2g^2 + 3gh - 2h^2} \cdot \dfrac{g^2 + 3gh - 4h^2}{3g^2 - 7gh + 2h^2}$

17. $\dfrac{5x^2 - 5}{3x^2 - 16x + 5} \cdot \dfrac{x^2 - 9x + 20}{5x + 5}$

18. $\dfrac{4j^2 + 3j - 1}{2j^2 - 18} \cdot \dfrac{2j - 6}{j^2 + 5j + 4}$

19. $\dfrac{A^2 - 5A + 6}{3A - 6} \cdot \dfrac{9A - 9}{5A - 15}$

20. $\dfrac{6k - 30}{2k^2 - 11k + 5} \cdot \dfrac{3k - 3}{10k - 10}$

21. $\dfrac{R}{5} \div \dfrac{R}{10}$

22. $\dfrac{a^2}{6} \div \dfrac{a}{12}$

23. $\dfrac{rs}{t} \div \dfrac{r}{s}$

24. $\dfrac{3}{a} \div \dfrac{aA}{6}$

25. $\dfrac{5B}{3B - 3} \div \dfrac{15B^2}{6}$

26. $\dfrac{22B^3}{3b} \div \dfrac{11bB^2}{3B + 6}$

27. $\dfrac{x^2 - 4}{7y} \div \dfrac{x - 2}{4y^2}$

28. $\dfrac{v}{w + 3} \div \dfrac{10w}{w^2 - 9}$

29. $\dfrac{p^2 - 9q^2}{2p - 4q} \div \dfrac{p^2 + 5pq + 6q^2}{p - 2q}$

30. $\dfrac{M^2 - 4N^2}{M^2 + 3MN - 4N^2} \div \dfrac{3M - 6N}{5M - 5N}$

31. $\dfrac{4x^2 - 9}{x - 7} \div \dfrac{2x + 3}{x^2 - 8x + 7}$

32. $\dfrac{\theta^2 + 3\theta - 10}{4\theta - 5} \div \dfrac{\theta - 2}{16\theta^2 - 25}$

33. $\dfrac{2X^2 + 5X - 3}{X^3 + 3X^2 + 2X} \div \dfrac{4X^2 - 1}{X^3 + 2X^2}$

34. $\dfrac{3w^2 - w - 2}{w^5 - 4w^3} \div \dfrac{3w^2 + 5w + 2}{w^3 - 2w^2}$

35. $\dfrac{9t^2 - 12t + 4}{2t^2 + 3t - 5} \div \dfrac{3t^2 - 5t + 2}{2t^2 + 7t + 5}$

36. $\dfrac{4r^2 - 4rs + s^2}{5r^2 + 13rs - 6s^2} \div \dfrac{4r^2 - 6rs + 2s^2}{5r^2 - 17rs + 6s^2}$

37. $\dfrac{S^2 - 8S + 5}{S + 2} \div (S + 2)$

38. $\dfrac{x - 1}{(2x - 1)^2} \div \dfrac{1}{x - 1}$

39. $\dfrac{2z + 6}{z^2 - 7z + 12} \div \dfrac{4z + 12}{z^2 - 16}$

40. $\dfrac{5mn - 25n^2}{m^2 - 4mn - 5n^2} \div \dfrac{10m^2 - 10n^2}{3m + 3n}$

41. The centripetal force, F, of a body in a circular path is the product of its mass, m, and its acceleration, a. If $m = \dfrac{W}{g}$ and $a = \dfrac{v^2}{r}$, find F.

42. The modulus of elasticity, Y, of the material of a body is found by dividing the stress by the strain. If the stress equals $\frac{F}{A}$ and the strain equals $\frac{\Delta L}{L}$, find the formula for $\frac{A}{Y}$.

7.3   Lowest Common Denominator of Algebraic Fractions

To add or subtract two or more algebraic fractions with different denominators, it is necessary to convert all of the fractions to equivalent fractions with a common denominator. To convert to fractions with a common denominator means to find the lowest common denominator of the fractions and then to express each fraction as an equivalent fraction with the denominator equal to the lowest common denominator.

The lowest common denominator of two or more fractions is the smallest number which is a multiple of all the denominators.

Rule 1:   To find the L.C.D. (lowest common denominator):

Step 1:   Factor each denominator completely.

Step 2:   Take all the different factors that occur in the factorizations of Step 1 and write them as a product.

Step 3:   Raise each of the factors of the product of Step 2 to the highest power to which they occur in Step 1.

The L.C.D. is the result of Step 3.

Example 1:   Find the L.C.D. of $\frac{1}{9r^2st}$ and $\frac{u}{6rs^3}$.

Step 1: $9r^2st = 3^2r^2st$ **Factor.**

$6rs^3 = 2 \cdot 3rs^3$ **Factor.**

Step 2: $2 \cdot 3rst$ **Write as a product all different factors.**

Step 3: $2 \cdot 3^2r^2s^3t$ **Raise each factor to its highest expressed power.**

Therefore, the L.C.D. $= 2 \cdot 3^2r^2s^3t$

$= 18r^2s^3t$

Example 2:   Find the L.C.D. of $\frac{x-2}{x^2-6x-7}$ and $\frac{x+5}{x^2+3x+2}$.

Step 1: $x^2 - 6x - 7 = (x - 7)(x + 1)$ **Factor.**

$x^2 + 3x + 2 = (x + 2)(x + 1)$ **Factor.**

Step 2: $(x - 7)(x + 2)(x + 1)$ **Write as a product all different factors.**

Step 3: $(x - 7)(x + 2)(x + 1)$ **Raise each factor to its highest expressed power.**

Example 3:   Find the L.C.D. of $\frac{3q}{2p^2}, \frac{q^3}{4p^2-4}$ and $\frac{4}{p^2-p}$.

Step 1: $2p^2 = 2p^2$ **Factor.**

$4p^2 - 4 = 4(p^2 - 1) = 2^2(p + 1)(p - 1)$ **Factor.**

$p^2 - p = p(p - 1)$ **Factor.**

Step 2: $2p(p - 1)(p + 1)$ **Write as a product all different factors.**

Step 3: $2^2p^2(p - 1)(p + 1)$ **Raise each factor to its highest expressed power.**

Therefore, the L.C.D. $= 4p^2(p - 1)(p + 1)$.

Example 4: Find the L.C.D. of $\dfrac{5R}{R^2 + 6R + 9}$, $\dfrac{2}{R^2 - 9}$ and $\dfrac{R^2}{5R + 15}$.

Step 1: $R^2 + 6R + 9 = (R + 3)(R + 3) = (R + 3)^2$ **Factor.**

$R^2 - 9 \qquad\quad = (R + 3)(R - 3)$ **Factor.**

$5R + 15 \qquad\quad = 5(R + 3)$ **Factor.**

Step 2: $5(R + 3)(R - 3)$ **Write as a product all different factors.**

Step 3: $5(R + 3)^2(R - 3)$ **Raise each factor to the highest power to which it occurs.**

Therefore, the L.C.D. $= 5(R - 3)(R + 3)^2$

Rule 2: To convert two or more algebraic fractions to a common denominator:

Step 1: Find the L.C.D. of the fractions.

Step 2: Write an equivalent fraction with denominator equal to the L.C.D. for each of the given fractions.

Example 5: Convert to a common denominator: $\dfrac{t}{9i^2r}$ and $\dfrac{2t}{6ir^3}$.

Step 1: L.C.D. $= 18i^2r^3$

Step 2: $\dfrac{t}{9i^2r} = \dfrac{t}{9i^2r} \cdot \dfrac{2r^2}{2r^2} = \dfrac{2r^2t}{18i^2r^3}$ **Multiply the numerator and the denominator by $2r^2$ in the first fraction and $3i$ in the second fraction, so that the equivalent fractions will have denominators equal to the L.C.D.**

$\dfrac{2t}{6ir^3} = \dfrac{2t}{6ir^3} \cdot \dfrac{3i}{3i} = \dfrac{6it}{18i^2r^3}$

Example 6: Convert to a common denominator: $\dfrac{x - 2}{x^2 + 6x - 7}$ and $\dfrac{x + 5}{x^2 - 3x + 2}$.

Step 1: $x^2 + 6x - 7 = (x + 7)(x - 1)$

$x^2 - 3x + 2 = (x - 2)(x - 1)$

L.C.D. $= (x + 7)(x - 2)(x - 1)$

Step 2: $\dfrac{x - 2}{x^2 + 6x - 7} = \dfrac{(x - 2)}{(x + 7)(x - 1)} \cdot \dfrac{(x - 2)}{(x - 2)}$

$= \dfrac{(x - 2)^2}{(x + 7)(x - 1)(x - 2)}$

$= \dfrac{x^2 - 4x + 4}{(x + 7)(x - 1)(x - 2)}$

$\dfrac{x + 5}{x^2 - 3x + 2} = \dfrac{(x + 5)}{(x - 2)(x - 1)} \cdot \dfrac{(x + 7)}{(x + 7)}$

$= \dfrac{(x + 5)(x + 7)}{(x - 2)(x - 1)(x + 7)}$

$= \dfrac{x^2 + 12x + 35}{(x - 2)(x - 1)(x + 7)}$

EXERCISES 7.3

Find the L.C.D. of the following algebraic fractions and convert to equivalent fractions with common denominators:

1. $\dfrac{4}{15a^3b}$, $\dfrac{5}{12ab^2}$

2. $\dfrac{5y}{18x^3}$, $\dfrac{7x}{30x^2y}$

3. $\dfrac{3}{e^2 i^2}, \dfrac{-2}{eir}$

4. $\dfrac{2}{2\alpha^2\beta^3}, \dfrac{2}{3\alpha\beta}, \dfrac{3}{4\alpha^2\beta^4}$

5. $\dfrac{X-1}{X^2-9}, \dfrac{3X+2}{X^2-X-6}$

6. $\dfrac{h}{h^2-5h+6}, \dfrac{2h-1}{h^2-4}$

7. $\dfrac{1}{E^2-I^2}, \dfrac{1}{(E-I)^2}$

8. $\dfrac{1}{Z^2+Z}, \dfrac{1}{Z^2-1}$

9. $\dfrac{2e}{E^2-Ee}, \dfrac{E}{Ee-e^2}$

10. $\dfrac{ab}{3a-6}, \dfrac{2a}{a^2-2a}$

11. $\dfrac{3}{R^2-4}, \dfrac{2}{R^2-1}, \dfrac{1}{R^2-3R+2}$

12. $\dfrac{x}{x^2-9}, \dfrac{2}{x^2+x-6}, \dfrac{x}{x^2-4x+4}$

13. $\dfrac{3}{v-t}, \dfrac{s}{v+t}, \dfrac{t^2}{v-t}$

14. $\dfrac{\theta}{\theta-\phi}, \dfrac{\phi}{\theta+\phi}$

15. $\dfrac{5}{w^2-w-2}, \dfrac{2}{w^2-4}, \dfrac{w-2}{w^2-4w-5}$

16. $\dfrac{Z-1}{Z^2-25}, \dfrac{Z}{5Z-25}, \dfrac{Z-2}{10Z+50}$

17. $\dfrac{7E}{2IR-2I^2}, \dfrac{5I}{3R^2-3IR}$

18. $\dfrac{3}{r^2+3r+2}, \dfrac{2}{r^2-2r-3}$

19. One resistance is given as $\dfrac{1}{r_1}$ and another resistance is given as $\dfrac{1}{r_2}$. Find the common denominator of these parallel resistances.

20. The differential of a pulley is given as $\dfrac{2pr}{r-2}$. The differential of a second pulley is given as $\dfrac{2pr}{r-5}$. Find the common denominator of the two differentials.

7.4   Addition and Subtraction of Algebraic Fractions

Addition and subtraction of algebraic fractions involve the same principles as addition and subtraction of fractions in arithmetic. When the denominators are the same, the numerators are added and the sum is placed over the common denominator. When the denominators are different, it is necessary to express each fraction as an equivalent fraction with denominator equal to the L.C.D. in order to be able to add or subtract the fractions.

Rule: To add or subtract algebraic fractions:

Step 1:   Find the L.C.D.

Step 2:   If the fractions have a common denominator, go to Step 3. If the denominators are different, convert the fractions to equivalent fractions with denominators equal to the L.C.D.

Step 3:   Add or subtract the numerators and write the result over the common denominator.

Step 4:   Reduce to lowest terms, when possible.

Example 1: $\dfrac{6}{5xy} + \dfrac{3a}{5xy} - \dfrac{7a}{5xy} =$

Steps 1 and 2:   The fractions already have common denominators.

Step 3: $\dfrac{6 + 3a - 7a}{5xy} = \dfrac{6 - 4a}{5xy}$ Add and subtract the numerators and write the result over the common denominator.

Step 4: $\dfrac{6 - 4a}{5xy}$ is in lowest terms.

Example 2: $\dfrac{7}{UV} + \dfrac{W}{3V^2} - \dfrac{5}{3W}$

Step 1:   L.C.D. $= 3UV^2W$

Step 2: $\dfrac{7}{UV} = \dfrac{7}{UV} \cdot \dfrac{3VW}{3VW} = \dfrac{21VW}{3UV^2W}$ Convert to equivalent fractions with denominators equal to the L.C.D.

$\dfrac{W}{3V^2} = \dfrac{W}{3V^2} \cdot \dfrac{UW}{UW} = \dfrac{UW^2}{3UV^2W}$

$\dfrac{5}{3W} = \dfrac{5}{3W} \cdot \dfrac{UV^2}{UV^2} = \dfrac{5UV^2}{3UV^2W}$

Step 3: $\dfrac{21VW + UW^2 - 5UV^2}{3UV^2W}$

The above calculations may also be performed as follows:

$$\frac{7}{UV} + \frac{W}{3V^2} - \frac{5}{3W} = \frac{21VW}{3UV^2W} + \frac{UW^2}{3UV^2W} - \frac{5UV^2}{3UV^2W}$$

$$= \frac{21VW + UW^2 - 5UV^2}{3UV^2W}$$

Example 3: $\dfrac{E}{E-1} - \dfrac{2}{E^2-1} = \dfrac{E}{E-1} - \dfrac{2}{(E-1)(E+1)}$

Step 1: L.C.D. $= (E-1)(E+1)$

Step 2: $\dfrac{E}{(E-1)} - \dfrac{2}{(E-1)(E+1)} = \dfrac{E(E+1)-2}{(E-1)(E+1)}$

Step 3: $= \dfrac{E^2+E-2}{(E-1)(E+1)}$

Step 4: $= \dfrac{\overset{1}{\cancel{(E-1)}}(E+2)}{\underset{1}{\cancel{(E-1)}}(E+1)}$

$= \dfrac{E+2}{E+1}$

Example 4: $\dfrac{5}{x+2} - \dfrac{3}{x-1}$

Step 1: L.C.D. $= (x+2)(x-1)$

Step 2 : $\dfrac{5}{x+2} - \dfrac{3}{x-1} = \dfrac{5(x-1)}{(x+2)(x-1)} - \dfrac{3(x+2)}{(x+2)(x-1)}$

Step 3: $= \dfrac{5(x-1)-3(x+2)}{(x+2)(x-1)}$

$= \dfrac{5x-5-3x-6}{(x+2)(x-1)}$

$= \dfrac{2x-11}{(x+2)(x-1)}$

Example 5: $\dfrac{3}{r^2-r-2} - \dfrac{4}{r^2-3r+2} + \dfrac{1}{r^2-2r+1}$

$= \dfrac{3}{(r-2)(r+1)} - \dfrac{4}{(r-2)(r-1)} + \dfrac{1}{(r-1)^2}$

$= \dfrac{3(r-1)^2}{(r-2)(r+1)(r-1)^2} - \dfrac{4(r-1)(r+1)}{(r-2)(r+1)(r-1)^2} + \dfrac{1(r-2)(r+1)}{(r-2)(r+1)(r-1)^2}$

$= \dfrac{3(r^2-2r+1) - 4(r^2-1) + (r^2-r-2)}{(r-2)(r+1)(r-1)^2}$

$= \dfrac{3r^2-6r+3-4r^2+4+r^2-r-2}{(r-2)(r+1)(r-1)^2}$

$= \dfrac{-7r+5}{(r-2)(r+1)(r-1)^2}$

EXERCISES 7.4

Perform the indicated operations and simplify the answers:

1. $\dfrac{a-2}{12} + \dfrac{a+3}{18}$

2. $\dfrac{2r-s}{15} - \dfrac{r-s}{21}$

3. $\dfrac{A}{6a} + \dfrac{a}{10b}$

4. $\dfrac{1}{4R^2} - \dfrac{h}{12R^3}$

5. $\dfrac{2y}{7w^2} - \dfrac{3w}{14y}$

6. $\dfrac{4}{15M^3N} + \dfrac{5}{12MN^2}$

7. $\dfrac{5\theta}{18\phi^3} - \dfrac{7\phi}{30\phi^2\theta}$

8. $\dfrac{3\gamma}{\gamma - 1} - \gamma$

9. $b + \dfrac{2}{b - 3}$

10. $\dfrac{x + 3}{x - 2} + \dfrac{x - 1}{x - 5}$

11. $\dfrac{B - 3}{B + 1} - \dfrac{B}{B - 2}$

12. $\dfrac{3}{d - L} - \dfrac{2}{d + L}$

13. $\dfrac{2q}{q^2 - 9} + \dfrac{q}{q - 3}$

14. $\dfrac{I}{I - 1} - \dfrac{5I}{I^2 - 1}$

15. $\dfrac{s - 1}{s^2 - 9} + \dfrac{3s + 2}{s^2 - s - 6}$

16. $\dfrac{E}{E^2 - 5E + 6} - \dfrac{2E - 1}{E^2 - 4}$

17. $\dfrac{2r}{i^2 - ir} + \dfrac{3i}{ir - r^2}$

18. $\dfrac{\alpha\beta}{3\alpha - 6} - \dfrac{2\beta}{\alpha^2 - 2\alpha}$

19. $\dfrac{2z - 3}{z^2 - 5z + 6} - \dfrac{3}{z - 2}$ 20. $\dfrac{2f}{f^2 - 9f + 20} - \dfrac{5f - 2}{f - 5}$

21. $\dfrac{3}{R^2 - 4} - \dfrac{R}{R^2 - 1} + \dfrac{1}{R^2 + R - 2}$ 22. $\dfrac{w - 1}{w^2 - 25} - \dfrac{w}{5w - 25} + \dfrac{w - 2}{10w + 50}$

23. In compound interest problems where payments are uniform, the present worth factor, w, is equal to

$$\frac{1}{i} - \frac{1}{i(1 + i)^n}$$

Express w in another form by performing the subtraction above.

24. The weight, w, of a person is given by

$$\frac{GMm}{r^2} - \frac{mv^2}{r}$$

Perform the subtraction above to obtain another expression for w.

25. If $p = \dfrac{m}{d - L} - \dfrac{m}{d + L}$, simplify p.

7.5 Fractional Equations

To solve a fractional equation it is helpful to eliminate the denominators. This is accomplished by finding the L.C.D. of all the denominators in the equation and then multiplying both members of the equation by the L.C.D. When this multiplication is performed and both members of the equation are simplified, the resulting equation can be solved as those in Unit 4.

It is essential to substitute the solutions obtained back into the original equation in order to assure their validity. It is necessary to check for validity because extraneous roots might have been introduced by multiplying both sides by the L.C.D. This has occurred if a solution would make the value of the L.C.D. equal to zero.

Rule: To solve a fractional equation:

Step 1:   Find the L.C.D. of all the denominators in the equation.

Step 2:   Multiply both members of the equation by the L.C.D.

Step 3:   Simplify each member of the equation.

Step 4:   Solve the resulting equation.

Example 1:   Solve $\dfrac{t}{4} - \dfrac{t-2}{5} = \dfrac{3}{10}$.

Step 1:   L.C.D. = 20

Step 2: $(20)\left(\dfrac{t}{4} - \dfrac{t-2}{5}\right) = (20)\left(\dfrac{3}{10}\right)$

Multiply both members of the equation by the L.C.D.

Step 3: $(20) \cdot \dfrac{t}{4} - (20) \cdot \dfrac{(t-2)}{5} = (20) \cdot \dfrac{3}{10}$

$\overset{5}{(20)} \cdot \dfrac{t}{\underset{1}{4}} - \overset{4}{(20)} \cdot \dfrac{t-2}{\underset{1}{5}} = \overset{2}{(20)} \cdot \dfrac{3}{\underset{1}{10}}$

$$5t - 4(t - 2) = 2(3)$$

$$5t - 4t + 8 = 6$$

$$t + 8 = 6$$

Step 4: $t + 8 - 8 = 6 - 8$

$$t = -2$$

Check: $\dfrac{t}{4} - \dfrac{t-2}{5} = \dfrac{3}{10}$ **Check by substituting (-2) for every t.**

$$\dfrac{-2}{4} - \dfrac{-2-2}{5} = \dfrac{3}{10}$$

$$-\dfrac{1}{2} - \dfrac{-4}{5} = \dfrac{3}{10}$$

$$-\dfrac{5}{10} + \dfrac{8}{10} = \dfrac{3}{10}$$

$$\dfrac{3}{10} = \dfrac{3}{10}$$

Therefore, $t = -2$ satisfies the equation.

Example 2:   Solve $\dfrac{2}{w} + \dfrac{3}{w-3} = \dfrac{5}{w^2 - 3w}$ for w.

Step 1:   L.C.D. = $w(w - 3)$

Step 2: $w(w - 3)\left(\dfrac{2}{w} + \dfrac{3}{w-3}\right) = w(w - 3)\left(\dfrac{5}{w^2 - 3w}\right)$

Step 3: $w(w - 3) \cdot \dfrac{2}{w} + w(w - 3) \cdot \dfrac{3}{(w-3)} = w(w - 3)\dfrac{5}{w(w-3)}$

$\overset{1}{w}(w - 3) \cdot \dfrac{2}{\underset{1}{w}} + w\overset{1}{(w - 3)} \cdot \dfrac{3}{\underset{1}{(w-3)}} = \overset{1}{w}\overset{1}{(w-3)} \cdot \dfrac{5}{\underset{1}{w}\underset{1}{(w-3)}}$

$$(w - 3)(2) + (w)(3) = 5$$

$$2w - 6 + 3w = 5$$

$$5w - 6 = 5$$

Step 4: $5w - 6 + 6 = 5 + 6$

$$5w = 11$$

$$\frac{5w}{5} = \frac{11}{5}$$

$$w = \frac{11}{5}$$

Example 3: $\dfrac{2X - 4}{X - 3} = 3 + \dfrac{2}{X - 3}$

Step 1:   L.C.D. $= (X - 3)$

Step 2: $(X - 3)\left(\dfrac{2X - 4}{X - 3}\right) = (X - 3)\left(3 + \dfrac{2}{X - 3}\right)$

Step 3: $(X - 3) \cdot \dfrac{(2X - 4)}{(X - 3)} = (X - 3)(3) + (X - 3) \cdot \dfrac{2}{(X - 3)}$

$$2X - 4 = 3X - 9 + 2$$

$$2X - 4 = 3X - 7$$

Step 4: $2X - 4 - 3X = 3X - 7 - 3X$

$$-X - 4 = -7$$

$$-X - 4 + 4 = -7 + 4$$

$$-X = -3$$

$$X = 3$$

Check: $\dfrac{2X - 4}{X - 3} = 3 + \dfrac{2}{X - 3}$ **Check by substituting 3 for each X.**

$$\frac{2(3) - 4}{3 - 3} = 3 + \frac{2}{3 - 3}$$

$$\frac{6 - 4}{0} = 3 + \frac{2}{0}$$

Since division by zero is meaningless, there is no solution to the equation.

EXERCISES 7.5

Solve and check:

1. $\dfrac{1}{2x} = \dfrac{5}{4} - \dfrac{3}{x}$

2. $\dfrac{2}{a} + \dfrac{5}{3} = 6$

3. $\dfrac{1}{R} + \dfrac{3}{R} + \dfrac{5}{R} = 7$

4. $\dfrac{3}{G} + \dfrac{2}{G} + \dfrac{1}{G} = 1$

5. $\dfrac{2}{3p} + \dfrac{1}{3} = \dfrac{1}{15}$

6. $\dfrac{1}{w} + \dfrac{2}{w} = 3 - \dfrac{5}{w}$

7. $\dfrac{E-4}{3E} = \dfrac{3E-5}{4E}$

8. $\dfrac{x-1}{6x} = \dfrac{x+2}{9x} - \dfrac{2}{12}$

9. $\dfrac{r-2}{6r} = \dfrac{r+8}{9r}$

10. $\dfrac{2}{15A} = \dfrac{5}{10} - \dfrac{A-2}{3A}$

11. Given $Q = \dfrac{f}{F-f}$, solve for F.

12. $\dfrac{e-3}{e-2} = \dfrac{2e-4}{e+1} - 1$

13. Given $E = \dfrac{Z}{r+Z}$, solve for r.

14. $\dfrac{5x-1}{x-3} = 6 - \dfrac{x-3}{x-2}$

15. $\dfrac{i-1}{i^2-4} = \dfrac{3}{i+2}$

16. $\dfrac{8}{T+3} = \dfrac{5T-1}{T^2-9}$

17. $\dfrac{2}{b^2 - 9} = \dfrac{5}{b + 3} + \dfrac{4}{b - 3}$

18. $\dfrac{5}{4A - 8} = \dfrac{1}{A^2 - 4} - \dfrac{3}{A - 2}$

19. $\dfrac{5}{d^2 + d - 6} = \dfrac{2}{d^2 + 3d - 10}$

20. $\dfrac{5}{a^2 - 2a} = \dfrac{a}{a^2 - 3a + 2} - \dfrac{1}{a}$

21. $\dfrac{-3}{2A - 2} = \dfrac{2}{A^2 + 2A - 3} - \dfrac{3}{2A}$

22. $\dfrac{Z}{5z^2 - 20} = \dfrac{2}{10z} + \dfrac{1}{z^2 - 4}$

23. $\dfrac{2}{f^2 + f - 6} = \dfrac{3}{f^2 - f - 2} - \dfrac{1}{f^2 + 2f - 3}$

24. $\dfrac{5}{h^2 + 3h - 4} - \dfrac{2}{h^2 - 1} = \dfrac{3}{h^2 + 5h + 4}$

25. Given $\dfrac{R_g}{I} = \dfrac{R_s}{R_y + R_s}$, solve for R_g.

26. $\dfrac{3x - 7}{x - 4} + 4 = \dfrac{5}{x - 4}$

27. $\dfrac{1}{a - 3} - \dfrac{1}{a + 2} = \dfrac{7}{a^2 - a - 6}$

28. Given $I = \dfrac{nE}{Re + inR}$, solve for i.

29. Given $P = \dfrac{HD + 20}{100 + E}$, solve for D.

30. $\dfrac{1}{j + 1} + \dfrac{1}{j - 1} = \dfrac{4}{j^2 - 1}$

31. $\dfrac{1}{Q^2 - 4} = \dfrac{2}{Q^2 + 4Q + 4}$

32. Given $\dfrac{1}{C} = \dfrac{1}{C_1} + \dfrac{1}{C_2}$, solve for C.

33. Given $\dfrac{1}{R_T} = \dfrac{1}{R_1} + \dfrac{1}{R_2} + \dfrac{1}{R_3}$, solve for R_T.

34. The width of a rectangle is three fifths its length. If the perimeter of the rectangle is 32 meters, find the width of the rectangle.

35. A steel ring is heated and allowed to shrink over an aluminum cylinder. The formula

$$t_a s_a + t_s s_s = \dfrac{pD}{2}$$

relates the stresses caused by internal pressure where t_a is the thickness of the aluminum, t_s is the thickness of the steel ring, s_a is the stress on the aluminum, s_s is the stress on the steel, p is the internal pressure, and D is the diameter of the aluminum cylinder. Solve the formula for the stress on the aluminum.

Unit 7 Self-Evaluation

1. Reduce: $\dfrac{Q^2}{Q^2 - 5Q}$

2. Reduce: $\dfrac{3r_1 - 6r_2}{r_1{}^2 - 4r_2{}^2}$

3. Reduce: $\dfrac{X^2 - 5X + 4}{2X^2 + X - 3}$

4. Multiply: $\dfrac{3p}{p^2 - pq} \cdot \dfrac{5p - 5q}{6q}$

5. Divide: $\dfrac{I^2 - i^2}{3i} \div (I - i)$

6. Divide: $\dfrac{a^2 - 4b^2}{2a^2 - 4ab} \div \dfrac{5a + 10b}{15a^3}$

7. Subtract: $\dfrac{I}{IR - R^2} - \dfrac{R}{I^2 - IR}$

8. Add: $\dfrac{3}{u^2 - uv - 2v^2} + \dfrac{4}{u^2 + 3uv + 2v^2}$

9. Solve: $\dfrac{5w}{w - 3} = 8$

10. Solve: $\dfrac{r}{r^2 - 9} - \dfrac{1}{r + 5} = \dfrac{3}{r^2 + 8r + 15}$

Systems of Linear Equations

Unit 8 Objectives
1. To solve systems of linear equations in two variables by graphing.
2. To solve systems of linear equations algebraically.
3. To solve systems of linear equations in two variables using determinants.
4. To solve three equations in three unknowns using determinants.

unit 8

8.1 Solving Systems of Linear Equations by Graphing

Two linear equations in two variables is called a *system* of linear equations. To solve a system of linear equations is to find a single ordered pair, (x,y), that satisfies each equation. The equation $x + y = 10$ has an unlimited number of ordered pairs that satisfy the given equation. $(5,5)$, $(3,7)$, $(7,3)$, $(0,10)$, $(-2,12)$ are examples of ordered pairs that satisfy the given equation. That is, in each ordered pair, the sum of x and y is 10. The equation $2x - y = -4$ also has an unlimited number of ordered pairs which, when substituted in the equation, will yield a true statement. One method of obtaining the ordered pair that satisfies both equations is to graph the two equations on the same set of axes and determine their *point of intersection*. For example, to solve the system of linear equations $x + y = 10$ and $2x - y = -4$, each equation is graphed as follows:

Solve each equation for y:

$$x + y = 10 \qquad\qquad 2x - y = -4$$
$$y = -x + 10 \qquad\qquad -y = -2x - 4$$
$$\qquad\qquad\qquad y = 2x + 4$$

The line $x + y = 10$ has a y-intercept of 10 and the slope is -1.
The line $2x - y = -4$ has a y-intercept of 4 and the slope is 2.

The point of intersection is $(2,8)$. Therefore, $x = 2$ and $y = 8$ should satisfy each equation.

$$x + y = 10 \qquad\qquad 2x - y = -4$$
$$(2) + (8) = 10 \qquad\qquad 2(2) - (8) = -4$$
$$10 = 10 \qquad\qquad -4 = -4$$

Equations that have one solution are said to be *consistent*.

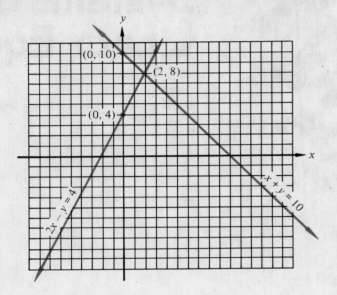

Figure 8.1

If two equations, when graphed, do not intersect at one point (i.e., if the lines are parallel), the system of equations is said to be *inconsistent*. Here there is no solution since no single ordered pair will satisfy each equation.

Two equations are said to be *dependent* if, when graphed, the two lines are the same line. There will be an infinite number of solutions, since every ordered pair that satisfies one equation will also satisfy the other equation.

Example 1: Graph to find the solution and tell whether the system is consistent, inconsistent, or dependent:

$$4x + 2y = 6$$
$$6x + 3y = -9$$

To graph, solve each equation for y:

$$4x + 2y = 6 \qquad\qquad 6x + 3y = -9$$
$$2y = -4x + 6 \qquad\qquad 3y = -6x - 9$$
$$y = -2x + 3 \qquad\qquad y = -2x - 3$$

Note that each line has a slope, -2, but the y-intercepts are not equal. Lines with the same slope are parallel.

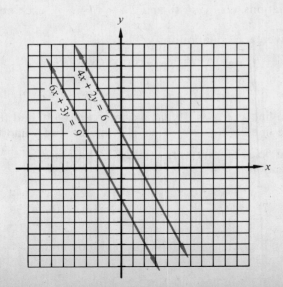

Figure 8.2

Since the lines are parallel, as seen in Figure 8.2, there is no solution, and the system is inconsistent.

Example 2: Graph to find the solution, and tell whether the system is consistent, inconsistent, or dependent:

$$2x + 3y = 12$$
$$-\frac{2}{3}x - y = -4$$

To graph, solve each equation for y:

$$2x + 3y = 12 \qquad\qquad -\frac{2}{3}x - y = -4$$

$$3y = -2x + 12 \qquad\qquad -y = \frac{2}{3}x - 4$$

$$y = -\frac{2}{3}x + 4 \qquad\qquad y = -\frac{2}{3}x + 4$$

The two equations have the same slope and the same y-intercept.

When graphed (Figure 8.3), the two lines are the same line. Therefore, the system is dependent. Every ordered pair that satisfies the first equation also satisfies the second equation.

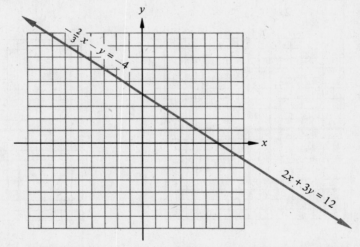

Figure 8.3

EXERCISES 8.1

Solve the following systems by graphing and tell whether they are consistent, inconsistent, or dependent. If consistent, state the solution.

1. $2x = 3y$
$x - y = 1$

2. $5x + y = -5$
$x + 2 = 0$

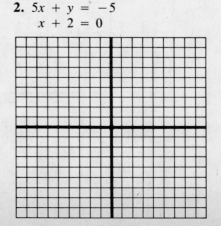

3. $3x - y = 2$
$y = 3x + 5$

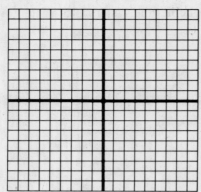

4. $y + 1 = 0$
$y = 2x - 1$

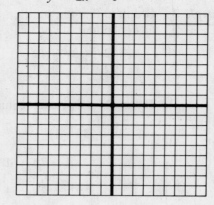

5. $y = x + 7$
$y = 2x + 11$

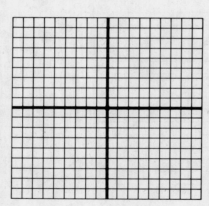

6. $3y = x - 6$
$y = \frac{1}{3}x - 2$

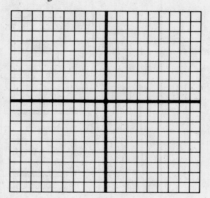

7. $5y - x = 0$
$5y + x = -10$

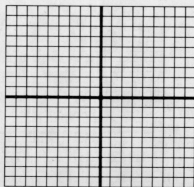

8. $x + y = 2$
$3x - 2y = 1$

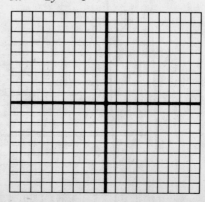

9. $2x - y = -1$
$\quad 4x - 2y = 2$

10. $\quad 3x - 7y = -7$
$\quad\quad -\frac{3}{7}x + y = 1$

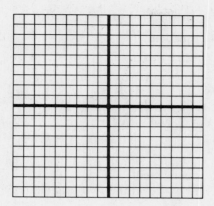

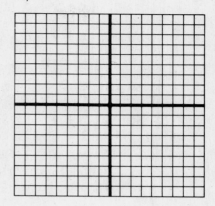

11. $y = x$
$\quad y = 6 - x$

12. $2y = 4x$
$\quad\quad x = 2 + y$

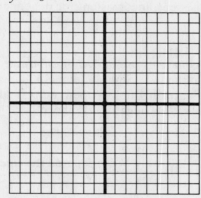

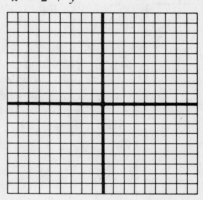

8.2   Solving Systems of Linear Equations Algebraically

It is not always practical to solve systems of equations graphically. For example, if the solution of two equations is $\left(\frac{9}{11}, \frac{7}{11}\right)$, the point would be difficult to determine graphically. Systems of linear equations may also be solved algebraically. Recall that in solving linear equations, the *same quantity* was added to or subtracted from both sides of an equation. This concept can be extended by adding or subtracting *equal quantities* to the given equal quantities.

Rule 1: To solve two linear equations by addition or subtraction:

Step 1:   Add or subtract the two equations to obtain an equivalent equation which contains only one variable. If neither variable is eliminated by adding or subtracting the two equations, multiply either (or both) equation(s) by a number (or numbers) that will make the numerical coefficients of one variable alike.

Step 2:   Solve the resulting linear equation in one variable.

Step 3:   Substitute the variable obtained in either of the given equations and solve for the remaining variable.

Example 1:   Find the solution of the following system using the addition or subtraction method. Tell whether the system is consistent, inconsistent, or dependent:

$$x + y = 3$$
$$x - y = 1$$

Step 1: $x + y = 3$ **Add the two equations.**
$$\underline{x - y = 1}$$
$$2x \quad\; = 4$$

Step 2: $2x = 4$ **Solve for x.**
$$x = 2$$

Step 3: $x + y = 3$ **Substitute $x = 2$, and solve for y.**
$$2 + y = 3$$
$$y = 1$$

The solution is $(2,1)$. The system is consistent.

To check, substitute $x = 2$ and $y = 1$ in the two original equations:

$$x + y = 3 \qquad x - y = 1$$
$$2 + 1 = 3 \qquad 2 - 1 = 1$$
$$3 = 3 \qquad\quad 1 = 1$$

Example 2:   Find the solution of the following system using the addition or subtraction method. Tell whether the system is consistent, inconsistent, or dependent:

$$x + 2y = 8$$
$$3x + \;\;y = 5$$

Step 1:   Note that if the two equations are added or subtracted, an equation in one variable is not obtained. Since adding or subtracting does not eliminate one of the variables unless their numerical coefficients are equal, or unless one is the negative of the other, multiply the first equation by -3.

$$-3(x + 2y) = -3(8) \rightarrow -3x - 6y = -24$$
$$3x + y = 5 \rightarrow \underline{3x + \;\;y = \quad\; 5} \quad \textbf{Add the equations.}$$
$$-5y = -19$$

$$y = \frac{19}{5} \quad \textbf{Solve for the variable.}$$

$$x + 2y = \quad 8 \quad \textbf{Substitute } y = \frac{19}{5} \textbf{ in one of the}$$
$$\textbf{given equations, and solve for } x.$$

$$x + 2\left(\frac{19}{5}\right) = 8$$

$$x + \frac{38}{5} = \frac{40}{5}$$

$$x = \frac{2}{5}$$

The solution is the ordered pair, $\left(\frac{2}{5}, \frac{19}{5}\right)$. The system is consistent.

Example 3:   Solve the given system algebraically and tell whether the system is consistent, inconsistent, or dependent:

$$3a - \;\;b = \;\;6$$
$$6a - 2b = 12$$

Multiplying the first equation by 2 and leaving the second equation as is:

$$2(3a - b) = 2(6) \rightarrow 6a - 2b = 12$$
$$6a - 2b = \;\;12 \;\rightarrow 6a - 2b = 12$$

Note that the first and second equations are now the same. Therefore, the two lines are dependent and every point on either line is a solution. If the two equations had been subtracted, the result would be $0 = 0$:

$$6a - 2b = 12$$
$$\underline{6a - 2b = 12}$$
$$0 = 0$$

The result, $0 = 0$, which is a true statement, denotes that the two equations are dependent.

Example 4: Solve the given system algebraically and tell whether the system is consistent, inconsistent, or dependent:

$$R + T = 8$$
$$-R - T = 7$$

$$R + T = 8 \quad \textbf{Add the two equations.}$$
$$\underline{-R - T = 7}$$
$$0 = 15$$

The statement $0 = 15$ is a false statement. Therefore, no ordered pair will satisfy the two given equations. The lines are parallel lines and the system is inconsistent.

Systems of linear equations may also be solved algebraically by the *substitution* method.

Rule 2: To solve two linear equations by substitution:

Step 1: Solve either of the given equations for one of the variables.

Step 2: Substitute the results in the other equation and solve.

Step 3: Substitute the variable found in either of the given equations and solve for the remaining variable.

Example 5: Solve the system using the substitution method.

$$2R - 3T = 3$$
$$R + T = 4$$

In choosing one of the two equations to solve for R or T, note that $R + T = 4$ may be solved more easily for R or T than the equation $2R - 3T = 3$.

Step 1:
$$R + T = 4$$
$$R = 4 - T$$

Step 2:
$$2R - 3T = 3$$
$$2(4 - T) - 3T = 3 \quad \textbf{Substitute } 4 - T \textbf{ for } R.$$
$$8 - 2T - 3T = 3 \quad \textbf{Solve for } T.$$
$$8 - 5T = 3$$
$$-5T = -5$$
$$T = 1$$

Step 3:
$$R + T = 4$$
$$R + 1 = 4 \quad \textbf{Substitute } T = 1.$$
$$R = 3 \quad \textbf{Solve for } R.$$

Since $R = 3$ and $T = 1$ satisfy each of the given equations, the system is consistent.

Check:
$$R + T = 4 \qquad\qquad 2R - 3T = 3$$
$$3 + 1 = 4 \qquad\qquad 2(3) - 3(1) = 3$$
$$4 = 4 \qquad\qquad\quad 6 - 3 = 3$$
$$3 = 3$$

Example 6: A board measuring 65 in. is cut into two pieces. The smaller piece is 1 in. longer than one third the length of the larger piece. What are the lengths of the two pieces?

Let x = the length of the smaller piece, and let y = the length of the larger piece.

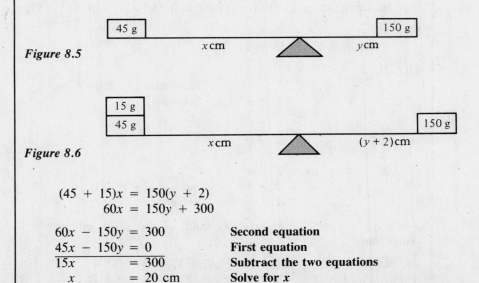

$$x + y = 65$$

$$x = \frac{1}{3}y + 1$$

Figure 8.4

$\frac{1}{3}y + 1 + y = 65$	**Substitute the second equation in the first equation and solve for y.**
$y + 3 + 3y = 195$	
$4y + 3 = 195$	
$4y = 192$	
$y = 48$	
$x + y = 65$	**Substitute $y = 48$ into first equation to obtain value for x.**
$x + 48 = 65$	
$x = 17$	

Therefore, the lengths of the two pieces of board are 48 in. and 17 in.

Check: The sum of the two pieces, 48 in. + 17 in. = 65 in. 1 added to $\frac{1}{3}$ of 48 = 17 in.

Example 7: A lever is balanced on a fulcrum with weights of 45 g at one end and 150 g at the other end (Fig. 8.5). The lever remains balanced when a 15-g weight is added to the 45-g weight if the 150-g weight is moved 2 cm farther from the fulcrum (Fig. 8.6). Find the original lengths of the arms of the lever.

Let x = the distance from the fulcrum to the 45-g weight, and let y = the distance from the fulcrum to the 150-g weight.

Figure 8.5

Figure 8.6

$$(45 + 15)x = 150(y + 2)$$
$$60x = 150y + 300$$

$60x - 150y = 300$	**Second equation**
$45x - 150y = 0$	**First equation**
$15x \qquad = 300$	**Subtract the two equations**
$x \qquad = 20$ cm	**Solve for x**

$$45x \quad\quad = 150y$$
$$45(20) = 150y$$
$$900 = 150y$$
$$6 = y$$
$$y = 6 \text{ cm}$$

Substitute $x = 20$ in one of the equations. Solve for y

Therefore, the 45-g weight is 20 cm from the fulcrum, and the 150-g weight is 6 cm from the fulcrum.

Check: $45(20) = 150(6)$
$$900 = 900$$

The method of solving two equations algebraically can be extended to solve three equations.

Rule 3:   To solve three linear equations in three unknowns algebraically:

Step 1:   Choose two of the three equations and eliminate any one of the three variables (as in Step 1 of Rule 1).

Step 2:   Choose a different pair of equations and eliminate the same variable eliminated in Step 1.

Step 3:   Solve the equations obtained from the results of Steps 1 and 2 by using Rule 1.

Step 4:   Substitute the two variables obtained in Step 3 in any one of the three given equations and solve for the third variable.

Example 8:   Solve:
$$2x + y + 4z = 12$$
$$3x + 3y + z = 0$$
$$5x - y - 3z = -2$$

Step 1:
$$2x + y + 4z = 12$$
$$\underline{5x - y - 3z = -2}$$
$$7x \quad\quad + z = 10$$

Choose the first and last equations and add to eliminate y.

Step 2:
$$-6x - 3y - 12z = -36$$
$$\underline{3x + 3y + \quad z = \quad 0}$$
$$-3x \quad\quad - 11z = -36$$

Choose the first two equations. Multiply the first equation by -3 and add the equations.

Step 3:
$$7x + \quad z = \quad 10$$
$$-3x - 11z = -36$$

$$77x + 11z = 110$$
$$\underline{-3x - 11z = -36}$$
$$74x \quad\quad = \quad 74$$
$$x = \quad 1$$

$$7x + z = 10$$
$$7(1) + z = 10$$
$$7 + z = 10$$
$$z = 3$$

Solve the two equations from Steps 1 and 2 for x and z using Rule 1.

Step 4:
$$2x + y + 4z = 12$$
$$2(1) + y + 4(3) = 12$$
$$2 + y + 12 = 12$$
$$y + 14 = 12$$
$$y = -2$$

Substitute $x = 1$ and $z = 3$ in the first equation and solve for y.

The solution is written in the form of an ordered triple, (x,y,z). Therefore, the solution is $(1, -2, 3)$.

EXERCISES 8.2

Solve the following systems of linear equations using the addition or subtraction method. Tell whether the systems are consistent, inconsistent or dependent. If consistent, state the solution.

1. $x + y = 11$
$x - y = 7$

2. $2x + y = 10$
$2x + 4y = 4$

3. $3p + 2q = 5$
$5p - 2q = 3$

4. $E + 2I = 6$
$3E + 2I = 6$

5. $3a - 7b = -11$
$3a - 5b = -8$

6. $3R_1 - 4R_2 = 1$
$2R_1 + 4R_2 = 7$

7. $4X_L + X_c = 12$
$X_c = 4$

8. $x + 5y = 14$
$x = -6$

9. $2X + Y = 3$
$-2X - Y = 6$

10. $2c - d = 6$
$3c + 2d = 9$

11. $3P_1 + 2P_2 = -4$
$9P_1 + 5P_2 = 8$

12. $4s - 3t = 5$
$8s - 6t = 10$

13. $7J - 2K = 8$
$5J - K = 3$

14. $7L_0 - 10L_1 = 24$
$11L_0 + 5L_1 = 2$

15. $2f + 3F = 1$
$3f + 7F = -1$

16. $11m + 5n = -7$
$-9m - 4n = 6$

17. $6E - 9e = 2$
$9E - 7e = -1$

18. $8\theta + 6\phi = 3$
$12\theta - 10\phi = 11$

19. $\frac{2}{3}Z_1 + \frac{5}{9}Z_2 = 6$
$6Z_1 + 5Z_2 = 1$

20. $\frac{1}{5}S - \frac{3}{10}T = 2$
$6S - 9T = 60$

21. $.02x + .03y = .40$
$.04x - .06y = 1.6$

22. $1.9x - .7y = -1.8$
$3.8x - .6y = -4.4$

23. $\frac{x}{3} - \frac{y}{12} = \frac{1}{6}$

$\frac{x}{8} - \frac{y}{24} = \frac{1}{6}$

24. $\frac{x}{10} - \frac{y}{15} = \frac{1}{5}$

$\frac{3x}{20} + \frac{y}{10} = \frac{1}{4}$

25. $\frac{x}{6} - \frac{5y}{10} = \frac{2}{5}$

$\frac{x}{18} + \frac{y}{3} = \frac{2}{3}$

Solve the following systems of linear equations using the substitution method. Tell whether the systems are consistent, inconsistent, or dependent. If consistent, state the solution.

26. $P + Q = 6$
$P + Q = -2$

27. $2L - 7D = 5$
$L - 6D = 0$

28. $3M - W = 0$
$M + 5W = -24$

29. $f - 3e = 4$
$2f - 4e = 4$

30. $7y - 2x = 3$
$x + y = 5$

31. $3a - 2b = 4$
$\frac{1}{2}a - \frac{1}{3}b = \frac{2}{3}$

32. $2L_0 - 6L_1 = 10$
$-L_0 - 2L_1 = -1$

33. $2h - 6k = -4$
$-3h + 7k = 22$

34. $\frac{1}{5}r - \frac{1}{5}s = 10$
$\phantom{\frac{1}{5}}r - \phantom{\frac{1}{5}}s = 2$

35. $2c - 6d = 10$
$c - 3d = 2$

Solve using any method.

36. The sum of two numbers is 87 and the difference between them is 21. Find the two numbers.

37. The sum of two voltages is 240 and their difference is 90. Find the voltages.

38. When the width of a rectangle is decreased by two cm and its length is doubled, its perimeter increases by 2 cm. Find the original dimensions of the rectangle if the original perimeter is 28 cm.

39. When a weight of 200 lb is placed on one end of a fulcrum and a weight of 360 lb is placed on the other end, a lever is balanced. If 100 lb is added to the 200 lb weight, the 360 lb weight must be moved $2\frac{1}{2}$ ft farther away to balance the lever. Find the original lengths of the two arms of the lever.

40. Two cement blocks have a total weight of 180 kg. One block weighs 30 kg less than twice the other. How much does each weigh?

41. Two rotating pulleys, P_1 and P_2, have the following relationship:

A small pulley is 8 cm in diameter while a larger pulley is 12 cm in diameter. If the smaller pulley is replaced with one which has a 10-cm diameter, the rpm of the larger pulley is increased by 10. Find the rpm of each of the original pulleys.

$$(\text{diameter of } P_1)(\text{rpm } P_1) = (\text{diameter of } P_2)(\text{rpm } P_2)$$

Solve the following systems of linear equations using Rule 3:

42.
$$2x + 4y + z = 6$$
$$-3x + 2y - z = -14$$
$$x + y + z = 1$$

43.
$$-x + 3y + 2z = 12$$
$$x - y + 3z = -7$$
$$x + 2y - 4z = 15$$

44.
$$5x - 3y + z = 18$$
$$2x + y - 3z = 14$$
$$-x + 3y - 2z = 0$$

45.
$$-2x - 4y - 5z = 1$$
$$3x + 6y + 5z = -4$$
$$4x - y + z = 16$$

8.3   Second Order Determinants

Solve the following equations algebraically:

$$a_1x + b_1y = c_1$$
$$a_2x + b_2y = c_2$$

Multiply the first equation by b_2 and the second equation by b_1.

$$a_1b_2x + b_1b_2y = b_2c_1$$
$$a_2b_1x + b_1b_2y = b_1c_2$$

Subtract $(a_1b_2 - a_2b_1)x = b_2c_1 - b_1c_2$

Solve for x. $x = \dfrac{b_2c_1 - b_1c_2}{a_1b_2 - a_2b_1}$

Instead of substituting the value of x in one of the original equations to find y, the equations may be solved again to determine y. Multiply the first equation by a_2 and the second equation by a_1.

$$a_1a_2x + a_2b_1y = a_2c_1$$
$$a_1a_2x + a_1b_2y = a_1c_2$$
$$\overline{(a_2b_1 - a_1b_2)y = a_2c_1 - a_1c_2}$$

$$y = \frac{a_2c_1 - a_1c_2}{a_2b_1 - a_1b_2} = \frac{a_1c_2 - a_2c_1}{a_1b_2 - a_2b_1}$$

Note that the denominators of x and y are the same, $a_1b_2 - a_2b_1$. The denominator may be expressed as

$$\begin{vmatrix} a_1 & b_1 \\ a_2 & b_2 \end{vmatrix}$$

An expression of this form is called a second order *determinant*. It has two rows, a_1b_1 and a_2b_2, and two columns, $\begin{matrix} a_1 \\ a_2 \end{matrix}$ and $\begin{matrix} b_1 \\ b_2 \end{matrix}$. a_1, a_2, b_1, and b_2 are called *elements* of the determinant. By definition,

$$\begin{vmatrix} a_1 & b_1 \\ a_2 & b_2 \end{vmatrix} = a_1b_2 - a_2b_1$$

The definition suggests the following: to evaluate a second order determinant, multiply the elements in the first diagonal and subtract the product of the elements in the second diagonal.

$$\begin{vmatrix} a_1 & b_1 \\ a_2 & b_2 \end{vmatrix} \begin{matrix} \text{second diagonal} \\ \text{first diagonal} \end{matrix} = a_1b_2 - a_2b_1$$

Using this definition, the numerators of x and y can also be written as determinants.

$$\begin{vmatrix} c_1 & b_1 \\ c_2 & b_2 \end{vmatrix} = b_2c_1 - b_1c_2$$

$$\begin{vmatrix} a_1 & c_1 \\ a_2 & c_2 \end{vmatrix} = a_1c_2 - a_2c_1$$

Note that in comparing the determinant for the numerator for x to the determinant for the denominator, the coefficients of x have been replaced by the constants. In comparing the determinant for the numerator of y to the determinant for the denominator, the coefficients of y have been replaced by the constants.

The use of determinants to solve systems of linear equations is known as *Cramer's Rule*.

Example 1:   Use determinants to solve

$$2x + 3y = 1$$
$$5x + 6y = 4$$

Recall that a_1 and a_2 were the coefficients of x, b_1 and b_2 were the coefficients of y, and c_1, c_2 were the constants. Therefore, by substitution,

$$x = \frac{\begin{vmatrix} c_1 & b_1 \\ c_2 & b_2 \end{vmatrix}}{\begin{vmatrix} a_1 & b_1 \\ a_2 & b_2 \end{vmatrix}} = \frac{\begin{vmatrix} 1 & 3 \\ 4 & 6 \end{vmatrix}}{\begin{vmatrix} 2 & 3 \\ 5 & 6 \end{vmatrix}} = \frac{(1)(6) - (4)(3)}{(2)(6) - (5)(3)} = \frac{6 - 12}{12 - 15} = \frac{-6}{-3} = 2$$

$$y = \frac{\begin{vmatrix} a_1 c_1 \\ a_2 c_2 \end{vmatrix}}{\begin{vmatrix} a_1 b_1 \\ a_2 b_2 \end{vmatrix}} = \frac{\begin{matrix} 2 & 1 \\ 5 & 4 \end{matrix}}{\begin{matrix} 2 & 3 \\ 5 & 6 \end{matrix}} = \frac{(2)(4) - (5)(1)}{-3} = \frac{8 - 5}{-3} = \frac{3}{-3} = -1$$

The solution is $(2, -1)$, and the system is consistent.

Check:
$$2x + 3y = 1 \qquad\qquad 5x + 6y = 4$$
$$2(2) + 3(-1) = 1 \qquad 5(2) + 6(-1) = 4$$
$$4 - 3 = 1 \qquad\qquad 10 - 6 = 4$$
$$1 = 1 \qquad\qquad\quad 4 = 4$$

Example 2: Use determinants to solve

$$2C - 3D = 4$$
$$-2C + 3D = 5$$

$$C = \frac{\begin{vmatrix} c_1 b_1 \\ c_2 b_2 \end{vmatrix}}{\begin{vmatrix} a_1 b_1 \\ a_2 b_2 \end{vmatrix}} = \frac{\begin{matrix} 4 & 3 \\ 5 & 3 \end{matrix}}{\begin{matrix} 2 & 3 \\ -2 & 3 \end{matrix}} = \frac{(4)(3) - (5)(-3)}{(2)(3) - (-2)(-3)} = \frac{(12) - (-15)}{6 - 6} = \frac{27}{0}$$

Since 27 divided by zero is undefined, there is no solution. The lines are parallel and the system is inconsistent.

Example 3: Use determinants to solve

$$4x + y = 12$$
$$\tfrac{2}{3}x + \tfrac{1}{6}y = 2$$

$$x = \frac{\begin{vmatrix} c_1 b_1 \\ c_2 b_2 \end{vmatrix}}{\begin{vmatrix} a_1 b_2 \\ a_2 b_2 \end{vmatrix}} = \frac{\begin{matrix} 12 & 1 \\ 2 & \tfrac{1}{6} \end{matrix}}{\begin{matrix} 4 & 1 \\ \tfrac{2}{3} & \tfrac{1}{6} \end{matrix}} = \frac{(12)\left(\tfrac{1}{6}\right) - (2)(1)}{(4)\left(\tfrac{1}{6}\right) - \left(\tfrac{2}{3}\right)(1)} = \frac{2 - 2}{\tfrac{2}{3} - \tfrac{2}{3}} = \frac{0}{0}$$

$\frac{0}{0}$ indicates that there are an infinite number of solutions. The system is dependent. Every point on the line is a solution.

EXERCISES 8.3

Use determinants to solve the following systems of linear equations. Tell whether the systems are consistent, inconsistent, or dependent. If consistent, state the solution.

1. $2x + y = 9$
$3x - 2y = -4$

2. $3x - 5y = -1$
$x = 3$

3. $4x - 2y = 8$
$3x + y = -4$

4. $X_0 - X = 5$
$-2X_0 + 2X = 3$

5. $5x - 2y = 8$
$3x + y = -4$

6. $9x + y = -5$
$\phantom{9x + {}}y = 4$

7. $R - 4S = 7$
$3R + 2S = 7$

8. $5a - 10b = -1$
$15a + 20b = 5$

9. $3p - 5q = 1$
$-9p + 15q = 1$

10. $11F - 22H = -5$
$-33F + 44H = 1$

11. $.07c - .03d = -.05$
$.2c + .4d = -1.6$

12. $.3r + .4s = -.07$
$.2r + .02s = .052$

13. $\frac{3}{5}T_1 + \frac{2}{3}T_2 = 1$
$9T_1 + 10T_2 = 15$

14. $\frac{5}{6}L_1 + \frac{1}{12}L_0 = 6$
$\frac{2}{3}L_1 - \frac{1}{4}L_0 = 1$

15. $4m + 6n = -7$
$-6m - 8n = 9$

16. $\frac{5}{7}e + \frac{2}{3}d - 1 = 0$
$15e + 14d - 21 = 0$

17. $\dfrac{x}{4} - \dfrac{y}{3} = -\dfrac{1}{2}$

$\dfrac{x}{6} - \dfrac{y}{7} = -\dfrac{1}{3}$

18. $\dfrac{5x}{7} + \dfrac{2y}{3} = \dfrac{2}{21}$

$-\dfrac{3x}{5} - \dfrac{7y}{10} = \dfrac{1}{5}$

19. $215x - 786y = 571$
$-321x + 851y = -530$

20. $23x + 42y - 99 = 0$
$13x - 8y + 71 = 0$

21. $8K - 16M = -19$
$32K + 32M = 8$

22. $\dfrac{2}{3}W + \dfrac{5}{2}Z = -\dfrac{1}{6}$

$-8W - 30Z = \dfrac{1}{2}$

23. The sum of two numbers is 37, and one number is 7 more than one half the other. Find the two numbers.

24. The perimeter of a rectangle is 56 cm. The length is 1 cm more than twice the width. Find the dimensions of the rectangle.

25. A grocer wishes to mix candy selling for $5.20 a kg with candy selling for $3.20 a kg to get a mixture that he wishes to sell at $4.00 a kg. How many kilograms each of $5.20 candy and $3.20 candy should he use to get 30 kg of the mixture?

26. Mr. Burns invests a total of $12,000. He receives $1390 interest from investing part of the money at 11% for one year and the remainder at 13%. How much was invested at each rate?

27. The total weight of two packages to be mailed is 28 kg. The lighter of the packages weighs 6 kg less than the other. How much does each package weigh?

28. The total price of a house and lot was $54,000. Four times the price of the lot was $6000 less than the price of the house. Find the price of the house.

8.4   Third Order Determinants

A third order determinant has three rows and three columns. By definition

$$\begin{vmatrix} a_1 & b_1 & c_1 \\ a_2 & b_2 & c_2 \\ a_3 & b_3 & c_3 \end{vmatrix} = a_1 \begin{vmatrix} b_2 & c_2 \\ b_3 & c_3 \end{vmatrix} - a_2 \begin{vmatrix} b_1 & c_1 \\ b_3 & c_3 \end{vmatrix} + a_3 \begin{vmatrix} b_1 & c_1 \\ b_2 & c_2 \end{vmatrix}$$

Note that three second order determinants are obtained. The first is obtained by crossing out the row and column that contain a_1:

$$\begin{vmatrix} a_1 & b_1 & c_1 \\ a_2 & b_2 & c_2 \\ a_3 & b_3 & c_3 \end{vmatrix}$$

The second is obtained by crossing out the row and column that contain a_2:

$$\begin{vmatrix} a_1 & b_1 & c_1 \\ a_2 & b_2 & c_2 \\ a_3 & b_3 & c_3 \end{vmatrix}$$

The third is obtained by crossing out the row and column that contain a_3:

$$\begin{vmatrix} a_1 & b_1 & c_1 \\ a_2 & b_2 & c_2 \\ a_3 & b_3 & c_3 \end{vmatrix}$$

The second order determinant that remains after deleting the row and column that contain an element is called the *minor* of the element.

The minor of a_1 is $\begin{vmatrix} b_2 & c_2 \\ b_3 & c_3 \end{vmatrix}$.

The minor of a_2 is $\begin{vmatrix} b_1 & c_1 \\ b_3 & c_3 \end{vmatrix}$.

The minor of a_3 is $\begin{vmatrix} b_1 & c_1 \\ b_2 & c_2 \end{vmatrix}$.

Using the correct signs, a third order determinant may be evaluated by using the elements of any row or column with their respective minors. To evaluate

$$\begin{vmatrix} a_1 & b_1 & c_1 \\ a_2 & b_2 & c_2 \\ a_3 & b_3 & c_3 \end{vmatrix}$$

using the second row, first consider the elements in row 2:

a_2 is in column *1*, row *2*, and $1 + 2$ is an *odd* number. Therefore, use the negative of a_2.

b_2 is in colunm *2*, row *2*, and $2 + 2$ is an *even* number. Therefore, use b_2.

c_2 is in column *3*, row *2*, and $3 + 2$ is an *odd* number. Therefore, use the negative of c_2.

$$\begin{vmatrix} a_1 & b_1 & c_1 \\ a_2 & b_2 & c_2 \\ a_3 & b_3 & c_3 \end{vmatrix} = -a_2 \begin{vmatrix} b_1 & c_1 \\ b_3 & c_3 \end{vmatrix} + b_2 \begin{vmatrix} a_1 & c_1 \\ a_3 & c_3 \end{vmatrix} - c_2 \begin{vmatrix} a_1 & b_1 \\ a_3 & b_3 \end{vmatrix}$$

Verify that this is equivalent to the given definition for a third order determinant.

Example 1:   Evaluate $\begin{vmatrix} 2 & 1 & -2 \\ 3 & -5 & -3 \\ 4 & 6 & -1 \end{vmatrix}$ using the elements in column two and their respective minors.

First, consider the elements in column two:

1 is in column *2*, row *1*, and $1 + 2$ is odd. Use the negative of 1.
-5 is in column *2*, row *2*, and $2 + 2$ is even. Use -5.
6 is in column *2*, row *3*, and $2 + 3$ is odd. Use the negative of 6.

$$\begin{vmatrix} 2 & 1 & -2 \\ 3 & -5 & -3 \\ 4 & 6 & -1 \end{vmatrix} = -1 \begin{vmatrix} 3 & -3 \\ 4 & -1 \end{vmatrix} - 5 \begin{vmatrix} 2 & -2 \\ 4 & -1 \end{vmatrix} - 6 \begin{vmatrix} 2 & -2 \\ 3 & -3 \end{vmatrix}$$

$$= -1(9) - 5(6) - 6(0)$$
$$= -9 - 30$$
$$= -39$$

Example 2:   Evaluate $\begin{vmatrix} 6 & 3 & 5 \\ -2 & 4 & -3 \\ -1 & 2 & -6 \end{vmatrix}$ using the elements in row three and their respective minors.

-1 is in column *1*, row *3*, and $1 + 3$ is even. Use -1.
2 is in column *2*, row *3*, and $2 + 3$ is odd. Use the negative of 2.
-6 is in column *3*, row *3*, and $3 + 3$ is even. Use -6.

$$\begin{vmatrix} 6 & 3 & 5 \\ -2 & 4 & -3 \\ -1 & 2 & -6 \end{vmatrix} = -1 \begin{vmatrix} 3 & 5 \\ 4 & -3 \end{vmatrix} - 2 \begin{vmatrix} 6 & 5 \\ -2 & -3 \end{vmatrix} - 6 \begin{vmatrix} 6 & 3 \\ -2 & 4 \end{vmatrix}$$

$$= -1(-29) - 2(-8) - 6(30)$$
$$= 29 + 16 - 180$$
$$= -135$$

To solve three equations in three unknowns, such as

$$a_1x + b_1y + c_1z = d_1$$
$$a_2x + b_2y + c_2z = d_2$$
$$a_3x + b_3y + c_3z = d_3$$

the algebraic method of elimination by addition is used. Cramer's Rule is also applicable to systems of three equations in three unknowns. The solution is written as follows:

$$x = \frac{\begin{vmatrix} d_1 & b_1 & c_1 \\ d_2 & b_2 & c_2 \\ d_3 & b_3 & c_3 \end{vmatrix}}{\begin{vmatrix} a_1 & b_1 & c_1 \\ a_2 & b_2 & c_2 \\ a_3 & b_3 & c_3 \end{vmatrix}} \qquad y = \frac{\begin{vmatrix} a_1 & d_1 & c_1 \\ a_2 & d_2 & c_2 \\ a_3 & d_3 & c_3 \end{vmatrix}}{\begin{vmatrix} a_1 & b_1 & c_1 \\ a_2 & b_2 & c_2 \\ a_3 & b_3 & c_3 \end{vmatrix}} \qquad z = \frac{\begin{vmatrix} a_1 & b_1 & d_1 \\ a_2 & b_2 & d_2 \\ a_3 & b_3 & d_3 \end{vmatrix}}{\begin{vmatrix} a_1 & b_2 & c_1 \\ a_2 & b_2 & c_2 \\ a_3 & b_3 & c_3 \end{vmatrix}}$$

Note that all denominators are the same. The coefficients of x, y, and z are used in that order in the determinant. In comparing the determinant for the numerator of x to the determinant for the denominator of x, the column of coefficients of x has been replaced by the column of constants. In comparing the determinant for the numerator of y to the determinant for the denominator of y, the column of coefficients of y has been replaced by the column of constants. In a similar comparison of z, the column of coefficients of z has been replaced by the column of constants.

Example 3:   Using determinants, solve

$$\begin{aligned} x + y + z &= 6 \\ 2x + 3y - z &= 5 \\ 6x - 2y - 3z &= -7 \end{aligned}$$

$$x = \frac{\begin{vmatrix} d_1 & b_1 & c_1 \\ d_2 & b_2 & c_2 \\ d_3 & b_3 & c_3 \end{vmatrix}}{\begin{vmatrix} a_1 & b_1 & c_1 \\ a_2 & b_2 & c_2 \\ a_3 & b_3 & c_3 \end{vmatrix}} = \frac{\begin{vmatrix} 6 & 1 & 1 \\ 5 & 2 & -1 \\ -7 & -2 & -3 \end{vmatrix}}{\begin{vmatrix} 1 & 1 & 1 \\ 2 & 3 & -1 \\ 6 & -2 & -3 \end{vmatrix}} = \frac{-33}{-33} = 1$$

$$y = \frac{\begin{vmatrix} a_1 & d_1 & c_1 \\ a_2 & d_2 & c_2 \\ a_3 & d_3 & c_3 \end{vmatrix}}{\begin{vmatrix} a_1 & b_1 & c_1 \\ a_2 & b_2 & c_2 \\ a_3 & b_3 & c_3 \end{vmatrix}} = \frac{\begin{vmatrix} 1 & 6 & 1 \\ 2 & 5 & -1 \\ 6 & -7 & -3 \end{vmatrix}}{\begin{vmatrix} 1 & 1 & 1 \\ 2 & 3 & -1 \\ 6 & -2 & -3 \end{vmatrix}} = \frac{-66}{-33} = 2$$

$$z = \frac{\begin{vmatrix} a_1 & b_1 & d_1 \\ a_2 & b_2 & d_2 \\ a_3 & b_3 & d_3 \end{vmatrix}}{\begin{vmatrix} a_1 & b_1 & c_1 \\ a_2 & b_2 & c_2 \\ a_3 & b_3 & c_3 \end{vmatrix}} = \frac{\begin{vmatrix} 1 & 1 & 6 \\ 2 & 3 & 5 \\ 6 & -2 & -7 \end{vmatrix}}{\begin{vmatrix} 1 & 1 & 1 \\ 2 & 3 & -1 \\ 6 & -2 & -3 \end{vmatrix}} = \frac{-99}{-33} = 3$$

Check:
$$\begin{aligned} x + y + z &= 6 & 2x + 3y - z &= 5 & 6x - 2y - 3z &= -7 \\ 1 + 2 + 3 &= 6 & 2(1) + 3(2) - 3 &= 5 & 6(1) - 2(2) - 3(3) &= -7 \\ 6 &= 6 & 2 + 6 - 3 &= 5 & 6 - 4 - 9 &= -7 \\ & & 5 &= 5 & -7 &= 7 \end{aligned}$$

The solution is written as an ordered triple, (x,y,z). Therefore, the solution is $(1,2,3)$.

EXERCISES 8.4

Evaluate the following third order determinants.

1. $\begin{vmatrix} 2 & 3 & 1 \\ 2 & 1 & 4 \\ 2 & 3 & 1 \end{vmatrix}$

2. $\begin{vmatrix} -5 & 3 & 2 \\ 4 & -2 & -3 \\ 0 & 1 & -4 \end{vmatrix}$

3. $\begin{vmatrix} 2 & 1 & -2 \\ 3 & 4 & -30 \\ 2 & 8 & -2 \end{vmatrix}$

4. $\begin{vmatrix} 0 & 3 & -2 \\ 0 & -4 & 1 \\ 0 & 6 & 4 \end{vmatrix}$

5. $\begin{vmatrix} 3 & 2 & -4 \\ 1 & -1 & 5 \\ 3 & 2 & -4 \end{vmatrix}$

6. $\begin{vmatrix} 0 & -7 & 3 \\ 7 & 2 & 0 \\ 0 & 3 & 4 \end{vmatrix}$

7. $\begin{vmatrix} 10 & 14 & 2 \\ 4 & -20 & -8 \\ 1 & 16 & -30 \end{vmatrix}$

8. $\begin{vmatrix} 1 & 1 & 1 \\ 1 & 1 & 1 \\ 1 & 1 & 1 \end{vmatrix}$

9. $\begin{vmatrix} a_1 & b_2 & b_1 \\ a_2 & b_1 & b_2 \\ a_3 & b_1 & b_2 \end{vmatrix}$

10. $\begin{vmatrix} -.02 & -.4 & -.6 \\ .3 & .3 & .2 \\ 1.1 & .01 & .5 \end{vmatrix}$

Solve using determinants:

11.
$$3x + 2y + z = 3$$
$$2x - y + z = 5$$
$$5x + y - 3z = -2$$

12.
$$x + y + z = 5$$
$$-2x + 4y - z = -4$$
$$3x - 5y + 2z = 9$$

13.
$$4D + E - 3F = 10$$
$$-D + 2E + 2F = -8$$
$$D - 3E - 4F = 9$$

14.
$$5E - 10F + 20G = 13$$
$$10E + 5F + 10G = 9$$
$$2E + 2F + 3G = 2$$

15.
$$7R + 2S - 3T + 4 = 0$$
$$-5R - S + T - 6 = 0$$
$$2S + T - 2 = 0$$

16.
$$3R_0 - 2R_1 + R_2 = 4$$
$$5R_0 + R_1 + 6R_2 = 11$$
$$-3R_0 + 5R_1 + 2R_2 = -1$$

17.
$$9a - 4b - c = -22$$
$$7a - 2c = 11$$
$$-2a + b + c = 33$$

18.
$$6X + 2Y - 4Z - 12 = 0$$
$$-8X + 8Y - 2Z + 17 = 0$$
$$10X - 2Y + 8Z + 18 = 0$$

19.
$$12W + 14X - 16Y + 102 = 0$$
$$-15W + 13X + 18Y - 24 = 0$$
$$22W + 32X + 19Y - 84 = 0$$

20.
$$5M + 2N - P = 11$$
$$-4M - 2N + 4P = 0$$
$$7M + 6N - 6P = -13$$

21.
$$1.2I_1 + 3.2I_2 + .6I_3 = -1.4$$
$$.64I_2 - .31I_3 = -.95$$
$$2.3I_1 - .42I_2 + .5I_3 = 3.22$$

22.
$$\frac{2}{3}T_1 - 4T_2 + T_3 = -\frac{19}{6}$$
$$4T_1 - \frac{4}{9}T_2 + 4T_3 = -\frac{1}{3}$$
$$\frac{1}{2}T_1 - T_2 + \frac{1}{2}T_3 = -\frac{3}{4}$$

23. A piece of plywood measuring 96 in. is to be cut into three different lengths. The sum of the lengths of the smallest and the largest pieces is 4 in. less than twice the length of the other piece. The largest piece is to be 16 in. less in length than the sum of the lengths of the other two pieces. Find the lengths of each piece of plywood cut.

24. Three steel beams weigh a total of 2000 kg. The two smaller beams equal the weight of the largest beam, and four times the weight of the smallest beam equals the total weight of the other two beams. Find the weight of each beam.

25. The perimeter of a triangle is 30 cm. Twice the length of the smallest side is 6 cm less than the sum of the lengths of the other sides, and the largest side is 6 cm less in length than the sum of the other two sides of the triangle. Find the lengths of each side of the triangle.

26. A total of 72 screws, sizes G, H, and K, are needed to repair a machine. The sum of screws sizes G and H is 12 less than the number of size K needed. The number of size G is 2 less than the number of size H needed. Find the number of each size of screw needed to repair the machine.

Unit 8 Self-Evaluation

Solve the following linear systems, and tell whether they are consistent, inconsistent or dependent. If consistent, state the solution.

1. Graph: $2x - 3y = 6$
   $x + y = -7$

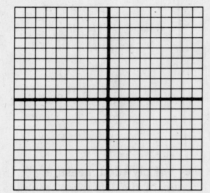

2. Solve by addition or subtraction method:

   $5E - 4F = 7$
   $3E - 5F = 12$

3. Solve using the substitution method:

   $2m - n = 6$
   $-3m + n = -12$

4. Solve using determinants:

   $8W + 4V = -1$
   $-12W - 8V = 5$

5. Solve using determinants and check algebraically:

$$2x - 3y - 4z = -5$$
$$5x - 2y - z = -14$$
$$6x + y + 3z = -15$$

Solve Exercises 6 through 10 using any method.

6. $6a - 9b = 7$
$9a + 3b = 5$

7. $\frac{7}{8}R - \frac{2}{3}T = \frac{1}{2}$
$-21R + 16T = -12$

8. $6.2D - 1.7L = 1.13$
$2.3D + .7L = .02$

9. $3p + 2q = 4$
$-6p - 4q = -1$

10. The sum of two numbers is 31, and one number is 7 more than $\frac{1}{5}$ the other. Find the numbers.

Exponents and Scientific Notation

Unit 9 Objectives
1. To simplify positive integral exponents.
2. To simplify zero exponents.
3. To simplify negative exponents.
4. To simplify fractional exponents.
5. To express numbers in scientific notation.
6. To multiply and divide numbers expressed in scientific notation.

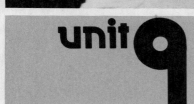

Exponents

Exponents offer a convenient way of expressing repeated factors. When repeated factors occur in a formula, scientists and technologists have opted to use exponents in order to simplify the formula.

For example, the minimum length of runway required by an airplane, given a jet-assisted take-off from a carrier deck, is given by s, where $s = \dfrac{v^2}{2a}$. This is simpler than expressing s without exponents, where $s = \dfrac{vv}{2a}$.

The amount of angular twist, θ, in twisting a round shaft is $\dfrac{2\tau L}{\pi n r^4}$. If θ were written without exponents, then it would equal $\dfrac{2\tau L}{\pi n r r r r}$. To find the number of times r occurs as a factor in the second formula, one would have to count the r's. However, by the use of exponents, the four r's are noted at a glance.

The impedance Z, given the resistance R, the inductive reactance ψ_L, and the capacitive reactance ψ_C, is equal to

$$\sqrt{R^2 + (\psi_L - \psi_C)^2}$$

The formula for the impedance written without exponents would appear in a more complicated form, as follows:

$$Z = \sqrt{R \cdot R + (\psi_L - \psi_C)(\psi_L - \psi_C)}$$

9.1 Positive Exponents

When a number is multiplied by itself several times, the product may be represented by means of exponents. If 2 is multiplied by itself 3 times, then the product may be written as 2^3. The 2 is called the base and the 3 is called the exponent; 2^3 is read as "2 to the third power" or "two cubed."

As introduced in Unit 1, an *exponent* is a number written on the upper right of another number called the *base* and represents how many times the base is used as a factor.

Definition: $a^n = \underbrace{a \cdot a \cdot a \cdot \ldots \cdot a}_{n \text{ times}}$

It is read "*a* to the *n*th power."

The product of the base number is referred to as the *power*. Since $8 = 2^3$, 8 is a power of 2; more specifically, 8 is the third power of 2.

In Sections 3.6 and 3.7, rules were given for multiplication and division of monomials with like bases. These rules are stated below as the first two of five Laws of Exponents.

Law   I: $a^m \cdot a^n = a^{m+n}$

Law   II: $\dfrac{a^m}{a^n} = \begin{cases} a^{m-n} \text{ if } m > n \\ \dfrac{1}{a^{n-m}} \text{ if } n > m \end{cases}$

Law III: $(ab)^n = a^n b^n$

Law IV: $\left(\dfrac{a}{b}\right)^n = \dfrac{a^n}{b^n}$

Law   V: $(a^m)^n = a^{mn}$

The definition of exponents may be used to simplify $(2x)^3$. $(2x)^3 = (2x)(2x)(2x) = 2 \cdot 2 \cdot 2 \cdot x \cdot x \cdot x = 2^3 x^3$. However, it is easier to use Law III,

$$(ab)^n = a^n b^n \; . \; (2x)^3 = 2^3 x^3 \text{ or } 8x^3.$$

Example 1: $(xy)^4 = x^4 y^4$ **Law III**

Example 2: $(3a)^5 = 3^5 a^5 = 243a^5$ **Law III**

To find the power of a quotient, the definition may again be applied. $\left(\dfrac{3}{4}\right)^5 = \dfrac{3}{4} \cdot \dfrac{3}{4} \cdot \dfrac{3}{4} \cdot \dfrac{3}{4} \cdot \dfrac{3}{4} = \dfrac{3^5}{4^5}$. According to Law IV, to find the power of a quotient, raise the numerator to the given power and the denominator to the given power. $\left(\dfrac{3}{4}\right)^5 = \dfrac{3^5}{4^5} \text{ or } \dfrac{243}{1024}$.

Example 3:   Simplify $\left(\dfrac{5}{X}\right)^3$

$\left(\dfrac{5}{X}\right)^3 = \dfrac{5^3}{X^3} = \dfrac{125}{X^3}$ **Law IV**

To find the power of a power, again apply the definition of an exponent. $(5^2)^3 = (5^2)(5^2)(5^2) = 5^{2+2+2} = 5^6$. Law V states that to raise a power to a power, multiply the powers. Therefore, according to Law V, $(5^2)^3 = 5^{2 \cdot 3} = 5^6$.

Example 4:   Simplify $(a^4)^6$

$(a^4)^6 = a^{4 \cdot 6} = a^{24}$ **Law V**

Example 5:   Simplify $(b^2 c^5)^3$.

$$\begin{aligned} (b^2 c^5)^3 &= (b^2)^3 \cdot (c^5)^3 &&\textbf{Law III} \\ &= b^6 \cdot c^{15} &&\textbf{Law V} \\ &= b^6 c^{15} \end{aligned}$$

or $(b^2 c^5)^3 = b^{2 \cdot 3} c^{5 \cdot 3} = b^6 c^{15}$

Example 6: Simplify $\left(\dfrac{r^3}{s^5}\right)^2$

$$\left(\frac{r^3}{s^5}\right)^2 = \frac{(r^3)^2}{(s^5)^2} \qquad \textbf{Law IV}$$

$$= \frac{r^6}{s^{10}} \qquad \textbf{Law V}$$

or $\qquad \left(\dfrac{r^3}{s^5}\right)^2 = \dfrac{r^{3\cdot 2}}{s^{5\cdot 2}} = \dfrac{r^6}{s^{10}}$

Example 7: Simplify $\left(\dfrac{10^2 s^5}{r}\right)^3\left(\dfrac{r^2 s^2}{10^4}\right)^4$

$$\left(\frac{10^2 s^5}{r}\right)^3 \cdot \left(\frac{r^2 s^2}{10^4}\right)^4 = \frac{10^6 s^{15}}{r^3} \cdot \frac{r^8 s^8}{10^{16}} \qquad \textbf{Laws III, IV, and V}$$

$$= \frac{s^{23} r^5}{10^{10}} \qquad \textbf{Laws I and II}$$

Example 8: Tension in the cord of a conical pendulum with a weight of W pounds and l foot lengths of cord is given by

$$T = \frac{Wl}{g}(2\pi n)^2$$

If the weight is numerically equal to the length cubed, find the tension.

$$T = \frac{Wl}{g}(2\pi n)^2$$

$$= \frac{l^3 l}{g}(2\pi n)^2 \qquad \textbf{Substituting } w = l^3$$

$$= \frac{l^{3+1}}{g}(2\pi n)^2 \qquad \textbf{Law I}$$

$$= \frac{l^4}{g} \cdot 2^2 \pi^2 n^2 \qquad \textbf{Law III}$$

$$= \frac{l^4}{g} 4\pi^2 n^2$$

$$= \frac{4l^4 \pi^2 n^2}{g} \qquad \textbf{Regrouping}$$

EXERCISES 9.1

Simplify:

1. $x^2 \cdot x^7$

2. $a \cdot a^5$

3. $10^2 \cdot 10^5$

4. $10^3 \cdot 10 \cdot 10^7$

5. $3^2 \cdot 3^2 \cdot 3^2$

6. $5^4 \cdot 5^4$

7. $b \cdot b^3 \cdot b^3$

8. $y^5 \cdot y^5 \cdot y^5$

9. $\dfrac{l^8}{l^3}$

10. $\dfrac{\pi^6}{\pi^4}$

11. $\dfrac{10^5}{10^3}$ **12.** $\dfrac{10^8}{10^7}$

13. $\dfrac{A^2}{A}$ **14.** $\dfrac{(-5)^5}{(-5)^4}$

15. $\dfrac{\omega^4}{\omega^5}$ **16.** $\dfrac{t}{t^3}$

17. $\dfrac{10^2}{10^3}$ **18.** $\dfrac{10}{10^5}$

19. $\dfrac{n^3}{n^9}$ **20.** $\dfrac{d^2}{d^{10}}$

21. $(vw)^3$ **22.** $(r_1 r_2)^4$

23. $(6T)^2$ **24.** $(5ir)^3$

25. $\left(\dfrac{Z}{R}\right)^2$ **26.** $\left(\dfrac{V_0}{V_f}\right)^7$

27. $\left(\dfrac{f_s}{3}\right)^4$ **28.** $\left(\dfrac{9}{Z_a}\right)^2$

29. $(Q^2)^4$ **30.** $(g^3)^3$

31. $(10^7)^6$ **32.** $(10^4)^4$

33. $(5^4)^3$ **34.** $(3^6)^2$

35. $(5i^3)^2$ **36.** $(2P^4)^4$

37. $(\theta^3\phi)^4$ **38.** $(u^2 v^4 w)^3$

39. $\left(\dfrac{G^2}{H^3}\right)^4$ **40.** $\left(\dfrac{i^3}{4}\right)^2$

41. $\left(\dfrac{9X_1^2}{X_2^3}\right)^3$ **42.** $\left(\dfrac{\psi}{3\psi^2}\right)^4$

43. $\left(\dfrac{10^3}{10^2}\right)^4$

44. $\left(\dfrac{10}{10^5}\right)^2$

45. $(i^2e^3)(ie)$

46. $(R^3S^3)(R^5S^2)$

47. $\dfrac{p^2q^3}{p^5q}$

48. $\dfrac{v_0v^3}{v_0^2v^2}$

49. $\dfrac{10^3i}{10i}$

50. $\dfrac{10^5d^3}{10^8d^4}$

51. $(7\alpha^2\beta)^3(3^2\alpha\beta^3)^3$

52. $(5fh^3)^5(5^2f^{\,6}hj)^3$

53. $\dfrac{(11\pi^5)^3}{(5\rho^3\pi)^2}$

54. $\dfrac{(7\gamma^5)^2}{(3\gamma\mu)^6}$

55. $[(5)(10^2)]^4$

56. $[(7)(10^5)]^3$

57. Find the expression for the area of a rectangle whose length is the square of the width.

58. The volume of a sphere is $V = \dfrac{\pi d}{6}$, where d is the diameter. Find the volume of a sphere whose radius is $3r^2$.

59. Simplify the formula giving the intensity of a wave:
$$I = 2vd(\pi - fd)^2$$

60. Simplify the formula for the tension of a stretching string:
$$T = \dfrac{M(2Lf)^2}{L}$$

9.2   Zero Exponent

Consider the quotient of two powers, where the same power appears in the numerator as in the denominator. To simplify that expression, Law II of Exponents may be applied.

Example 1:   Simplify $\frac{a^4}{a^4}$

$$\frac{a^4}{a^4} = a^{4-4} = a^0 \quad \textbf{Law II}$$

But from division, it is known that when a non-zero number is divided by itself, it equals one. That is,

$$\frac{a^4}{a^4} = 1$$

Thus,

$$a^0 = \frac{a^4}{a^4} \text{ and } \frac{a^4}{a^4} = 1$$

Therefore,

$$a^0 = 1$$

The exponent 0 is meaningless when applying the definition of an exponent because a number multiplied by itself 0 times does not make sense.

In order to give some meaning to the exponent 0 and still be consistent with Law IV of Exponents and the definition of division, any non-zero number to the zero exponent is defined to be equal to 1.

Definition: $a^0 = 1$ for any non-zero number a.

Example 2: $5^0 = 1$

Example 3: $(3x)^0 = 1$

Example 4: $\left(\frac{4}{a}\right)^0 = 1$

Example 5: $7 \cdot 10^0 = 7 \cdot 1 = 7$ **Because $10^0 = 1$**

Laws I through V of Exponents are extended to hold true for zero exponents as well as for positive whole numbers.

Example 6: $b^3 \cdot b^0 = b^{3+0} = b^3$ **Law I**

Example 7: $\frac{r^7}{r^0} = r^{7-0} = r^7$ **Law II**

Example 8: $\frac{m^0}{m^2} = \frac{1}{m^{2-0}} = \frac{1}{m^2}$ **Law II**

Example 9: $(4a)^0 = 1$ **By definition of the zero exponent**
or
$(4a)^0 = 4^0 a^0 = 1 \cdot 1 = 1$ **Law III**

Example 10: $\left(\frac{p}{q}\right)^0 = 1$ **By definition of the zero exponent**

or

$\left(\frac{p}{q}\right)^0 = \frac{p^0}{q^0} = \frac{1}{1} = 1$ **Law IV**

Example 11: $(h^5)^0 = h^{5 \cdot 0} = h^0 = 1$ **Law V**

Example 12: $(k^0)^3 = k^{0 \cdot 3} = k^0 = 1$ **Law V**

Example 13: $(R_i - R_f)^0 = 1$ **By definition of zero exponent**

EXERCISES 9.2

Simplify:

1. T^0

2. 7^0

3. 10^0

4. $(-1)^0$

5. $(-21)^0$

6. $\left(\dfrac{1}{4}\right)^0$

7. $2q^0$

8. $5M^0$

9. $(6u)^0$

10. $(10ei)^0$

11. $10^0 \cdot 10^4$

12. $10^0 \cdot 10^0$

13. $\dfrac{(2\rho)^5}{3\rho^0}$

14. $\dfrac{(5Z_1Z_2)^0}{(3Z_3)^3}$

15. $\dfrac{10^0}{10^8}$

16. $\dfrac{10}{10^0}$

17. $\dfrac{(5T)^2}{10m^0}$

18. $\dfrac{(3\theta\phi^2)^0}{(3\phi)^4}$

19. $R_1^2 \cdot R_1^3 \cdot R_1^0$

20. $\alpha^3\beta^0\gamma^3$

21. $(R - 2S)^0$

22. $(6a^2 - a)^0$

23. $\dfrac{1}{(2m - n)^0}$

24. $\dfrac{\omega - 2}{(\omega\sigma)^0}$

25. $(9r^2 - 3r + 1)^0$

26. $\left(\dfrac{3i - 5i^2}{6}\right)^0$

9.3   Negative Exponents

In Section 9.1, the Laws of Exponents were applied to powers involving positive integers or whole numbers as exponents. In Section 9.2, the Laws of Exponents were extended to powers involving the zero exponent. In this section, a base raised to a negative integer will be defined in such a way that the Laws of Exponents will hold true for negative exponents.

Example 1:   Simplify $T^{-3} \cdot T^3$

$$T^{-3} \cdot T^3 = T^{(-3)+3} \quad \textbf{Law I}$$
$$= T^0$$
$$= 1 \qquad \textbf{Definition of zero exponent}$$

Recalling from Unit 3 that the product of the reciprocal of a number times the number equals one, the following is true:

$$\frac{1}{T^3} \cdot T^3 = 1$$

By coupling this result with Example 1,

$$T^{-3} \cdot T^3 = 1 = \frac{1}{T^3} \cdot T^3$$

Therefore,

$$T^{-3} = \frac{1}{T^3}$$

Definition: A non-zero base a raised to a negative exponent $-n$ is equivalent to $\frac{1}{a^n}$; that is,

$$a^{-n} = \frac{1}{a^n}$$

Example 2: $8^{-2} = \frac{1}{8^2}$

Example 3: $10^{-1} = \frac{1}{10^1}$

Example 4: $x^{-7} = \frac{1}{x^7}$

With this definition, the restrictions on Law II of Exponents may be removed as follows:

LAW III: $\dfrac{a^m}{a^n} = a^{m-n}$

Example 5: $\dfrac{x^4}{x^6} = x^{4-6} = x^{-2} = \dfrac{1}{x^2}$

Consider the case in which a negative exponent occurs in the denominator, such as $\frac{1}{a^{-3}}$.

$$\frac{1}{a^{-3}} = \frac{1}{\left(\dfrac{1}{a^3}\right)} \qquad \textbf{By definition of the negative exponent}$$

$$= 1 \cdot \frac{a^3}{1} \qquad \textbf{By definition of division}$$

$$= a^3 \qquad \textbf{Multiplying}$$

Therefore, a rule can be formulated to remove negative exponents whether they occur in the numerator or the denominator.

Rule 1: Any power may be changed from the numerator to the denominator or from the denominator to the numerator *within the same term*, without altering the value of the term, by reversing the sign of the exponent.

Example 6: $w^5x^{-3} = \dfrac{w^5}{x^3}$

Example 7: $\dfrac{a^{-4}}{b^{-3}} = \dfrac{b^3}{a^4}$

Example 8: $\dfrac{1}{2x^{-1}} = \dfrac{x^1}{2} = \dfrac{x}{2}$ (*Note:* The exponent applies only to x.)

Example 9: $\left(\dfrac{r}{s}\right)^{-2} = \dfrac{r^{-2}}{s^{-2}}$ **Law IV**

$\quad\quad\quad\quad\quad = \dfrac{s^2}{r^2}$ **Rule 1 above**

Example 10: $(3x^4)^{-5} = 3^{-5} \cdot x^{-20}$ **Laws III and V**

$\quad\quad\quad\quad\quad = \dfrac{1}{3^5} \cdot \dfrac{1}{x^{20}}$ **Definition of negative exponent**

$\quad\quad\quad\quad\quad = \dfrac{1}{3^5x^{20}}$ **Multiplying**

Example 11: $\left(\dfrac{x^4}{y^3}\right)^{-2} = \dfrac{x^{-8}}{y^{-6}}$ **Laws IV and V**

$\quad\quad\quad\quad\quad = \dfrac{y^6}{x^8}$ **Definition of negative exponent**

Example 12: $2a^{-3} + y^{-2}$ is an expression involving two terms. Each term should be simplified separately.

The first term, $2a^{-3} = \dfrac{2}{a^3}$.

The second term, $y^{-2} = \dfrac{1}{y^2}$.

Adding both terms together, $2a^{-3} + y^{-2} = \dfrac{2}{a^3} + \dfrac{1}{y^2} = \dfrac{2y^2 + a^3}{a^3y^2}$

Example 13: The total reaction force of the air against a plane at the bottom of a vertical loop is

$$F = m(v^2r^{-1} + g).$$

Simplify the above expression for F.

$F = m(v^2r^{-1} + g)$

$\quad = mv^2r^{-1} + mg$ **Multiplying**

$\quad = \dfrac{mv^2}{r} + mg$ **Eliminating negative exponents**

EXERCISES 9.3

Simplify and give answers without negative exponents.

1. g^{-2} **2.** H^{-5} **3.** 10^{-4} **4.** 3^{-2}

5. $\dfrac{1}{f^{-3}}$ **6.** $\dfrac{1}{b^{-1}}$ **7.** $\dfrac{1}{2^{-5}}$ **8.** $\dfrac{3}{5^{-2}}$

9. $4K^{-1}$ **10.** $-2\,j^{-3}$ **11.** $(5r_0)^{-1}$ **12.** $(-3S)^{-2}$

13. $\dfrac{1}{8c^{-9}}$ **14.** $\dfrac{1}{3k^{-1}}$ **15.** $\dfrac{2}{(3\theta)^{-2}}$ **16.** $\dfrac{-1}{(5p)^{-3}}$

17. $10^{-7} \cdot 10^{7}$ **18.** $10^{5} \cdot 10^{-6}$ **19.** $\dfrac{10}{10^{-4}}$ **20.** $\dfrac{(-2)^{-3}}{(-2)}$

21. $\dfrac{i^{-2}}{i^{-3}}$ **22.** $\dfrac{Q^{-6}}{Q^{-1}}$ **23.** $\dfrac{10^{0}}{10^{-1}}$ **24.** $\dfrac{10^{-2}}{10^{-8}}$

25. $\dfrac{u^{-2}v}{w^{-1}}$ **26.** $\dfrac{P^{-1}Q^{-2}}{R}$ **27.** $\dfrac{10^{4} \cdot 10^{-3}}{10^{-5}}$ **28.** $\dfrac{10^{0} \cdot 10^{-7}}{10^{-2}}$

29. $T^{-1}fd^{-2}$ **30.** $9h^{-1}j^{3}$ **31.** $(L^{-2})^{-1}$ **32.** $(5y^{3})^{-3}$

33. $(43g^{2})^{-4}$ **34.** $(R_0^{2}R_1^{-3})^{-5}$ **35.** $\left(\dfrac{x^{2}}{z^{3}}\right)^{-2}$ **36.** $\left(\dfrac{L^{2}d}{m^{3}}\right)^{-2}$

37. $\left(\dfrac{2c^{-1}}{5E}\right)^{-3}$ **38.** $\left(\dfrac{6\theta^{3}}{7n^{-2}s}\right)^{-2}$ **39.** $(-3J_x^{-2}J_y)^{-3}(2J_x\,J_y^{-3})^{2}$

40. $(5s^{2})^{-1}(25s^{3})$ **41.** $4f_1^{-2}(3f_1^{-3})^{2}$ **42.** $3^{-2}R^{-3}(2R^{-3})^{2}$

43. $\dfrac{(mv)^{-1}}{m^{-3}v}$ **44.** $\dfrac{3(A^{2}p)^{-2}}{A^{-1}p^{-2}}$ **45.** $\dfrac{(2g^{-2}h)^{-3}}{(5g^{2}h^{-1})^{2}}$

46. $\dfrac{3^{-1}\gamma(2\mu^{-3})^2}{(3\gamma^2\mu^{-3})^{-3}}$

47. $I^{-1} - R^{-1}$

48. $t_0 - t^{-2}$

49. $2\alpha^{-2} + 3\beta^{-1}$

50. $5\sigma^0 - 2\epsilon^{-3}$

51. $(R_1^{-1} + R_2^{-1})^{-1}$ is a common electricity expression. Simplify it so that it has no parenthesis or negative exponents.

52. The rate at which energy is transported may be written

$$W = \frac{1}{(2m)^{-1}(\pi fA)^{-2}}.$$

Expess in simplest form without negative exponents.

53. The mirror equation is $f^{-1} = p^{-1} + q^{-1}$. Express it without negative exponents.

54. The equivalence resistance of the armature and field windings of an electric motor is $R = \dfrac{(R_a + R_f)^{-1}}{(R_a R_f)^{-1}}$. Express it without negative exponents.

55. The energy density of the electric field of a capacitor is

$$W = 2^{-1}\epsilon_0 \left(\frac{V}{d}\right)^2.$$

Express it in simplest form without negative exponents.

56. Resistance between the ends of a conductor is given by

$$R_1 = l_1^2 l_2^{-2} m_1^{-1} m_2 R_2.$$

Express it without negative exponents.

57. The expansion ratio of a steam engine is

$$E = \left(\frac{d}{D}\right)^{-2} (xC_n)^{-1}$$

Express it without negative exponents.

9.4   Fractional Exponents

The Laws of Exponents apply also to fractional exponents. All five Laws hold true when the exponent is a fraction.

Example 1:   Simplify $w^{1/3} \cdot w^{1/3}$

$$w^{1/3} \cdot w^{1/3} = w^{(1/3 + 1/3)} \quad \textbf{Law I}$$
$$= w^{2/3}$$

Example 2:   Simplify $\dfrac{a^{5/6}}{a^{1/6}}$

$$\frac{a^{5/6}}{a^{1/6}} = a^{(5/6 - 1/6)} \quad \textbf{Law II}$$
$$= a^{4/6}$$
$$= a^{2/3}$$

Example 3:   Simplify $(rs)^{2/3}$

$$(rs)^{2/3} = r^{2/3} \cdot s^{2/3} \quad \textbf{Law III}$$

Example 4:   Simplify $\left(\dfrac{v}{w}\right)^{1/8}$

$$\left(\frac{v}{w}\right)^{1/8} = \frac{v^{1/8}}{w^{1/8}} \quad \textbf{Law IV}$$

Example 5:   Simplify $(t^{1/5})^{2/3}$

$$(t^{1/5})^{2/3} = t^{(1/5)(2/3)} \quad \textbf{Law V}$$
$$= t^{2/15}$$

Example 6: Simplify $\dfrac{r^0 s^{1/3} t^{-1/2}}{r^{-1/4} s^2 t^{-1/4}}$

$$\frac{r^0 s^{1/3} t^{-1/2}}{r^{-1/4} s^2 t^{-1/4}} = \frac{1 \cdot s^{1/3} r^{1/4} t^{1/4}}{s^2 t^{1/2}} \qquad r^0 = 1$$

Change all negative exponents to positive exponents using Rule 1 of Section 9.3.

$$= \frac{r^{1/4}}{s^{2-1/3} t^{1/2-1/4}} \qquad \textbf{Law II}$$

$$= \frac{r^{1/4}}{s^{5/3} t^{1/4}}$$

Example 7: In hydraulics, the time required to lower the surface of a prismatic reservoir by means of a suppressed weir is

$$t = 3A(5b)^{-1}(IH_2^{-1/2} - IH_1^{-1/2}).$$

Express t in simpler form with no negative exponents.

$$t = 3A(5b)^{-1}(IH_2^{-1/2} - IH_1^{-1/2})$$

$$= 3A5^{-1}b^{-1}(IH_2^{-1/2} - IH_1^{-1/2}) \quad \textbf{Law III}$$

$$= \frac{3A}{5b}\left(\frac{I}{H_2^{1/2}} - \frac{I}{H_1^{1/2}}\right) \quad \textbf{Changing negative exponents}$$

EXERCISES 9.4

Simplify and express with positive exponents only:

1. $t^{1/4} t^{1/4}$

2. $BB^{1/2}$

3. $\dfrac{e^{4/5}}{e^{1/5}}$

4. $\dfrac{k^{7/8}}{k^{2/3}}$

5. $\dfrac{R^{1/3}}{R^{2/3}}$

6. $\dfrac{w^{1/4}}{w^{1/2}}$

7. $(S_a S_b)^{1/4}$

8. $(ei)^{1/2}$

9. $\left(\dfrac{2}{Z}\right)^{3/8}$

10. $\left(\dfrac{r}{3}\right)^{1/3}$

11. $(F^{1/2})^{1/2}$

12. $(c^{1/4})^{1/2}$

13. $\dfrac{r^{-1/3} S^2}{r^{1/2} S^{-1/4}}$

14. $\dfrac{P^{-3/4} Q^{-1/3}}{PQ^{-2/3}}$

15. $(j^{-1/4}h^{1/2})^{-4}$ **16.** $(E^{-2}I^{1/4})^{-1/2}$

17. The length of an object in motion with respect to an observer is

$$L = L_0 \left[1 - \left(\frac{v}{c} \right)^2 \right]^{1/2}.$$

Express in simpler form.

18. The angular velocity of a pendulum is given by

$$\omega = \left(\frac{g}{l} \right)^{1/2}.$$

Express in simpler form.

19. The plate resistance of a vacuum tube is

$$r = 2^2 6^{-1} K^{-2/3} I^{1/3} D^{-1}$$

Express it in simpler form without negative exponents.

20. The voltage drop in an A.C. circuit is

$$V = [V_R^2 + (V_L - V_C)^2]^{1/2}$$

Find V in terms of V_L if V_C is one half of V_L, and V_R is twice V_L.

9.5   Scientific Notation

In the scientific and technical world, very large and very small numbers are often encountered. For example, radio communications are carried out by electromagnetic waves traveling at a speed of 300,000,000 meters per second, the speed of light; the allowance made when shrinking a 2-inch collar on a shaft is 0.0008 inch; a liquid fuel rocket engine delivers 1,500,000 pounds of thrust in each of its five jets.

Numbers such as the ones above are more conveniently expressed as a product of a number between 1 and 10 and a power of 10. Numbers expressed in this form are said to be in *scientific notation*.

Example 1: Express 300,000,000 in scientific notation.

$$300,000,000 = 3 \cdot 100,000,000$$
$$= 3 \cdot 10^8$$

Example 2: Express 0.0008 in scientific notation.

$$0.0008 = \frac{8}{10,000}$$
$$= \frac{8}{10^4}$$
$$= 8 \cdot 10^{-4}$$

Familiarization with the powers of 10 is essential when expressing numbers in scientific notation. Consider the following powers of 10:

$$10^5 = 10 \cdot 10 \cdot 10 \cdot 10 \cdot 10 = 100,000$$
$$10^4 = 10 \cdot 10 \cdot 10 \cdot 10 = 10,000$$
$$10^3 = 10 \cdot 10 \cdot 10 = 1,000$$
$$10^2 = 10 \cdot 10 = 100$$
$$10^1 = 10 = 10$$
$$10^0 = 1 = 1$$
$$10^{-1} = \frac{1}{10} = 0.1$$
$$10^{-2} = \frac{1}{10^2} = \frac{1}{100} = 0.01$$
$$10^{-3} = \frac{1}{10^3} = \frac{1}{1000} = 0.001$$
$$10^{-4} = \frac{1}{10^4} = \frac{1}{10,000} = 0.0001$$

The following conclusions can be drawn:

Rule 1: 10^n, where n is a positive integer, is equal to a 1 followed by n zeros.

Example 3: Express 10^6 without exponents.

10^6 is written as the number 1 followed by 6 zeros; that is,

$$10^6 = 1,000,000$$

Rule 2: 10^n, where n is a negative integer, is equal to a decimal point, followed by $|n| - 1$ zeros followed by 1.

Example 4: Express 10^{-7} without exponents.

10^{-7} is written as a decimal point followed by $|7| - 1$ or 6 zeros followed by 1.

$$10^{-7} = 0.0000001$$

Now consider the number 1,500,000. In order to express it in scientific notation, it has to be written as a product of a number between 1 and 10 and a power of 10. 1.5 is a number between 1 and 10 which, when multiplied by 10^6, will equal 1,500,000. Therefore,

$$1,500,000 = 1.5 \cdot 10^6$$

is in scientific notation.

Rule 3:   To express a number in scientific notation:

Step 1:   Move the decimal place in order to have one non-zero digit to the left of the decimal.

Step 2:   If the decimal point is moved n places to the left, multiply the number obtained in Step 1, by 10^n. If the decimal point is moved n places to the right multiply the number obtained in Step 1 by 10^{-n}.

Example 5:   Express 245 in scientific notation.
    The decimal point is moved two places to the left in order to have one non-zero digit to the left of the decimal. Therefore 2.45 is multiplied by 10^2.

$$245 = (2.45)(10^2)$$

Example 6:   Express 0.00253 in scientific notation.
    The decimal point is moved three places to the right in order to have one non-zero digit to the left of the decimal. Therefore 2.53 is multiplied by 10^{-3}.

$$0.00253 = (2.53)(10^{-3})$$

Scientific notation is also useful to indicate the number of significant figures.

Example 7: $1.003 \cdot 10^{-2}$ has four significant figures because there are four digits in the first factor.

Example 8: $6.3 \cdot 10^4$ has two significant figures because there are two digits in the first factor.

Definition: The *significant figures* in a number expressed in scientific notation are indicated by the number of digits in the first factor.

Example 9:   Express 0.00302 in scientific notation and indicate the number of significant figures.

$$0.00302 = 3.02 \cdot 10^{-3}$$

0.00302 has three significant figures.

EXERCISES 9.5

Express the following numbers in scientific notation:

1. 286

2. 305

3. 18,000

4. 4500

5. 0.041

6. 0.472

7. 23.87

8. 298.5

9. 0.000842

10. 0.0079

11. 2,300,000

12. 54,000,000

13. 0.00000573

14. 0.000068

Express without exponents:

15. $3.7 \cdot 10^5$

16. $9 \cdot 10^7$

17. $2.85 \cdot 10^{-4}$

18. $8.1 \cdot 10^{-8}$

19. $7 \cdot 10^{-1}$

20. $2.00 \cdot 10^{-5}$

Express in scientific notation and indicate the number of significant figures:

21. 302,000

22. 0.00813

23. 0.00040

24. 788,500

25. 6700

26. 800,000

27. 0.0004

28. 0.0306

29. In chemistry, Avogadro's number is $6.02472 \cdot 10^{23}$. Express it in ordinary notation.

30. The modulus of elasticity for copper is $18 \cdot 10^6$. Express it in ordinary notation.

31. The speed of sound in air is 331.4 meters per second. Round it off to two significant figures and express it in scientific notation.

32. The density of the earth is 5520 kg/m^3. Round it off to one significant figure and express it in scientific notation.

33. The mass of an electron is 0.0005486 amu. Round it off to three significant figures and express it in scientific notation.

34. In electricity, 1 coulomb equals 0.0002788. Round it off to two significant figures and express it in scientific notation.

9.6 Calculations in Scientific Notation

Calculations involving multiplication and division are greatly simplified by expressing the numbers in scientific notation.

Example 1: $(23{,}000)(0.00002) = (2.3 \cdot 10^4)(2 \cdot 10^{-5})$
$$= (2.3)(2)(10^4)(10^{-5})$$
$$= 4.6 \cdot 10^{-1}$$

Rule: To multiply or divide numbers in scientific notation.

Step 1: Multiply or divide the first factors of the numbers in scientific notation. Round off to the correct number of significant figures, if the numbers are approximate numbers. Express the product or quotient in scientific notation.

Step 2: Multiply or divide the powers of 10, using the laws of exponents.

Step 3: The answer in scientific notation is the product of the results of Steps 1 and 2.

The rules given in Unit 1 for rounding off the results of multiplication and division of approximate numbers also apply when these numbers are expressed in scientific notation.

Example 2: Divide: $0.0045 \div 300$.

$$\frac{0.0045}{300} = \frac{4.5 \cdot 10^{-3}}{3 \cdot 10^2}$$

Step 1: $4.5 \div 3 = 1.5$

Step 2: $\dfrac{10^{-3}}{10^2} = \dfrac{1}{10^2 10^3} = \dfrac{1}{10^5} = 10^{-5}$

Step 3: $1.5 \cdot 10^{-5}$

Therefore, $0.0045 \div 300 = 1.5 \cdot 10^{-5}$
The above steps may be performed simultaneously, as follows:

$$\frac{0.0045}{300} = \frac{4.5 \cdot 10^{-3}}{3 \cdot 10^2} = 1.5 \cdot 10^{-3-2} = 1.5 \cdot 10^{-5}$$

Example 3: $\dfrac{(0.030)(3800)}{0.0050} = \dfrac{(3 \cdot 10^{-2})(3.8 \cdot 10^3)}{5 \cdot 10^{-3}}$

Step 1: $\dfrac{3.8 \cdot 3}{5} = \dfrac{11.4}{5} = 2.3$

Step 2: $\dfrac{10^3 10^{-2}}{10^{-3}} = 10^{3+(-2)-(-3)} = 10^4$

Therefore,

$$\frac{(3 \cdot 10^{-2})(3.8 \cdot 10^3)}{5 \cdot 10^{-3}} = 2.3 \cdot 10^4$$

Example 4: A flywheel on a gasoline motor is 45.8 cm in diameter. The velocity is $70.8 \, \dfrac{m}{sec}$. Calculate the acceleration given by $a = \dfrac{v^2}{r}$. Since the diameter is given in centimeters and the velocity is in meters per second, the diameter has to be changed to meters.

$$\text{diameter} = 45.8 \text{ cm} \cdot \frac{1 \text{ m}}{100 \text{ cm}} = 0.458 \text{ m}$$

The acceleration formula requires the radius. Since the radius is one half of the diameter, then $r = \frac{1}{2}$ (diameter) $= \frac{1}{2}$ (0.458 m) = 0.229 m. Therefore,

$$a = \frac{v^2}{r} = \frac{\left(70.8 \dfrac{m}{sec}\right)^2}{0.229 \, m} = \frac{(7.08 \cdot 10)^2}{2.29 \cdot 10^{-1}} \frac{m}{sec^2}$$

$$= \frac{7.08^2 \cdot 10^2}{2.29 \cdot 10^{-1}} \frac{m}{sec^2}$$

$$= 2.19 \cdot 10^4 \; \frac{m}{sec^2}$$

Example 5: Multiply 42,570,000 by 9,500,000 using a calculator.

$$42{,}570{,}000 = 4.25 \cdot 10^7$$
$$9{,}500{,}000 = 9.5 \; \cdot 10^6$$

The key $\boxed{EXP}$ or $\boxed{EE}$ on your calculator is used for numbers in scientific notation.

Enter	Press	Enter	Press	Enter	Press	Enter
4.257	$\boxed{EE}$	7	$\boxed{\times}$	9.5	$\boxed{EE}$	6

Press	Display
$\boxed{=}$	4.04416 14

$$(2{,}570{,}000)(9{,}500{,}000) = 4.04416 \cdot 10^{14}$$

Example 6: Divide .869000 by .000914 using a calculator.

$$.869000 = 8.69 \cdot 10^{-1}$$
$$.000914 = 9.14 \cdot 10^{-4}$$

Enter	Press	Enter	Press	Press	Enter	Press
8.69	$\boxed{EE}$	1	$\boxed{+/-}$	$\boxed{\div}$	9.14	$\boxed{EE}$

Enter	Press	Press	Display
4	$\boxed{+/-}$	$\boxed{=}$	9.5076586 02

$$.869000 \div .000914 = 9.5076586 \cdot 10^2$$

EXERCISES 9.6

Perform the following calculations using scientific notation:

1. $(0.0036)(0.0021)$

2. $(18{,}000)(76{,}000{,}000)$

3. $\dfrac{0.0036}{18{,}000{,}000}$

4. $\dfrac{70{,}000{,}000}{0.00008}$

5. $\dfrac{(75{,}000)(260{,}000)}{(1{,}300)(200{,}000{,}000)}$

6. $\dfrac{(5{,}000{,}000)(0.00012)}{(2{,}000)(0.00006)}$

7. $\dfrac{(15{,}000{,}000)(0.0017)(1{,}300)}{(0.00026)(0.00034)(500)}$

8. $\dfrac{(0.00016)(0.00018)}{(0.032)(0.000054)}$

9. $\dfrac{(14{,}000{,}000)(0.270)(0.0150)}{(810{,}000{,}000)(0.000028)(0.0045)}$

10. A wrought iron bar 0.05 by 0.12 meter in cross section and 2 meters long will shorten by ΔL when subjected to a compression load of 2500 newtons, where

$$\Delta L = \frac{(2500 \text{ newtons})(2m)}{(0.05m)(0.12m)\left(18.3 \cdot 10^{10}\,\dfrac{\text{newtons}}{m^2}\right)}$$

Evaluate ΔL.

11. The factor of safety involved in using a steel cable to support a given load is:

$$\frac{(120,000)(3.14)(0.02)}{1500}$$

Evaluate the factor of safety.

12. The decrease in volume of 7.5 liters of water under a pressure of $1.5 \cdot 10^7$ dynes is

$$\frac{(1.5)(7.5)(10^5)}{(0.21)(10^6)} \text{ cm}^3$$

Find the volume decrease.

13. The capacitance reactance of a capacitor of 50μf capacitance when an alternating current of 60 amperes is impressed on it is

$$\frac{1}{(2)(3.14)(60)(50)(10^{-6})} \text{ ohms}$$

Find the capacitance reactance.

Perform the following calculations using scientific notation on a calculator:

14. $(876,400)(.0789)$

15. $\dfrac{97,428,568}{8522}$

16. $\dfrac{(.00000489)(.00889)}{.000432}$

17. $\dfrac{(4,286)(61,000)(48,000)}{53,000,000}$

Unit 9 Self-Evaluation

In Problems 1 through 8, simplify the expressions and express answers in positive exponents:

1. $\dfrac{R^2}{R^7}$

2. $\left(-\dfrac{3}{st}\right)^3$

3. $3x^0$

4. $(5xy^2)^{-2}$

5. $\dfrac{9x^2y}{z} \cdot \dfrac{y}{15x^3z^3}$

6. $\dfrac{3a^{-2}b}{c^{-3}} \div (6a^5c^{-2})^{-2}$

7. $m^0 - n^{-4}$

8. $B^{\frac{1}{2}}B^{\frac{1}{4}}$

Express Problems 9 and 10 in scientific notation:

9. 352,000

10. 0.0078

Perform the calculations in Problems 11 and 12 using scientific notation and rounding off answers to the correct number of significant digits.

11. $\dfrac{136,000}{0.042}$

12. $\dfrac{(13,000.0)(0.028)}{(1.04)}$

Roots and Radicals

Unit 10 Objectives
1. To find roots of given numbers.
2. To simplify radicals.
3. To multiply radicals.
4. To divide radicals and to rationalize monomial denominators.
5. To add and subtract radicals.
6. To multiply binomials containing radicals.
7. To rationalize denominators that are binomials.

10.1 Roots

Definition 1. The *square root* of a number is one of two equal factors. It is denoted by the radical sign, $\sqrt{}$

Example 1: Find the square root of 4.

$$\sqrt{4} = 2, \text{ because } 2^2 = 2 \cdot 2 = 4$$

However, (-2) squared will also yield 4:

$$(-2)^2 = (-2)(-2) = 4$$

Therefore,

$$\sqrt{4} = 2 \text{ or } (-2)$$

Example 2: Find the square root of 25.

$$5^2 = 5 \cdot 5 = 25 \text{ and}$$
$$\text{Since} \quad (-5)^2 = (-5)(-5) = 25,$$
$$\sqrt{25} = 5 \text{ or } (-5)$$

Any positive number has two square roots, one positive and one negative. The positive root is called the *principal square root*. In this unit, only principal square roots will be considered.

Example 3: Find the principal square root of 36.

Since $6^2 = 6 \cdot 6 = 36$ and 6 is positive, 6 is the principal square root of 36.

Example 4: Find the square root of (-9).

The square root of (-9) is not a real number because no real number multiplied by itself will yield a negative nine. The numbers 3 and (-9) appear to be square roots of (-9) but they are not so, because

$$3^2 = 3 \cdot 3 = 9 \quad \text{and} \quad (-3)^2 = (-3)(-3) = 9$$

The square root of a negative number cannot be expressed as a real number. Only the square roots of positive numbers can be expressed as real numbers.

Definition 2.   The *cube root* of a number is one of three equal factors. It is denoted by $\sqrt[3]{}$

Example 5:   Find the cube root of 27.

Since $3^3 = 3 \cdot 3 \cdot 3 = 27$, $\sqrt[3]{27} = 3$.
    (-3) appears to be a cube root of 27. However, $(-3)^3 = (-3)(-3)(-3) = -27$. Hence, (-3) is not a cube root of 27. Therefore, 3 is the only number that when cubed equals 27.

Example 6:   Find the cube root of (-64).

Since $(-4)^3 = (-4)(-4)(-4) = -64,$
$$\sqrt[3]{-64} = -4.$$

4 appears to be a cube root of (-64). However, $4^3 = 4 \cdot 4 \cdot 4 = 64$. Hence, 4 is not a cube root of (-64). Therefore, (-4) is the only number that, when cubed, equals (-64).

Any number, whether positive or negative, has only one cube root. The cube root of a positive number is positive, and the cube root of a negative number is negative.

Rule: To summarize:

1. The even root of a positive number has two solutions, but only the positive one will be considered here.
2. The even root of a negative number is not a real number.
3. The odd root of a positive number is positive.
4. The odd root of a negative number is negative.

Definition 3.   The *nth root* of a number is one of n equal factors. It is denoted by $\sqrt[n]{}$. n is called the *index*. The number under the radical sign is the *radicand*.

Example 7:   Find the fourth root of 81.

Since $3^4 = 81$, $\sqrt[4]{81} = 3$.

Example 8:   Find the sixth root of (-1).

There is no real number that when raised to the sixth power will yield (-1). Therefore, $\sqrt[6]{-1}$ is not a real number.

Example 9:   Find the fifth root of 32.

Since $2^5 = 32$, $\sqrt[5]{32} = 2$

Example 10:   Find the cube root of (-8).

Since $(-2)^3 = (-2)(-2)(-2) = -8$, $\sqrt[3]{-8} = -2$.

Since variables represent numbers, the preceding definitions also apply to taking the nth root of variables. Assuming all variables to be non-negative numbers, note the following examples:

$$\sqrt{a^2} = a, \text{ since } (a)^2 = a \cdot a = a^2$$

$$\sqrt{b^4} = b^2, \text{ since } (b^2)^2 = b^2 \cdot b^2 = b^4$$

$$\sqrt[3]{27y^3} = 3y, \text{ since } (3y)^3 = 3y \cdot 3y \cdot 3y = 27y^3$$

$$\sqrt[4]{16m^8} = 2m^2, \text{ since } (2m^2)^4 = 2m^2 \cdot 2m^2 \cdot 2m^2 \cdot 2m^2 = 16m^8$$

$$\sqrt[3]{a^6} = a^2, \text{ since } (a^2)^3 = a^2 \cdot a^2 \cdot a^2 = a^6$$

In the last example, note that $\sqrt[3]{a^6} = a^2$. $a^{6/3}$ is also equal to a^2. Therefore $\sqrt[3]{a^6} = a^{6/3}$, since they are equal to a^2.

Definition 4. $\sqrt[n]{a^m} = a^{m/n} = (\sqrt[n]{a})^m$

Example 11: Using the definition $\sqrt[n]{a^m} = a^{m/n}$, express the following in exponential form:

$$\sqrt[3]{a^4} = a^{4/3}$$

$$\sqrt{y^7} = y^{7/2}$$

$$\sqrt[5]{7^2} = 7^{2/5}$$

$$\sqrt{x} = x^{1/2}$$

Example 12: Express the following in radical form:

$$a^{3/4} = \sqrt[4]{a^3}$$

$$6^{1/2} = \sqrt{6}$$

$$x^{3/8} = \sqrt[8]{x^3}$$

The index of the radical can often be reduced by expressing a radical in exponential form.

Example 13: Reduce the index of the radicals.

$$\sqrt[6]{a^3} = a^{3/6} = a^{1/2} = \sqrt{a}$$

$$\sqrt[12]{2^4} = 2^{4/12} = 2^{1/3} = \sqrt[3]{2}$$

$$\sqrt{x^2} = x^{2/2} = x$$

To evaluate $8^{2/3}$, take the cube root of 8^2 or take the cube root of 8 and square it.

$$\sqrt[3]{8^2} = \sqrt[3]{64} = 4 \text{ or } (\sqrt[3]{8})^2 = (2)^2 = 4.$$

Example 14: Evaluate

$$125^{2/3} = (\sqrt[3]{125})^2 = (5)^2 = 25$$

$$16^{3/4} = (\sqrt[4]{16})^3 = (2)^3 = 8$$

$$4^{-1/2} = \frac{1}{4^{1/2}} = \frac{1}{\sqrt{4}} = \frac{1}{2}$$

EXERCISES 10.1

Find the following principal roots; assume all variables to be non-negative numbers.

1. $\sqrt{49}$

2. $\sqrt{9}$

3. $\sqrt[3]{8}$

4. $\sqrt[3]{125}$

5. $\sqrt[4]{81}$

6. $\sqrt[4]{16}$

7. $\sqrt[3]{-27}$

8. $\sqrt[3]{-1}$

9. $\sqrt[3]{-125}$

10. $\sqrt[5]{-32}$

11. $\sqrt[4]{1}$

12. $\sqrt{100}$

13. $\sqrt{-4}$

14. $\sqrt{-25}$

15. $\sqrt[6]{64}$

16. $\sqrt[8]{1}$

17. $\sqrt{121}$

18. $\sqrt{144}$

19. $\sqrt[3]{-1000}$

20. $\sqrt[3]{-64}$

21. $-\sqrt{64}$

22. $-\sqrt{81}$

23. $-\sqrt[3]{-1}$

24. $-\sqrt[3]{-8}$

25. $\sqrt{p^2}$

26. $\sqrt{m^4}$

27. $\sqrt{D^6}$

28. $\sqrt[3]{f^3}$

29. $\sqrt[3]{n^6}$

30. $\sqrt[4]{e^4}$

31. $\sqrt[3]{c^3}$

32. $-\sqrt[3]{-a^3}$

33. $-\sqrt{T^{10}}$

34. $\sqrt[5]{-x^5}$

35. $\sqrt[4]{y^{12}}$

36. $-\sqrt{B^6}$

37. $\sqrt[3]{W^9}$

38. $\sqrt{r^8}$

39. $\sqrt{t^{12}}$

40. $\sqrt[3]{t^{12}}$

41. $\sqrt{16p^2}$

42. $\sqrt{25r^4}$

43. $\sqrt[3]{8d^6}$

44. $\sqrt[4]{16a^4}$

45. $\sqrt{a^2b^4c^6}$

46. $\sqrt[3]{x^3y^6z^3}$

47. $\sqrt{9B^2C^4}$

48. $\sqrt{4W^8}$

Express the following in exponential form:

49. $\sqrt{b}$

50. $\sqrt[3]{d}$

51. $\sqrt{V^3}$

52. $\sqrt[4]{a^5}$

53. $\sqrt[3]{T^2}$

54. $\sqrt[7]{m^2}$

Express the following in radical form:

55. $r^{1/3}$

56. $t^{1/5}$

57. $W^{2/3}$

58. $S^{1/4}$

59. $2^{2/3}$

60. $5^{2/7}$

Reduce the index of the radical in the following:

61. $\sqrt[10]{a^5}$

62. $\sqrt[8]{n^6}$

63. $\sqrt[8]{5^2}$

64. $\sqrt[4]{7^2}$

65. $\sqrt[7]{B^7}$

66. $\sqrt[3]{R^3}$

67. $\sqrt{3^6}$

68. $\sqrt{2^2}$

Evaluate:

69. $\left(-\dfrac{1}{8}\right)^{2/3}$

70. $\left(\dfrac{1}{8}\right)^{-2/3}$

71. $81^{3/4}$ **72.** $32^{3/5}$

73. $\left(\dfrac{81}{16}\right)^{3/4}$ **74.** $16^{-1/2}$

75. $\left(\dfrac{125}{-27}\right)^{-2/3}$ **76.** $(16)^{-3/4}$

10.2 Simplifying Radicals

$$\sqrt{4 \cdot 100} = \sqrt{400} = 20$$
$$\sqrt{4} \cdot \sqrt{100} = 2 \cdot 10 = 20$$

Therefore, $\sqrt{4 \cdot 100} = \sqrt{4}\sqrt{100}$

Definition: $\sqrt[n]{ab} = \sqrt[n]{a}\sqrt[n]{b}$, if a and b are both zero or positive numbers.

This definition may be used to express irrational numbers in a simpler form. Recall that irrational numbers were defined in Unit 3 to be numbers that could not be expressed as the quotient of two integers, $\dfrac{a}{b}$, $b \neq 0$.

The square root of 27 is an irrational number. It is approximately equal to 5.196. If exacting work does not require the decimal form, $\sqrt{27}$ may be expressed simply as $3\sqrt{3}$ by using the above definition: $\sqrt{27} = \sqrt{9 \cdot 3} = \sqrt{9}\sqrt{3} = 3\sqrt{3}$.

Example 1: Simplify $\sqrt{12}$

$$\begin{aligned}\sqrt{12} &= \sqrt{4 \cdot 3}\\ &= \sqrt{4} \cdot \sqrt{3}\\ &= 2\sqrt{3}\end{aligned}$$

2, 4, and 6 are all factors of 12. Since 4 is a perfect square, 12 is factored as $4 \cdot 3$.

Example 2: Simplify $\sqrt{a^5}$

$$\begin{aligned}\sqrt{a^5} &= \sqrt{a^4 \cdot a}\\ &= \sqrt{a^4}\sqrt{a}\\ &= a^2\sqrt{a}\end{aligned}$$

a^2, a^3, and a^4 are all factors of a^5. Since a^4 is the largest square, a^5 is factored as $a^4 \cdot a$.

Example 3: Simplify $2\sqrt[3]{16b^4}$

$$\begin{aligned}2\sqrt[3]{16b^4} &= 2\sqrt[3]{8b^3 \cdot 2b}\\ &= 2\sqrt[3]{8b^3} \cdot \sqrt[3]{2b}\\ &= 2 \cdot 2b \cdot \sqrt[3]{2b}\\ &= 4b\sqrt[3]{2b}\end{aligned}$$

Since 8 is 2^3, 16 is factored as $8 \cdot 2$. b^4 is factored as the cube, $b^3 \cdot b$.

Example 4: Simplify $\sqrt{\dfrac{3}{4}}$

$$\begin{aligned}\sqrt{\dfrac{3}{4}} &= \sqrt{\dfrac{1}{4} \cdot 3}\\[6pt] &= \sqrt{\dfrac{1}{4}}\sqrt{3}\\[6pt] &= \dfrac{1}{2}\sqrt{3}\end{aligned}$$

EXERCISES 10.2

Simplify the following radicals:

1. $\sqrt{18}$

2. $\sqrt{27}$

3. $\sqrt{50}$

4. $\sqrt{28}$

5. $\sqrt{48}$

6. $\sqrt{200}$

7. $\sqrt{192}$

8. $\sqrt{108}$

9. $-\sqrt{72}$

10. $-\sqrt{125}$

11. $\sqrt{720}$

12. $\sqrt{450}$

13. $\sqrt{p^7}$

14. $\sqrt{r^3}$

15. $-\sqrt{x^9}$

16. $-\sqrt{y^{11}}$

17. $\sqrt{ab^3}$

18. $\sqrt{r^3 t^5}$

19. $\sqrt{8m^3}$

20. $\sqrt{24d^9}$

21. $\frac{3}{2}\sqrt{12}$

22. $-\frac{5}{2}\sqrt{164}$

23. $-\frac{1}{3}\sqrt{162}$

24. $\frac{1}{x}\sqrt{x^3}$

25. $\frac{a}{b}\sqrt{a^3 b^5}$

26. $\sqrt[3]{375}$

27. $\sqrt[3]{d^5}$

28. $\sqrt[4]{32}$

29. $\sqrt[4]{Y^4}$

30. $\sqrt[3]{256M^7}$

31. $\sqrt[5]{D^2 E^6}$

32. $-\sqrt{98 A^2 B^3 C}$

33. $-2K\sqrt{8K^8}$

34. $3\sqrt{\dfrac{m}{9}}$

35. $\sqrt[3]{\dfrac{F^4}{8}}$

36. $\sqrt[3]{-27}$

37. $-\sqrt[3]{-8X^6}$

38. $\sqrt{WT^2}$

39. $\sqrt[3]{L^3 P^5}$

Evaluate for the variable in the left member of each equation:

40. $v = \sqrt{\dfrac{k}{m}}\, s$; $k = 9$, $s = 2.5$, $m = .01$

41. $T = 2\sqrt{\dfrac{m}{k}}$; $m = .1$, $k = 40$

42. $t = \sqrt{\dfrac{2d}{a}}$; $d = 288$, $a = .16$

43. $v = \sqrt{rg}$; $g = 9.8$, $r = 0.2$

44. $E_a = \sqrt{E_b{}^2 + E_c{}^2}$; $E_b = .5$, $E_c = 1.2$

45. $Z = \sqrt{R^2 - (X_L - X_C)^2}$; $R = 17$, $X_L = 50$, $X_C = 35$

10.3 Multiplication of Radicals

In the preceeding section, it was noted that $\sqrt[n]{ab} = \sqrt[n]{a}\sqrt[n]{b}$. By the symmetric property of equality, $\sqrt[n]{a}\,\sqrt[n]{b} = \sqrt[n]{ab}$.

Rule: Radicals with like indexes may be multiplied. The product is a radical with the same index, and a radicand which is the product of the radicands being multiplied.

Example 1: $\sqrt{3}\sqrt{11} = \sqrt{3 \cdot 11} = \sqrt{33}$

Example 2: $-\sqrt[3]{55}\sqrt[3]{3} = -\sqrt[3]{55 \cdot 3} = -\sqrt[3]{165}$

Example 3: $\sqrt[4]{6}\,\sqrt[4]{8} = \sqrt[4]{48} = \sqrt[4]{16}\,\sqrt[4]{3} = 2\sqrt[4]{3}$

Example 4: $\begin{aligned} 3\sqrt{5} \cdot 2\sqrt{7} &= (3 \cdot 2)\sqrt{5}\sqrt{7} \\ &= 6\sqrt{5 \cdot 7} \\ &= 6\sqrt{35} \end{aligned}$

Example 5: $\begin{aligned} 2\sqrt{5}(\sqrt{3} - 3\sqrt{10}) &= (2\sqrt{5})(\sqrt{3}) - (2\sqrt{5})(3\sqrt{10}) \\ &= 2\sqrt{15} - 6\sqrt{50} \\ &= 2\sqrt{15} - 6\sqrt{25 \cdot 2} \\ &= 2\sqrt{15} - 6\sqrt{25}\sqrt{2} \\ &= 2\sqrt{15} - 6 \cdot 5\sqrt{2} \\ &= 2\sqrt{15} - 30\sqrt{2} \end{aligned}$

Radicals cannot be added or subtracted unless they have the same index and the same radicand. Use the distributive property to multiply.

EXERCISES 10.3

Multiply the following radicals. Simplify whenever possible.

1. $\sqrt{7} \cdot \sqrt{13}$

2. $-\sqrt{6} \cdot -\sqrt{5}$

3. $\sqrt{2} \cdot \sqrt{11}$

4. $\sqrt{3} \cdot \sqrt{13}$

5. $-\sqrt{5} \cdot \sqrt{3}$

6. $-\sqrt{27} \cdot -\sqrt{3}$

7. $\sqrt[4]{8} \cdot \sqrt[4]{3}$

8. $\sqrt[4]{2} \cdot \sqrt[4]{6}$

9. $\sqrt{6} \cdot \sqrt{8}$

10. $\sqrt{50} \cdot \sqrt{3}$

11. $-\sqrt{5a} \cdot \sqrt{10b}$

12. $\sqrt{3x} \cdot \sqrt{5x}$

13. $\sqrt[3]{a} \cdot \sqrt[3]{2a}$

14. $\sqrt{7y} \cdot \sqrt{5y}$

15. $\sqrt{rs} \cdot \sqrt{r^2s^3}$

16. $\sqrt[5]{6m} \cdot \sqrt[5]{6m^3}$

17. $\sqrt{3m^2} \cdot \sqrt{6mn}$

18. $\sqrt{8uv} \cdot \sqrt{14u^2v}$

19. $2\sqrt{7} \cdot 3\sqrt{6}$

20. $-3\sqrt{11} \cdot 2\sqrt{3}$

21. $-5\sqrt{10} \cdot 2\sqrt{2}$

22. $-6\sqrt{6} \cdot -3\sqrt{3}$

23. $3x\sqrt{5x} \cdot 2x\sqrt{7x}$

24. $ab\sqrt[4]{a^3} \cdot b\sqrt[4]{ab}$

25. $\sqrt{2} \cdot \sqrt{3} \cdot \sqrt{7}$

26. $\sqrt{5} \cdot \sqrt{6} \cdot \sqrt{8}$

27. $\sqrt[3]{ab} \cdot \sqrt[3]{ab^2} \cdot \sqrt[3]{b}$

28. $-\sqrt{3} \cdot \sqrt{5} \cdot \sqrt{10}$

29. $2\sqrt{6} \cdot 3\sqrt{5} \cdot 2\sqrt{8}$

30. $3r\sqrt{rs} \cdot \sqrt{2rs} \cdot 3\sqrt{2rs}$

31. $\sqrt[6]{8PQ^4} \cdot 3\sqrt[6]{8P^5Q^3}$

32. $5\sqrt[3]{3CD} \cdot 2\sqrt[3]{6C^2D}\sqrt[3]{-3C^2D^2}$

33. $\sqrt{2}(\sqrt{6} - \sqrt{3})$

34. $\sqrt{5}(\sqrt{2} + \sqrt{7})$

35. $3\sqrt{3}(2\sqrt{5} - \sqrt{2})$

36. $-6\sqrt{11}(\sqrt{11} + 5\sqrt{6} - \sqrt{10})$

37. The voltage of a heater is equal to the square root of the product of the power measured in watts times the resistance measured in ohms. Determine the voltage of a heater that has 25 ohms of resistance and uses power at the rate 2200 watts.

10.4 Division of Radicals

Note that $\dfrac{\sqrt{16}}{\sqrt{4}} = \dfrac{4}{2} = 2$ and $\sqrt{\dfrac{16}{4}} = \sqrt{4} = 2$

Definition. If a and b are both positive numbers or a is zero,

$$\frac{\sqrt[n]{a}}{\sqrt[n]{b}} = \sqrt[n]{\frac{a}{b}}$$

This definition may be used to divide or to simplify radicals.
Examples of division:

Example 1: $\dfrac{\sqrt{21}}{\sqrt{3}} = \sqrt{\dfrac{21}{3}} = \sqrt{7}$

Example 2: $\dfrac{\sqrt{x^3}}{\sqrt{x}} = \sqrt{\dfrac{x^3}{x}} = \sqrt{x^2} = x$

Example 3: $\dfrac{6\sqrt{24}}{3\sqrt{8}} = 2\sqrt{3}$

Simplifying a fraction whose denominator contains one term and a radical by writing the fraction as an equivalent fraction whose denominator is free of radicals is called *rationalizing the denominator*.

Rule: Eliminate the radical in the denominator by multiplying both numerator and denominator by a number that will yield a rational denominator. Write the fraction as an equivalent fraction with the rational denominator.

Example 1: Simplify $\dfrac{\sqrt{2}}{\sqrt{3}}$.

$$\frac{\sqrt{2}}{\sqrt{3}} \cdot \frac{\sqrt{3}}{\sqrt{3}} = \frac{\sqrt{6}}{\sqrt{9}} = \frac{\sqrt{6}}{3}$$

Rationalizing the denominator gives an equivalent fraction that, if used in a technical problem, is easier to put into decimal form.

$$\frac{\sqrt{2}}{\sqrt{3}} = \frac{1.414}{1.732} \doteq 0.816$$

$$\frac{\sqrt{6}}{3} = \frac{2.449}{3} \doteq 0.816$$

Example 2: Simplify $\dfrac{\sqrt{7}}{\sqrt{8}}$

$$\frac{\sqrt{7}}{\sqrt{8}} \cdot \frac{\sqrt{2}}{\sqrt{2}} = \frac{\sqrt{14}}{\sqrt{16}} = \frac{\sqrt{14}}{4}$$

Note that multiplying $\sqrt{8}$ by $\sqrt{8}$ also rationalizes the denominator. Multiplying by the smaller number, $\sqrt{2}$, avoids having to simplify the numerator after rationalizing.

Example 3: Simplify $\dfrac{\sqrt[3]{x}}{\sqrt[3]{y}}$

$$\frac{\sqrt[3]{x}}{\sqrt[3]{y}} \cdot \frac{\sqrt[3]{y^2}}{\sqrt[3]{y^2}} = \frac{\sqrt[3]{xy^2}}{\sqrt[3]{y^3}} = \frac{\sqrt[3]{xy^2}}{y}$$

EXERCISES 10.4

Simplify:

1. $\dfrac{\sqrt{20}}{\sqrt{2}}$

2. $\dfrac{\sqrt{15}}{\sqrt{3}}$

3. $\dfrac{\sqrt{21}}{\sqrt{7}}$

4. $\dfrac{\sqrt{39}}{\sqrt{13}}$

5. $\dfrac{\sqrt[3]{55}}{\sqrt[3]{5}}$

6. $\dfrac{\sqrt[3]{63}}{\sqrt[3]{7}}$

7. $\dfrac{\sqrt{48}}{\sqrt{6}}$

8. $\dfrac{\sqrt{x^3}}{\sqrt{x}}$

9. $\dfrac{\sqrt[4]{12a^2b}}{\sqrt[4]{6ab}}$

10. $\dfrac{3\sqrt{8p}}{\sqrt{2}}$

11. $\dfrac{\sqrt{22a^3bc^2}}{\sqrt{2abc}}$

12. $\dfrac{\sqrt[3]{27m}}{\sqrt[3]{3m}}$

13. $\dfrac{\sqrt{2}}{\sqrt{7}}$

14. $\dfrac{-\sqrt{3}}{\sqrt{5}}$

15. $\dfrac{\sqrt{5}}{\sqrt{18}}$

16. $\dfrac{\sqrt{6}}{-\sqrt{11}}$

17. $\dfrac{\sqrt{18}}{\sqrt{12}}$

18. $\dfrac{\sqrt[3]{2}}{\sqrt[3]{y}}$

19. $\dfrac{\sqrt[3]{rs^2}}{\sqrt[3]{s}}$

20. $\dfrac{\sqrt{s}}{\sqrt{t^3}}$

21. $\dfrac{3}{\sqrt{7}}$

22. $\dfrac{20}{\sqrt{15}}$

23. $\dfrac{3\sqrt{8}}{\sqrt{12}}$

24. $\dfrac{\sqrt{5d}}{\sqrt{3c}}$

25. $\dfrac{1}{\sqrt[4]{r}}$

26. $\dfrac{3}{\sqrt{t^3}}$

27. $\dfrac{\sqrt{6m^3}}{\sqrt{3m}}$

28. $\dfrac{\sqrt{x^2}}{\sqrt{4y}}$

29. $\sqrt{\dfrac{24}{3p}}$

30. $\sqrt[4]{\dfrac{32}{4z}}$

31. The amperage of an electric circuit is equal to the square root of the quotient of the power in watts and the resistance. Find the amperage when the power is 220 watts and the resistance is 150 ohms.

10.5   Addition and Subtraction of Radicals

In previous sections, similar terms, such as $2a^2$ and $3a^2$, were added or subtracted by using the distributive law, $2 \cdot a^2 + 3 \cdot a^2 = (2 + 3)a^2 = 5a^2$. Similar terms containing radicals may also be added by using the distributive property. Terms with radicals are similar if the radicals have the same index and the same radicand.

Example 1: $\qquad 3\sqrt{5} + 4\sqrt{5} = (3 + 4)\sqrt{5} \qquad\qquad = 7\sqrt{5}$

Example 2: $\qquad 2\sqrt{x} - 7\sqrt{x} = (2 - 7)\sqrt{x} \qquad\qquad = -5\sqrt{x}$

Example 3: $\qquad \sqrt[3]{6} + 3\sqrt[3]{6} - 2\sqrt[3]{6} = (1 + 3 - 2)\sqrt[3]{6} \qquad = 2\sqrt[3]{6}$

Example 4: $\sqrt{x} - \sqrt{y} + 8\sqrt{x} - 3\sqrt{y} = (1 + 8)\sqrt{x} + (-1 - 3)\sqrt{y} = 9\sqrt{x} - 4\sqrt{y}$

Rule: To add or subtract similar radical terms, add or subtract their coefficients. The sum or difference obtained is the coefficient of the similar radical.

Example 5: $\sqrt{50} + 3\sqrt{18} - \sqrt{8}$

Since the terms do not contain similar radicals, they may not be added in their present form. Simplify each term, and inspect again for similar terms.

$$\sqrt{50} + 3\sqrt{18} - \sqrt{8} = 5\sqrt{2} + 9\sqrt{2} - 2\sqrt{2}$$
$$= (5 + 9 - 2)\sqrt{2}$$
$$= 12\sqrt{2}$$

EXERCISES 10.5

1. $3\sqrt{7} + 2\sqrt{7}$

2. $5\sqrt{11} - 4\sqrt{11}$

3. $3\sqrt[3]{m} - \sqrt[3]{m}$

4. $\sqrt[4]{3r} + \sqrt[4]{3r}$

5. $3\sqrt{2d} - \sqrt{8d}$

6. $2\sqrt{45} + \sqrt{20} - 3\sqrt{80}$

7. $5\sqrt{2} - 3\sqrt{3} + \sqrt{2} + \sqrt{3}$

8. $\sqrt{25} - 3\sqrt{32} + \sqrt{49}$

9. $2\sqrt{p} + 3\sqrt{25p}$

10. $\sqrt{3q} - \sqrt{12q}$

11. $\sqrt{3} + \sqrt{18} - 3\sqrt{12}$

12. $\sqrt{4t^3} - t\sqrt{25t} + t\sqrt{t}$

13. $3\sqrt{2} + \sqrt{\dfrac{1}{2}} - 2\sqrt{\dfrac{1}{8}}$

14. $\sqrt{\dfrac{3}{5}} + \dfrac{2}{15}\sqrt{15} + \sqrt{\dfrac{20}{3}}$

15. $5\sqrt{\dfrac{2}{3}} - \dfrac{1}{2}\sqrt{24} + 3\sqrt{\dfrac{3}{2}}$

16. $\sqrt{.04m} - \sqrt{.25m}$

17. $.3\sqrt{.09x} + .02\sqrt{.16x}$

18. $10\sqrt{\dfrac{9}{10}} + 5\sqrt{\dfrac{8}{5}} - 4\sqrt{\dfrac{5}{8}}$

19. $\sqrt{b^3} - \sqrt{d^2e} + b\sqrt{b} + 3d\sqrt{e}$

20. $3\sqrt{112} - 2\sqrt{175} - 2\sqrt{44}$

21. $5\sqrt[3]{16} - 2\sqrt[3]{54}$

22. $\sqrt[4]{16xy} + \sqrt[4]{81xy} - \sqrt[4]{xy}$

10.6   Multiplication of Binomials Containing Radicals

Recall multiplication of two binomials:

$$(a - 2)(a + 5) = a \cdot a + (5a - 2a) + (-2 \cdot 5) = a^2 + 3a - 10$$

Binomials containing radicals can be multiplied in the same manner.

Example 1: $(\sqrt{a} - 2)(\sqrt{a} + 5) = \sqrt{a} \cdot \sqrt{a} + (5\sqrt{a} - 2\sqrt{a}) + (-2 \cdot 5)$
$$= a + 3\sqrt{a} - 10$$

Example 2: $(3\sqrt{2} - \sqrt{5})(\sqrt{2} - \sqrt{5})$

$$= (3\sqrt{2} \cdot \sqrt{2}) + [(-\sqrt{5} \cdot \sqrt{2}) - (3\sqrt{2} \cdot \sqrt{5})] + (-\sqrt{5} \cdot -\sqrt{5})$$
$$= 3\sqrt{4} + (-\sqrt{10} - 3\sqrt{10}) + \sqrt{25}$$
$$= 3 \cdot 2 + (-4\sqrt{10}) + 5$$
$$= 6 - 4\sqrt{10} + 5$$
$$= 11 - 4\sqrt{10}$$

EXERCISES 10.6

Multiply and simplify answers when possible.

1. $(\sqrt{5} - 2)(\sqrt{5} + 6)$

2. $(7 - \sqrt{3})(4 - \sqrt{3})$

3. $(2\sqrt{2} + 6)(3\sqrt{2} - 5)$

4. $(2 + 3\sqrt{7})(5 - \sqrt{7})$

5. $(9 - \sqrt{3})(9 + \sqrt{3})$

6. $(2\sqrt{5} - 6)(2\sqrt{5} + 6)$

7. $(2\sqrt{3} + \sqrt{5})(3\sqrt{3} - \sqrt{5})$

8. $(5\sqrt{x} + 3)(11\sqrt{x} - 7)$

9. $(2 - \sqrt{6})^2$

10. $(\sqrt{8} - 3)^2$

11. $(\sqrt{B} + 1)(\sqrt{B} + 2)$

12. $(4 - \sqrt{T})(9 + \sqrt{T})$

13. $(\sqrt{x} + \sqrt{y})(\sqrt{x} - \sqrt{y})$

14. $(\sqrt{a} - b)^2$

15. $(3\sqrt{R} + 2)(2\sqrt{R} - 5)$

16. $(W - \sqrt{6})(W + 3\sqrt{6})$

17. $(4\sqrt{P} - \sqrt{M})(4\sqrt{P} + \sqrt{M})$

18. $(11\sqrt{7} - 2\sqrt{5})(8\sqrt{7} + 2\sqrt{5})$

19. $(\sqrt[3]{2} + 5)(\sqrt[3]{2} - 6)$

20. $(2 - \sqrt[3]{P})(9 + \sqrt[3]{P})$

10.7   Rationalizing Binomial Denominators

To rationalize a binomial denominator where at least one term contains a radical, such as $\dfrac{7}{\sqrt{2} - 3}$, it is necessary to free the denominator of radicals. Recall that, when two binomials with identical first terms and identical last terms with opposite signs were multiplied, the product was a binomial; for example $(x - 3)(x + 3) = x^2 - 9$. Similarly, $(\sqrt{2} - 3)(\sqrt{2} + 3) = 2 - 9 = -7$. Thus we see that by multiplying the denominator by the same two terms, but with the sign of the last term the opposite, the denominator is rationalized. $\sqrt{2} + 3$ is called the *conjugate* of $\sqrt{2} - 3$.

Rule: To rationalize a binomial denominator where one or both terms contain a radical, multiply the denominator by the conjugate of the given denominator.

Example 1: Rationalize $\dfrac{7}{\sqrt{2} - 3}$

$$\frac{7}{\sqrt{2} - 3} \cdot \frac{\sqrt{2} + 3}{\sqrt{2} + 3} = \frac{7(\sqrt{2} + 3)}{\sqrt{4} - 9}$$

$$= \frac{7(\sqrt{2} + 3)}{2 - 9}$$

$$= \frac{7(\sqrt{2} + 3)}{-7}$$

$$= -(\sqrt{2} + 3)$$

Example 2: Rationalize $\dfrac{\sqrt{2} - \sqrt{3}}{\sqrt{2} + \sqrt{3}}$

$$\frac{(\sqrt{2} - \sqrt{3})}{(\sqrt{2} + \sqrt{3})} \cdot \frac{(\sqrt{2} - \sqrt{3})}{(\sqrt{2} - \sqrt{3})} = \frac{\sqrt{4} - 2\sqrt{6} + \sqrt{9}}{\sqrt{4} - \sqrt{9}}$$

$$= \frac{2 - 2\sqrt{6} + 3}{2 - 3}$$

$$= \frac{5 - 2\sqrt{6}}{-1}$$

$$= 2\sqrt{6} - 5$$

EXERCISES 10.7

Rationalize:

1. $\dfrac{5}{\sqrt{3} - 1}$

2. $\dfrac{11}{\sqrt{5} + 3}$

3. $\dfrac{9}{7 - \sqrt{2}}$

4. $\dfrac{1}{12 - \sqrt{11}}$

5. $\dfrac{\sqrt{2}}{\sqrt{7}+3}$

6. $\dfrac{\sqrt{3}}{5-\sqrt{2}}$

7. $\dfrac{2\sqrt{6}-5}{2\sqrt{6}+5}$

8. $\dfrac{2+\sqrt{5}}{3-\sqrt{5}}$

9. $\dfrac{3+\sqrt{7}}{3+\sqrt{2}}$

10. $\dfrac{3\sqrt{5}-\sqrt{2}}{2\sqrt{5}+\sqrt{2}}$

11. $\dfrac{W-\sqrt{5}}{2W-\sqrt{5}}$

12. $\dfrac{3+\sqrt{f}}{2-\sqrt{f}}$

13. $\dfrac{9+2\sqrt{e}}{7+3\sqrt{e}}$

14. $\dfrac{\sqrt{M}+6}{\sqrt{M}-2}$

15. $\dfrac{\sqrt{6}+\sqrt{2}}{\sqrt{6}-\sqrt{2}}$

16. $\dfrac{\sqrt{7}-3}{\sqrt{7}+3}$

17. $\dfrac{\sqrt{H}}{\sqrt{H}-1}$

18. $\dfrac{\sqrt{X}-\sqrt{Y}}{\sqrt{X}+\sqrt{Y}}$

19. $\dfrac{\sqrt{3R}-4\sqrt{2S}}{\sqrt{3R}+\sqrt{2S}}$

20. $\dfrac{6\sqrt{3}-2\sqrt{2}}{5\sqrt{3}-2\sqrt{2}}$

Unit 10 Self-Evaluation

---

1. Find the principal root of $\sqrt{64d^2}$.

2. Express $e^{7/5}$ in radical form.

3. Express $\sqrt{a}$ in exponential form.

4. Reduce the index of $\sqrt[6]{y^3}$.

Simplify:

5. $\dfrac{2}{3}\sqrt{27}$ **6.** $a\sqrt[3]{a^5}$ **7.** $-3\sqrt[3]{-125}$

Perform the indicated operations. Simplify answers whenever possible.

8. $\sqrt{6}\sqrt{3}$ **9.** $\sqrt[4]{y^3}\sqrt[4]{y^2}$

10. $\sqrt{3}(\sqrt{5} - 2\sqrt{7})$ **11.** $(2\sqrt{2} - \sqrt{7})(3\sqrt{2} + 2\sqrt{7})$

12. $(\sqrt{3} - 9)^2$ **13.** $\dfrac{\sqrt[3]{9}}{\sqrt[3]{3}}$

14. $\dfrac{a}{\sqrt[4]{a}}$

15. $\dfrac{\sqrt{5}}{\sqrt{2}+5}$

16. $4\sqrt{5}+\sqrt{5}-3\sqrt{5}$

17. $2\sqrt{25}-5\sqrt{32}+2\sqrt{49}$

18. $p\sqrt{\dfrac{1}{p}}-6\sqrt{p}$

Quadratic Equations

Unit 11 Objectives
1. To solve quadratic equations by factoring.
2. To solve quadratic equations by taking the square root of both members of the equation.
3. To solve quadratic equations by completing the square.
4. To solve quadratic equations using the quadratic formula.
5. To graph quadratic functions and to determine the nature of the zeros of a quadratic by using the discriminant.
6. To solve quadratic equations having solutions that are complex numbers.

unit 11

11.1 Solving Quadratic Equations by Factoring

Quadratic equations are equations in which the highest exponent of the unknown is 2. For example, $ax^2 + bx + c = 0$, $a \neq 0$, is a second degree, or quadratic, equation in one unknown, x. The variables a, b, and c represent constants.

Just as linear equations were transformed into equivalent equations in order to determine the solution by inspection, quadratic equations may be transformed into equivalent equations to determine solutions. The rules used to transform linear equations are also used, when needed, to transform a quadratic equation into an equivalent equation.

Rule 1: The same polynomial added to, or subtracted from, both members of an equation produces an equivalent equation.

Rule 2: Both members of an equation may be multiplied or divided by the same non-zero polynomial to produce an equivalent equation.

The following property of zero is helpful in solving a quadratic equation by factoring:

Property of Zero: The product of two or more factors is zero if and only if at least one of the factors is zero. That is, if a and b are real numbers, and if $ab = 0$, then $a = 0$ or $b = 0$.

To solve the quadratic equation $x^2 - 2x - 15 = 0$ by factoring, inspect the left member to see if it can be factored, using one of the methods of factoring polynomials from Unit 5:

$$x^2 - 2x - 15 = 0$$
$$(x - 5)(x + 3) = 0$$

Since the product of two factors is zero if and only if at least one of the factors is zero, the next step is to determine what value of the variables will make the factors zero.

$$x - 5 = 0 \quad \text{or} \quad x + 3 = 0$$
$$x = 5 \quad \text{or} \quad x = -3$$

Therefore, solutions of the given equation are $x = 5$ or $x = -3$. That is, when these solutions are substituted in the original equation, a true statement is obtained.

Rule 3:   To solve quadratic equations by factoring:

Step 1:   Write the equation as an equivalent equation equal to zero.

Step 2:   Factor the left member.

Step 3:   Set each factor equal to zero and solve.

Example 1:   Solve $3f^2 = f$

$$3f^2 - f = 0 \qquad\qquad \textbf{Step 1}$$
$$f(3f - 1) = 0 \qquad\qquad \textbf{Step 2}$$
$$f = 0 \quad \text{or} \quad 3f - 1 = 0 \quad \textbf{Step 3}$$
$$f = 0 \quad \text{or} \quad 3f = 1$$
$$f = 0 \quad \text{or} \quad f = \frac{1}{3}$$

Check: $3f^2 = f \qquad\qquad 3f^2 = f$

$$3(0)^2 = 0 \qquad\qquad 3\left(\frac{1}{3}\right)^2 = \frac{1}{3}$$

$$0 = 0 \qquad\qquad 3\left(\frac{1}{9}\right) = \frac{1}{3}$$

$$\frac{1}{3} = \frac{1}{3}$$

Therefore, $f = 0$ or $f = \frac{1}{3}$

Example 2:   Solve $6R^2 - 11R = 10$

$$6R^2 - 11R - 10 = 0 \qquad\qquad\qquad \textbf{Step 1}$$
$$(3R + 2)(2R - 5) = 0 \qquad\qquad\qquad \textbf{Step 2}$$
$$3R + 2 = 0 \quad \text{or} \quad 2R - 5 = 0 \quad \textbf{Step 3}$$
$$3R = -2 \quad \text{or} \quad 2R = 5$$
$$R = -\frac{2}{3} \quad \text{or} \quad R = \frac{5}{2}$$

Example 3:   A rectangular piece of sheet metal is 3 cm longer than it is wide. The area of the piece of sheet metal is 28 cm². Find its dimensions.

Let x = width of the piece of sheet metal, and let $x + 3$ = its length (Fig. 11.1).

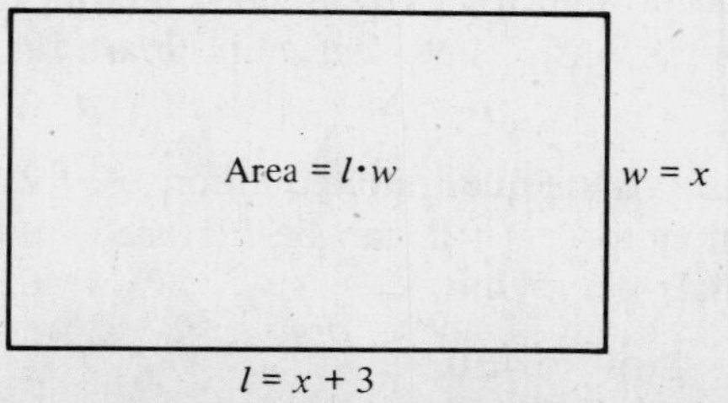

Area = $l \cdot w$ $w = x$

$l = x + 3$

Figure 11.1

$$x(x + 3) = 28 \qquad \text{Length times width = area}$$
$$x^2 + 3x - 28 = 0 \qquad \text{Step 1}$$
$$(x - 4)(x + 7) = 0 \qquad \text{Step 2}$$
$$x - 4 = 0 \quad \text{or} \quad x + 7 = 0 \qquad \text{Step 3}$$
$$x = 4 \quad \text{or} \quad x = -7$$

Since the width of the piece of sheet metal, cannot equal -7, the width is 4 cm. The length $= x + 3 = 4 + 3 = 7$ cm.

Check: $A = l \cdot w = 7 \cdot 4 = 28$ cm.²

Example 4: A border of uniform width is to be built around a rectangular garden. The garden is 20 meters by 18 meters, and the area of the border is 123 square meters. Find the width of the border.

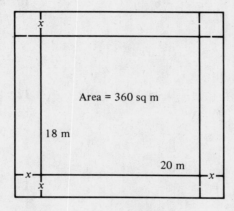

18 m

Area = 360 sq m

20 m

Figure 11.2

Area of the garden and walk − Area of the garden = Area of the walk

$$(20 + 2x)(18 + 2x) - (20)(18) = 123$$
$$360 + 76x + 4x^2 - 360 = 123$$
$$4x^2 + 76x - 123 = 0$$
$$(2x + 41)(2x - 3) = 0$$
$$x = -\frac{41}{2} \quad \text{or} \quad x = \frac{3}{2}$$

Since the width of the border can not be a negative number, the width of the border is $1\frac{1}{2}$ meters.

EXERCISES 11.1

Solve the following quadratic equations by factoring and check:

1. $x^2 - 2x = 0$

2. $2y^2 - y = 0$

3. $m^2 - 25 = 0$

4. $p^2 - 9 = 0$

5. $R^2 + 6R + 8 = 0$

6. $W^2 + 7W + 10 = 0$

7. $T^2 - 6T - 16 = 0$

8. $d^2 + 7d - 18 = 0$

9. $q^2 - 5q = -6$

10. $b^2 - 4b = -4$

11. $6a^2 + 11a = 35$

12. $12I^2 - 24I = -9$

13. $4y^2 = 25$

14. $64 - 9Q^2 = 0$

15. $16b^2 + 25 = 40b$

16. $9D^2 + 64 = 48D$

17. $7E^2 = E$

18. $.04p^2 - .17p - .15 = 0$

19. $2n^2 + 24n + 70 = 0$

20. $3e^2 + 3e - 36 = 0$

21. $30h^2 + 47h = -14$

22. $32t^2 - 52t = 45$

23. One number is 6 more than another number and their product is 160. Find the numbers.

24. The altitude of a triangle is 5 cm longer than the length of the base. Find the length of the base if the area is 7 cm.^{2}

25. Two times the square of a number, plus 15 is equal to 11 times the number. Find the number.

26. The sum of a positive number and its square is 12. Find the number.

27. The side of one square is 4 cm greater than the side of another, and the sum of their areas is 170 cm.^{2} How long is the side of each square?

28. The sum of two numbers is 13, and the sum of their squares is 85. Find the two numbers.

29. Three times the square of a certain number plus 8 is 25 times the number. What is the number?

30. Find the width of a uniform border that is to be placed around a 10 m by 14 m rectangular garden if the area of the border is to be 52 m.^{2}

31. A concrete walk of uniform width is to be built around a pool that is 60 ft by 40 ft. If the area of the walk is 416 ft^2, find the width of the walk.

32. $s = 16t^2$ is an equation for the distance, s, an object falls from a height. t represents the time in seconds the object is falling. If an object is dropped from a height of 900 meters, how long will it take the object to reach the ground?

33. $N = \dfrac{n(n-3)}{2}$ is the formula for finding the number of diagonals of a polygon of n sides. How many sides has a polygon having 20 diagonals?

34. $I^2R + IE = 15,200$ is a formula for finding the current, I, in a circuit with resistance, R, and voltage, E. Find the current in the circuit if the resistance is 8 ohms and the applied voltage is 408 volts.

11.2   Solving Equations of the Form $x^2 = a$

In the preceding section, the quadratic equation $x^2 = 25$ was solved by writing the equation as an equivalent equation equal to zero. The left member was then factored.

$$x^2 = 25$$
$$x^2 - 25 = 0$$
$$(x - 5)(x + 5) = 0$$

Since the product of two factors is zero if, and only if, at least one of the factors is zero, $x = 5$ or $x = -5$.

The equation $x^2 = 25$ may also be solved by taking the square root of the left and right members of the equation.

$$x^2 = 25$$
$$x = \pm\sqrt{25}$$
$$x = \pm 5$$
$$x = 5 \quad \text{or} \quad x = -5$$

Example 1: Solve the equation $x^2 = 49$

$$x^2 = 49$$
$$x = \pm\sqrt{49}$$ **Take the square root of both members of**
$$x = \pm 7$$ **the equation.**
$$x = 7 \quad \text{or} \quad x = -7$$

Check: $x^2 = 49$ $x^2 = 49$
 $7^2 = 49$ $(-7)^2 = 49$
 $49 = 49$ $49 = 49$

Therefore, $x = 7$ or $x = -7$

Example 2: Solve the equation $(x - 2)^2 = 7$

$$(x - 2)^2 = 7$$ **Take the square root of both**
$$x - 2 = \pm\sqrt{7}$$ **members of the equation.**
$$x = 2 + \sqrt{7} \quad \text{or} \quad x = 2 - \sqrt{7}$$ **Solve each equation for x.**

EXERCISES 11.2

Solve each of the following equations by taking the square root of both members of the equation.

1. $m^2 = 9$ **2.** $d^2 = 16$

3. $B^2 = 7$ **4.** $A^2 = 5$

5. $(x - 3)^2 = 25$ **6.** $(y - 2)^2 = 100$

7. $(T + 1)^2 = 144$ **8.** $(R + 3)^2 = 64$

9. $(c - 5)^2 = 6$ **10.** $(p + 3)^2 = 8$

11. $(f + 10)^2 = 32$ **12.** $(R - 9)^2 = 48$

13. $(d - 11)^2 = 11$ **14.** $(e + 7)^2 = 7$

15. $(2a + 3)^2 = 4$ **16.** $(5a - 1)^2 = 9$

17. $(3y - 4)^2 = 5$ **18.** $(7Q - 2)^2 = 7$

19. $(4r + 8)^2 = 32$ **20.** $(5S - 15)^2 = 60$

11.3   Solving Quadratic Equations by Completing the Square

Every quadratic equation will not be expressed in a form such that it may be solved simply by factoring or by taking the square root of each side. The equation $x^2 + 8x + 14 = 0$ cannot be factored simply. In the preceding section, equations of the form $(x + 4)^2 = 2$ were solved by taking the square root of each side. Examine the equation $x^2 + 8x + 14 = 0$ to see how it can be transformed into the equivalent equation $(x + 4)^2 = 2$.

First, subtract 14 from both members of the equation.

$$x^2 + 8x + 14 - 14 = 0 - 14$$
$$x^2 + 8x = -14$$

Note that if $\left(\frac{1}{2} \cdot 8\right)^2$ or 16 were added to the left member of the equation, it could be factored as $(x + 4)^2$. Therefore, the next step is to add 16 to both members of the equation in order to factor the left member.

$$x^2 + 8x + 16 = -14 + 16$$
$$(x + 4)^2 = 2$$

The equation may now be solved as in the preceding section. Take the square root of both members of the equation, and solve for x.

$$x + 4 = \sqrt{2} \quad \text{or} \quad x + 4 = -\sqrt{2}$$
$$x = -4 + \sqrt{2} \quad \text{or} \quad x = -4 - \sqrt{2}$$

This method of solving a quadratic equation is called *completing the square*.

Rule: To solve a quadratic equation by completing the square, use the following steps:

Step 1: Transform the equation into an equivalent equation with the constant term on the right side of the equation.

Step 2: If the coefficient of the squared term is not 1, divide each term by the coefficient of the squared term.

Step 3: Add the square of $\frac{1}{2}$ of the coefficient of the first degree term to both members of the equation.

Step 4: Factor the left member of the equation.

Step 5: Take the square root of each member of the equation.

Step 6: Solve the equation for the unknown.

Example 1: Solve the equation $2x^2 + 5x - 12 = 0$ by completing the square.

Step 1: $2x^2 + 5x = 12$ **Add 12 to each member of the equation.**

Step 2: $x^2 + \frac{5}{2}x = 6$ **Divide each member by 2.**

Step 3: $x^2 + \frac{5}{2}x + \frac{25}{16} = 6 + \frac{25}{16}$ $\left(\frac{1}{2}\text{ of }\frac{5}{2}\right)^2 = \frac{25}{16}$. **Add $\frac{25}{16}$ to each member.**

Step 4: $\left(x + \frac{5}{4}\right)^2 = \frac{121}{16}$ **Factor the left member.**

Step 5: $x + \frac{5}{4} = \frac{11}{4}$ or $x + \frac{5}{4} = -\frac{11}{4}$ **Take the square root of each member.**

Step 6: $x + \frac{5}{4} = \frac{11}{4}$ or $x + \frac{5}{4} = -\frac{11}{4}$

$x = -\frac{5}{4} + \frac{11}{4}$ $x = -\frac{5}{4} - \frac{11}{4}$ **Solve each equation for x.**

$x = \frac{3}{2}$ $x = -4$

Therefore $x = \frac{3}{2}$ or $x = -4$.

Check:

$$2x^2 + 5x - 12 = 0 \qquad\qquad 2x^2 + 5x - 12 = 0$$

$$2\left(\frac{3}{2}\right)^2 + 5\left(\frac{3}{2}\right) - 12 = 0 \qquad 2(-4)^2 + 5(-4) - 12 = 0$$

$$\frac{9}{2} + \frac{15}{2} - 12 = 0 \qquad\qquad 32 - 20 - 12 = 0$$

$$12 - 12 = 0 \qquad\qquad 12 - 12 = 0$$

$$0 = 0 \qquad\qquad 0 = 0$$

Example 2: Solve the equation $3m^2 + 7m - 5 = 0$ by completing the square.

Step 1: $3m^2 + 7m = 5$ **Add 5 to each member.**

Step 2: $m^2 + \frac{7}{3}m = \frac{5}{3}$ **Divide both members by 3.**

Step 3: $m^2 + \frac{7}{3}m + \frac{49}{36} = \frac{5}{3} + \frac{49}{36}$ $\left(\frac{1}{2}\text{ of }\frac{7}{3}\right)^2 = \frac{49}{36}$. **Add $\frac{49}{36}$ to each member.**

Step 4: $\left(m + \frac{7}{6}\right)^2 = \frac{109}{36}$ **Factor the left member.**

Step 5: $m + \frac{7}{6} = \pm\sqrt{\frac{109}{36}}$ **Take the square root of each member.**

Step 6: $m = -\frac{7}{6} \pm \frac{\sqrt{109}}{6}$ **Solve for m.**

EXERCISES 11.3

Solve each of the following equations for the given unknown by completing the square.

1. $t^2 + 6t - 7 = 0$

2. $9R^2 - 6R + 1 = 0$

3. $y^2 + y - 1 = 0$

4. $2d^2 + 3d - 14 = 0$

5. $12E^2 - 2E - 1 = 0$

6. $t^2 + 6t - 27 = 0$

7. $2m^2 + m - 15 = 0$

8. $16a^2 - 24a + 9 = 0$

9. $6b^2 + 13b - 5 = 0$

10. $3W^2 + 2W - 7 = 0$

11. $5D^2 + 3D - 6 = 0$

12. $x^2 + 5x + 1 = 0$

11.4 The Quadratic Formula

The method of completing the square can be generalized to obtain the quadratic formula. Solve the general equation $ax^2 + bx + c = 0$, by completing the square. a, b, and c represent constants, $a \neq 0$.

Step 1:   Subtract c from each member.

$$ax^2 + bx + c = 0$$
$$ax^2 + bx = -c$$

Step 2:   Divide both members by a.

$$x^2 + \frac{b}{a}x = -\frac{c}{a}$$

Step 3: $\left(\frac{1}{2} \text{ of } \frac{b}{a}\right)^2 = \frac{b^2}{4a^2}$. Add $\frac{b^2}{4a^2}$ to each member.

$$x^2 + \frac{b}{a}x + \frac{b^2}{4a^2} = -\frac{c}{a} + \frac{b^2}{4a^2}$$

Step 4:   Factor the left member.

$$\left(x + \frac{b}{2a}\right)^2 = \frac{b^2 - 4ac}{4a^2}$$

Step 5:   Take the square root of each member.

$$x + \frac{b}{2a} = \pm\frac{\sqrt{b^2 - 4ac}}{2a}$$

Step 6:   Solve for x.

$$x = -\frac{b}{2a} \pm \frac{\sqrt{b^2 - 4ac}}{2a}$$

$$x = \frac{-b \pm \sqrt{b^2 - 4ac}}{2a}$$

Rule: The Quadratic Formula states that

if $ax^2 + bx + c = 0$

then $x = \dfrac{-b \pm \sqrt{b^2 - 4ac}}{2a}$

Example 1:   Solve the equation $3x^2 + 7x - 5 = 0$ by using the quadratic formula.

From the given equation, $a = 3$, $b = 7$, $c = -5$,

$$x = \frac{-b \pm \sqrt{b^2 - 4ac}}{2a}$$

$$x = \frac{-7 \pm \sqrt{7^2 - 4(3)(-5)}}{2(3)} = \frac{-7 \pm \sqrt{49 + 60}}{6}$$

Substitute the values of *a*, *b*, and *c* in the quadratic equation.

$$x = \frac{-7 \pm \sqrt{109}}{6}$$

Example 2:   Solve the equation $5m^2 - 12m + 3 = 0$, using the quadratic equation.

$$a = 5, b = -12, c = 3$$

$$m = \frac{-b \pm \sqrt{b^2 - 4ac}}{2a}$$

$$m = \frac{-(-12) \pm \sqrt{(-12)^2 - 4(5)(3)}}{2(5)} = \frac{12 \pm \sqrt{84}}{10}$$

$$m = \frac{12 \pm 2\sqrt{21}}{10} = \frac{6 \pm \sqrt{21}}{5}$$

If an approximation of the values of m is needed, use the calculator.

Enter	Press	Enter	Press	Press	Press	Enter	Press	Display
12	$+$	84	$\sqrt{x}$	$=$	$\div$	10	$=$	2.1165151

which can be rounded to a given number of significant figures. To find the second value of m, follow the same procedure but press $-$ after 12 to get $m = .28348489$.

EXERCISES 11.4

Solve the following quadratic equations using the quadratic formula.

1. $E^2 - 6E - 4 = 0$ **2.** $m^2 + 12m + 30 = 0$

3. $I^2 + 8I + 4 = 0$ **4.** $5R^2 + 4R - 1 = 0$

5. $2r^2 + 3r - 1 = 0$ **6.** $i^2 - 4i + 3 = 0$

7. $2d^2 + 3d + 2 = 0$ **8.** $4p^2 + 5p = 1$

9. $3U^2 - 10U = -3$ **10.** $5b^2 = -9b - 3$

11. $y^2 + 2y = -6$ **12.** $10v^2 + 13v - 3 = 0$

13. $2w(w + 1) = 7$ **14.** $3r(r - 1) + 2(3r + 4) = 18$

15. $e + \dfrac{3}{5} = 3 + \dfrac{1}{e}$ **16.** $2s^2 - .4s = 2.2$

17. $\dfrac{Z + 3}{Z - 1} = \dfrac{Z - 1}{Z - 2} - 2$ **18.** $\dfrac{x + 2}{x + 1} - 4 = \dfrac{x + 1}{x + 3}$

19. $.3n^2 + .07n - .12 = 0$ **20.** $80K^2 - 32K - 48 = 0$

21. $\frac{3}{2}S^2 - 3S + \frac{1}{2} = 0$ **22.** $T + 2 = \frac{T^2}{3}$

23. $D(5D + 1) = (D + 3)^2$ **24.** $5 = \frac{12}{R^2} - \frac{3}{R}$

25. A square piece of sheet metal expands when heated. If the area changes by 6 square feet when the edge is 2 feet shorter than twice the original edge, what was the original length of one side of the piece?

11.5   Graph of $y = ax^2 + bx + c$

In Units 5 and 8, linear equations were graphed by finding two or more ordered pairs that satisfied the given equation. Quadratic equations of the form $y = ax^2 + bx + c$ can be graphed in a similar way. However, since this quadratic equation is not a straight line, but a curve called a *parabola*, it will be necessary to find more than two ordered pairs from the relation to sketch the equation.

Example 1:   Graph the equation, $y = x^2 + 2x - 8$.
Let $y = 0$ and solve for x.

$$
\begin{aligned}
0 &= x^2 + 2x - 8 \\
0 &= (x + 4)(x - 2) \\
x + 4 = 0 \quad &\text{or} \quad x - 2 = 0 \\
x = -4 \quad &\text{or} \quad x = 2.
\end{aligned}
$$

Therefore, -4 and 2 are called *zeros* of the equation. The curve will cross the x-axis at $(-4,0)$ and $(2,0)$.

To find more ordered pairs from this relation, let $x = -5, -3, -2, -1, 0, 1,$ and 3.

Points obtained are

$y = (-5)^2 + 2(-5) - 8 = 7$	$(-5,7)$
$y = (-3)^2 + 2(-3) - 8 = -5$	$(-3,-5)$
$y = (-2)^2 + 2(-2) - 8 = -8$	$(-2,-8)$
$y = (-1)^2 + 2(-1) - 8 = -9$	$(-1,-9)$
$y = (0)^2 + 2(0) - 8 = -8$	$(0,-8)$
$y = (1)^2 + 2(1) - 8 = -5$	$(1,-5)$
$y = (3)^2 + 2(3) - 8 = 7$	$(3,7)$

Plot the ordered pairs obtained and draw a smooth curve through the points. The point, $(-1,-9)$, is called the *vertex* of the parabola. The vertex can be obtained by $x = \frac{b}{2a}$. In the equation $y = x^2 + 2x - 8$, $x = \frac{-2}{2(1)} = -1$. Substitute $x = 1$ in the equation to get $y = 9$.

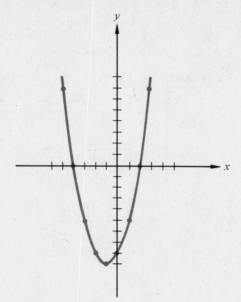

Figure 11.3

From Figure 11.3 it can be determined that the range of the relation is the set of real numbers greater than or equal to -9. The domain is the set of all real numbers. The relation is a function since each element of the domain is associated with one and only one element of the range.

Solving the quadratic equation $ax^2 + bx + c = 0$, $a \neq 0$, the solution was found to be

$$x = \frac{-b \pm \sqrt{b^2 - 4ac}}{2a}$$

The quantity $b^2 - 4ac$ is called the *discriminant*. From the following examples, conclusions concerning the zeros of quadratic equations can be determined from the discriminant.

Example 2: Solve the equation $3x^2 + 2x + 5 = 0$ using the quadratic formula.

$$x = \frac{-2 \pm \sqrt{2^2 - 4(3)(5)}}{2(3)} = \frac{-2 \pm \sqrt{-56}}{6} \qquad \textbf{Substitute } a = 3, b = 2, \text{ and } c = 5$$

Since the square root of a negative is not equal to any real number, the equation has no real solution.

By evaluating the discriminant $b^2 - 4ac$, this conclusion could have been determined immediately. $b^2 - 4ac = -56$. Since the discriminant is the quantity found under the square root symbol, it may be concluded that *if the discriminant is a negative number, the equation will have no real zeros.*

Example 3: Solve the equation $9m^2 - 30m + 25 = 0$ using the quadratic formula.

$$m = \frac{-(-30) \pm \sqrt{30^2 - 4(9)(25)}}{2(9)} \qquad \textbf{Substitute } a = 9, b = -30, \text{ and } c = 25 \text{ in the quadratic formula.}$$

$$m = \frac{30 \pm \sqrt{0}}{18}$$

$$m = \frac{30 + 0}{18} \quad \text{or} \quad m = \frac{30 - 0}{18}$$

$$m = \frac{5}{3} \quad \text{or} \quad m = \frac{5}{3}$$

Note that the equation has a double zero. By evaluating the discriminant $b^2 - 4ac = 30^2 - 4(9)(25) = 0$, we conclude that *if the discriminant is zero, the equation will have one (double) real zero.*

Example 4:   Solve the equation $6d^2 + 13d - 5 = 0$ using the quadratic formula.

$$d = \frac{-13 \pm \sqrt{13^2 - 4(6)(-5)}}{2(6)}$$

Substitute $a = 6$, $b = 13$, and $c = -5$ in the quadratic formula.

$$d = \frac{-13 \pm \sqrt{289}}{12}$$

$$d = \frac{-13 \pm 17}{12}$$

$$d = \frac{-13 + 17}{12} \quad \text{or} \quad d = \frac{-13 - 17}{12}$$

$$d = \frac{4}{12} \quad \text{or} \quad d = \frac{-30}{12}$$

$$d = \frac{1}{3} \quad \text{or} \quad d = \frac{-5}{12}$$

The discriminant $b^2 - 4ac = 13^2 - 4(6)(-5) = 289$. *If the discriminant is greater than zero, there are two unequal, real zeros.*

Rule: To determine the nature of the zeros of a quadratic equation $ax^2 + bx + c = 0$, evaluate the discriminant:

if $b^2 - 4ac = 0$, there is one real, double zero.
if $b^2 - 4ac < 0$, there is no real zero.
if $b^2 - 4ac > 0$, there are two unequal, real zeros.

EXERCISES 11.5

Without solving the equations, determine the nature of the zeros:

1. $I^2 + 2I - 15 = 0$

2. $5R^2 - 3R + 2 = 0$

3. $7E^2 - 10E - 5 = 0$

4. $x^2 - 4x + 4 = 0$

5. $4D^2 - 10D + 25 = 0$

6. $6m^2 - 31m + 35 = 0$

7. $121f^2 - 154f + 49 = 0$

8. $3b^2 + 2b + 7 = 0$

9. $6y^2 - 5y = 21$

10. $8e^2 = -14e - 3$

11. $.18M^2 - .09M - .02 = 0$

12. $\frac{1}{4}h^2 - \frac{3}{4}h - \frac{1}{2} = 0$

13. $81a^2 = 54a - 9$

14. $25T^2 - 10T = -1$

Graph the following functions:

15. $y = x^2 - 4$

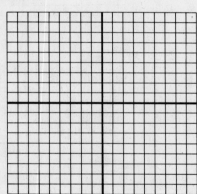

16. $y = x^2 + 2x - 3$

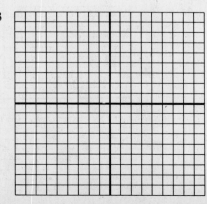

17. $y = x^2 - 4x$

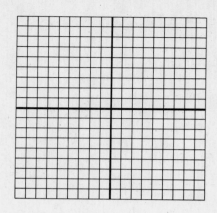

18. $y = x^2 - 2x + 2$

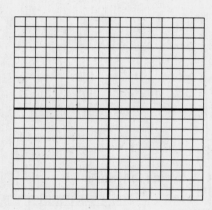

19. $y = -x^2 + 2x + 3$

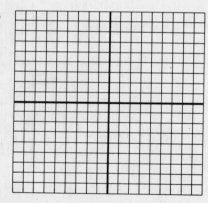

20. $y = 4x^2 - 4x - 3$

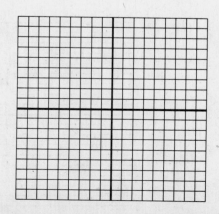

21. A company finds that the cost, y, of manufacturing x items of a certain product is

$$y = x^2 + 8x + 7$$

Graph the cost equation in a rectangular coordinate system.

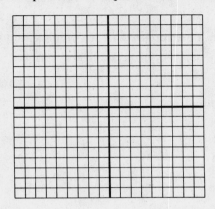

11.6 Complex Numbers

Use the quadratic formula to solve $x^2 - 2x + 2 = 0$. Substituting in the formula,
$$x = \frac{2 \pm \sqrt{-4}}{2}.$$

No real number, when squared, is equal to -4. Therefore, the equation $x^2 - 2x + 2 = 0$ has no real solution.

In order to be able to find solutions for this equation and similar equations, it is necessary to define the square root of a negative number.

Definition 1: The square root of -1 is j, an imaginary unit: $\sqrt{-1} = j$, $j^2 = -1$. (The $\sqrt{-1}$ is often denoted by the letter i. Since i is used in electrical problems to denote current, the letter j will be used in this section.)

Using the definition of $\sqrt{-1}$,
$$\sqrt{-4} = \sqrt{4}\sqrt{-1} = 2j.$$

The solution of the equation $x^2 - 2x + 2 = 0$ can therefore be expressed as

$$x = \frac{2 \pm 2j}{2} = 1 \pm j$$

The number $1 \pm j$, which is the sum of a real number and an imaginary number, is called a *complex number*. The equation $x^2 - 2x + 2 = 0$ has no real solution, but it has solutions that are complex numbers.

Definition 2: A complex number is a number of the form $a + bj$, where a and b are real numbers and $j = \sqrt{-1}$.

The following are examples of complex numbers:

$$3 + 2j$$
$$5 - 7j$$
$$-4 - 8j$$

Every real number may be expressed as a complex number, $a + jb$, with an imaginary component equal to zero.

$$4 = 4 + 0j$$
$$\frac{3}{7} = \frac{3}{7} + 0j$$
$$-11 = -11 + 0j$$

Every imaginary number may be expressed as a complex number with a real component equal to zero.

$$4j = 0 + 4j$$
$$-7j = 0 - 7j$$
$$\frac{5}{9}j = 0 + \frac{5}{9}j$$

Example 1: Simplify $\sqrt{-9}$

$$\sqrt{-9} = \sqrt{9}\sqrt{-1}$$
$$= 3j$$

Example 2: Simplify $\sqrt{-7}$

$$\sqrt{-7} = \sqrt{7}\sqrt{-1}$$
$$= \sqrt{7}\,j$$

Example 3: Simplify $1 + 3\sqrt{-16}$

$$1 + 3\sqrt{-16} = 1 + 3\sqrt{16}\sqrt{-1}$$
$$= 1 + (3 \cdot 4 \cdot j)$$
$$= 1 + 12j$$

Example 4: Solve the equation $3x^2 - 5x + 3 = 0$

$$x = \frac{5 \pm \sqrt{25 - 4(3)(3)}}{2(3)} = \frac{5 \pm \sqrt{25 - 36}}{6}$$

$$x = \frac{5 \pm \sqrt{-11}}{6} = \frac{5 \pm \sqrt{11}\,j}{6}$$

$$x = \frac{5}{6} \pm \frac{\sqrt{11}}{6}j$$

Definition 3: Two complex numbers, $a + bj$ and $c + dj$, are equal if, and only if, $a = c$ and $b = d$.

Definition 4: To add or subtract two complex numbers, find the sum, or difference, of their real parts and their imaginary parts.

$$(a + bj) + (c + dj) = (a + c) + (b + d)j$$
$$(a + bj) - (c + dj) = (a - c) + (b - d)j$$

Two complex numbers may be multiplied by using the distributive property, just as two binomials were multiplied.

$$(a + bj)(c + dj) = ac + adj + bcj + bdj^2$$
$$= ac + adj + bcj + bd(-1)$$
$$= (ac - bd) + adj + bcj$$
$$= (ac - bd) + (ad + bc)j$$

Definition 5: The product of two complex numbers

$$(a + bj)(c + dj) = (ac - bd) + (ad + bc)j.$$

Definition 6: The complex number $a - bj$ is called the *conjugate* of $a + bj$.

The product of a complex number and its conjugate is a complex number with an imaginary part equal to zero.

$$(a + bj)(a - bj) = a^2 - b^2j^2 = a^2 - b^2(-1) = a^2 + b^2$$

Since $j = \sqrt{-1}$, two complex numbers may be divided similarly to dividing irrational numbers. $\dfrac{a + bj}{c + dj}$ is multiplied by 1 in the form of $\dfrac{\text{the conjugate of the divisor}}{\text{the conjugate of the divisor}}$.

Definition 7: $\dfrac{a + bj}{c + dj} = \dfrac{a + bj}{c + dj} \cdot \dfrac{c - dj}{c - dj}$

Example 5: Add $(2 + 3j)$ and $(4 - 6j)$

$$\begin{aligned}(2 + 3j) + (4 - 6j) &= (2 + 4) + [(3 + (-6)]j \\ &= 6 - 3j\end{aligned}$$

Example 6: Subtract $(2 + 3j)$ from $(4 - 6j)$

$$\begin{aligned}(4 - 6j) - (2 + 3j) &= (4 - 2) + (-6 - 3)j \\ &= 2 - 9j\end{aligned}$$

Example 7: Multiply $(5 - 6j)$ by $(2 + 3j)$

$$\begin{aligned}(2 + 3j)(5 - 6j) &= [(2)(5) - (3)(-6)] + [(2)(-6) + (3)(5)]j \\ &= (10 + 18) + (-12 + 15)j \\ &= 28 + 3j\end{aligned}$$

Example 8: Divide $(2 + 3j)$ by $(5 - 6j)$

$$\frac{(2 + 3j)}{(5 - 6j)} \cdot \frac{(5 + 6j)}{(5 + 6j)} = \frac{[(2)(5) - (3)(6)] + [(2)(6) + (3)(5)]j}{(5)(5) - (6)(6)j}$$

$$= \frac{(10 - 18) + (12 + 15)j}{25 - 36j^2}$$

$$= \frac{-8 + 27j}{25 + 36}$$

$$= \frac{-8 + 27j}{61}$$

$$= -\frac{8}{61} + \frac{27}{61}j$$

EXERCISES 11.6

Simplify:

1. $\sqrt{-25}$

2. $\sqrt{-100}$

3. $4\sqrt{-121}$

4. $-6\sqrt{-64}$

5. $-\sqrt{-81}$

6. $-\frac{1}{2}\sqrt{-4}$

7. $\sqrt{-15}$ **8.** $\sqrt{-24}$

9. $1 + 2\sqrt{-64}$ **10.** $4 - 5\sqrt{-169}$

Solve the following quadratic equations. Express complex answers in the form of $a + bj$.

11. $x^2 - 2x + 3 = 0$ **12.** $y^2 - y + 5 = 0$

13. $T^2 - T + 10 = 0$ **14.** $3b^2 - 2b + 9 = 0$

15. $a^2 - 3a + 3 = 0$ **16.** $2D^2 + D + 3 = 0$

17. $4R^2 + 3R + 1 = 0$ **18.** $9x^2 + 2x + 4 = 0$

19. $7M^2 + M + 3 = 0$ **20.** $6W^2 + 5W + 2 = 0$

Perform the indicated operations for the following complex numbers:

21. $(2 + 7j) + (5 - 14j)$ **22.** $(15 - 3j) + (9 + 4j)$

23. $(6 - 2j) + (6 + 2j)$ **24.** $(-3 + 2j) + (3 - 2j)$

25. $(7 + 5j) + (-7 + 5j)$ **26.** $(-2 + j) - (-3 - j)$

27. $(16 - 11j) - (8 - 4j)$

28. $(-3 - 5j) - (2 + 5j)$

29. $(2 + j)(3 - 2j)$

30. $(1 - 3j)(1 - 7j)$

31. $(-4 + 3j)(2 + 6j)$

32. $(3 + 5j)(3 - 5j)$

33. $(6 - j)(6 + j)$

34. $\dfrac{2 - 7j}{2 - 3j}$

35. $\dfrac{5 - 2j}{5 + 2j}$

36. $\dfrac{3 - j}{-6 + 5j}$

37. $\dfrac{3j}{1 + 5j}$

38. $\dfrac{6}{4 + j}$

39. The voltage of a circuit is equal to the product of the current measured in amperes times the impedance measured in ohms. If the current is $(8 - 2j)$ amperes and the impedance is $(2 + 4j)$ ohms, find the voltage.

40. If the voltage of a circuit $(a + bj)$, the magnitude of the voltage is given as

$$v = \sqrt{a^2 + b^2}$$

Find the magnitude of the voltage $(19 - 15j)$ volts.

Unit 11 Self-Evaluation

Solve the following quadratic equations by factoring.

1. $b^2 + 5b - 14 = 0$

2. $6R^2 - 11R = 35$

3. $f^2 = 10f$

4. $49W^2 = -28W - 4$

Solve the following quadratic equations by taking the square root of both members of the equation.

5. $a^2 = 100$

6. $y^2 = 19$

7. $(T - 5)^2 = 9$

8. $(7p - 2)^2 = 18$

Solve the following quadratic equations by completing the square.

9. $r^2 + 4r - 12 = 0$

10. $6N^2 + 13N - 5 = 0$

11. $3x^2 - 2x - 3 = 0$

Solve the following quadratic equations using the quadratic formula.

12. $2E^2 + 7E - 1 = 0$

13. $6y^2 - y = 3$

14. $\dfrac{R - 3}{R + 2} - 5 = \dfrac{R + 1}{R - 3}$

15. $n^2 = 7n - 18$

Use the discriminant to determine the nature of the zeros of each of the following:

16. $64x^2 - 144x + 81 = 0$

17. $16c^2 - 7c - 2 = 0$

18. $10V^2 - V + 2 = 0$

Solve using any method:

19. The sum of a number and its reciprocal is $\frac{10}{3}$. Find the number.

20. The weight of one steel beam is 20 kg more than the weight of another beam. The product of the weight of the two beams is 22,400 kg. What is the weight of each beam?

21. The area of a rectangle is 208 m². The length is 2 m longer than three times the width. Find the length and the width of the rectangle.

22. The product of two conservative positive integers is 132. Find the integers.

Graph the following functions:

23. $y = 6x^2$

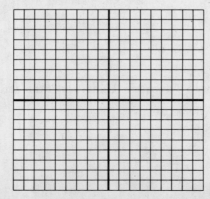

24. $y = x^2 - 9$

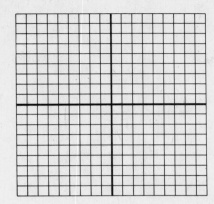

25. $y = -x^2 + 8x$

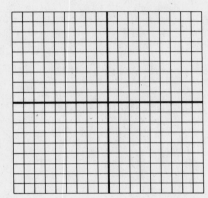

Perform the indicated operations for the following complex numbers:

26. $(8 + 2j) + (24 - 18j)$

27. $(6 - 13j) - (8 - 4j)$

28. $(9 + j)(4 - 7j)$

29. $(8 - 3j)(8 + 3j)$

30. $\dfrac{4 - 5j}{2 - 3j}$

Logarithms

Unit 12 Objectives
1. To graph exponential and logarithmic functions.
2. To solve logarithmic equations.
3. To use Table 2 to find logarithms.
4. To use Table 2 to find antilogarithms.
5. To use the properties of logarithms.
6. To compute using logarithms.
7. To find natural logarithms.

12.1 Exponential and Logarithmic Functions

The equation $y = a^x$, where a is a positive constant not equal to one, is called an *exponential function*.

Example 1: Graph:

(a) $y = 3^x$

Let $x = -2, -1, 0, 1, 2$ to get the ordered pairs $\left(-2, \frac{1}{9}\right)$, $\left(-1, \frac{1}{3}\right)$, $(0,1)$, $(1,3)$, $(2,9)$.

(b) $y = \left(\frac{1}{3}\right)^x$

Let $x = -2, -1, 0, 1, 2$ to get the ordered pairs $(-2,9), (-1,3)$, $(0,1)$, $\left(1, \frac{1}{3}\right)$, $\left(2, \frac{1}{9}\right)$.

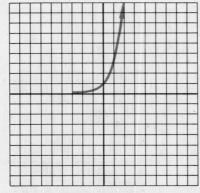

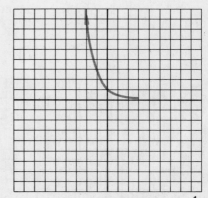

In this example the base a is 3. If $a > 1$, the graph rises (from left to right) and is called an *increasing function*.

In this example the base a is $\frac{1}{3}$. If $0 < a < 1$, the graph falls (from left to right) and is called a *decreasing function*.

$(0,1)$ is a point on both graphs.

371

The *inverse* of the exponential equation $y = a^x$ is $x = a^y$. The equation $x = a^y$, where $x < 0$ and a is a positive constant not equal to 1, is called a *logarithmic equation*.

Example 2:   Graph:

(a) $x = 3^y$ (b) $x = \left(\dfrac{1}{3}\right)^y$

The equations of these examples are the same as the equations used in Example 1 but with the x and y interchanged. Therefore, the ordered pairs from Example 1 can be used with the left and right members interchanged.

$\left(\dfrac{1}{9}, -2\right)$, $\left(\dfrac{1}{3}, -1\right)$, $(1,0)$, $(3,1)$, $(9,2)$. $(9, -2)$, $(3, -1)$, $(1,0)$, $\left(\dfrac{1}{3},1\right)$, $\left(\dfrac{1}{9},2\right)$.

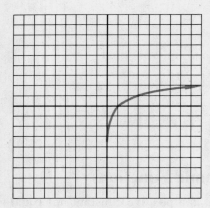   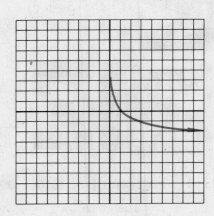

$(1,0)$ is a point on both graphs.

EXERCISES 12.1

Graph the exponential equations:

1. $y = 2^x$ **2.** $y = \left(\dfrac{1}{2}\right)^x$

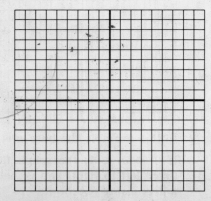   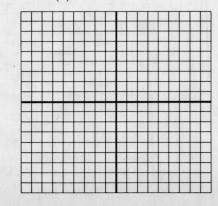

3. $y = 3^{x-1}$

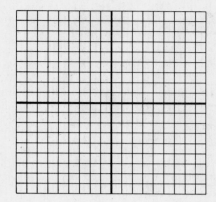

4. $y = \left(\dfrac{1}{3}\right)^{x-1}$

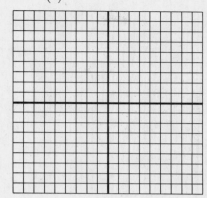

5. $y = 4^{x+1}$

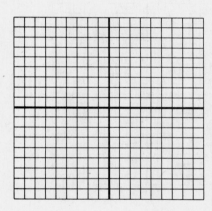

6. $y = \left(\dfrac{1}{4}\right)^{x+1}$

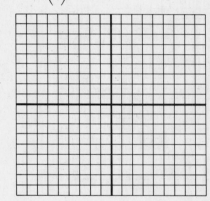

Graph the logarithmic equations:

7. $x = 2^y$

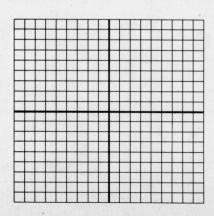

8. $x = \left(\dfrac{1}{2}\right)^y$

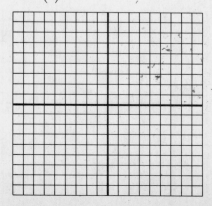

9. $x = 3^{y-1}$

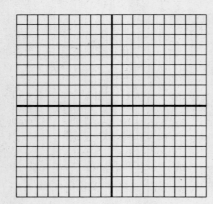

10. $x = \left(\dfrac{1}{3}\right)^{y-1}$

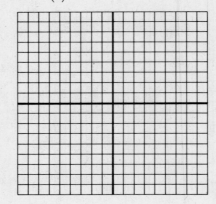

11. $x = 4^{y-1}$

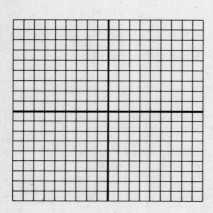

12. $x = \left(\dfrac{1}{4}\right)^{y-1}$

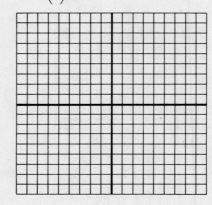

13. In a solution, the hydrogen ion concentrate and the pH are related by the equation:

$$H = 10^{-pH}$$

Graph this relation in a rectangular coordinate system, using the horizontal axis for the pH factor and the vertical axis for the hydrogen ion concentration.

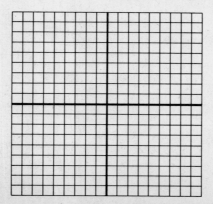

12.2 Logarithmic Equations

In technology, many involved calculations are encountered. Logarithms offer a method of performing these calculations. For example, a new truck costing $4000.00 is estimated to have a scrap value of $800.00 in ten years. To find the percentage depreciation, d, so that the truck is depreciated each year by a constant percentage of its book value:

$$d = 1 - \sqrt[10]{\frac{800}{4000}}$$
$$= 1 - \sqrt[10]{0.2}$$

In order to calculate d, the tenth root of 0.2 has to be found. The slide rule is limited to at most a three-place accuracy. Through the use of logarithms, a greater accuracy will be possible.

In this unit, logarithms are used to simplify extensive arithmetic computations and to find powers and roots. However, their use is not limited to simplifying calculations. In technical calculus, logarithms are used extensively for purposes other than calculations.

A logarithm is an equivalent way of writing an exponential expression. Consider the exponential expression $3^2 = 9$, where 3 is the base and 2 is the exponent. The logarithm of 9 to the base 3 is equal to the exponent to which the base needs to be raised to yield 9, that is, 2.

$\log_3 9 = 2$ because $3^2 = 9$

The generalized formal definition of a logarithm is as follows:

Definition: Where a is a positive constant not equal to 1, and $x > 0$

$x = a^y$ means $y = \log_a x$.

Example 1: Express $5^3 = 125$ in logarithmic form.

$5^3 = 125$ is equivalent to $\log_5 125 = 3$

Example 2: Express $\log_3 3 = 1$ in exponential form.

$\log_3 3 = 1$ is equivalent to $3^1 = 3$.

Equations involving logarithms can be solved if, after putting the equation in exponential form, both sides of the equation can be expressed as powers of the same base.

Rule: $a^r = a^s$ if and only if $r = s$.

Example 3: Solve: $\log_{12} 144 = x$

$\log_{12} 144 = x$ **Express the equation in exponential form.**
$12^x = 144$
$12^x = 12^2$
$x = 2$ **Express 144 as 12^2.**

Example 4:   Solve: $\log_3 \frac{1}{27} = x$

$$\log_3 \frac{1}{27} = x$$

$3^x = \frac{1}{27}$ **Express the equation in exponential form.**

$3^x = \frac{1}{3^3}$ **Express 27 as 3^3.**

$3^x = 3^{-3}$

$x = -3$

Example 5:   Solve: $\log_9 3 = x$

$9^x = 3$ **Express the equation in exponential form.**

$(3^2)^x = 3$ **Express 9 as 3^2.**

$3^{2x} = 3^1$

$2x = 1$

$x = \frac{1}{2}$

EXERCISES 12.2

Express in logarithmic form:

1. $4^2 = 16$ **2.** $3^0 = 1$

3. $10^3 = 1000$ **4.** $10^{-2} = \frac{1}{100}$

5. $2^{-1} = \frac{1}{2}$ **6.** $3^2 = 9$

7. $6^0 = 1$ **8.** $2^4 = 16$

Express in exponential form:

9. $\log_3 27 = 3$ **10.** $\log_2 32 = 5$

11. $\log_{10} 10 = 1$ **12.** $\log_{10} 1000 = 3$

13. $\log_2 \frac{1}{2} = -1$ **14.** $\log_3 \frac{1}{9} = -2$

15. $\log_{22} 1 = 0$ **16.** $\log_{1/2} 1 = 0$

Use the definition of logarithms to solve the following:

17. $\log_2 4 = x$

18. $\log_5 25 = x$

19. $\log_8 2 = x$

20. $\log_{16} 4 = x$

21. $\log_2 32 = x$

22. $\log_3 81 = x$

23. $\log_4 4 = x$

24. $\log_5 5 = x$

25. $\log_3 1 = x$

26. $\log_7 1 = x$

27. $\log_2 \frac{1}{8} = x$

28. $\log_3 \frac{1}{9} = x$

29. $\log_{10} 10,000 = x$

30. $\log_{10} 100 = x$

31. $\log_{10} 0.001 = x$

32. $\log_{10} 0.01 = x$

33. $\log_{10} \frac{1}{10} = x$

34. $\log_{10} \frac{1}{100} = x$

12.3   Common Logarithms

Logarithms are more widely used with the base 10 than with any other base. In fact, its use is so widespread that the number indicating the base of the logarithm is omitted and it is understood that it is in base 10. Logarithms in base 10 are called *common logarithms*.

Since the logarithm in base 10 of any number a is equal to the exponent to which 10 needs to be raised to yield a, it is convenient to be able to express a as a product of a number between 1 and 10 and a power of 10, that is, in scientific notation, which was introduced in Unit 9.

Consider the powers of 10 and their logarithmic equivalents:

$$
\begin{aligned}
10^3 &= 1000 & \log 1000 &= 3 \\
10^2 &= 100 & \log 100 &= 2 \\
10^1 &= 10 & \log 10 &= 1 \\
10^0 &= 1 & \log 1 &= 0 \\
10^{-1} &= 0.1 & \log 0.1 &= -1 \\
10^{-2} &= 0.01 & \log 0.01 &= -2 \\
10^{-3} &= 0.001 & \log 0.001 &= -3
\end{aligned}
$$

Notice that the logarithms of the perfect powers of 10 are integers.

Also, observe that the log 1 = 0 and the log 10 = 1. Therefore, the logarithm of any number between 1 and 10 will be between 0 and 1. For example, log 3 will be between 0 and 1. These values between 0 and 1 will be expressed in decimal form rounded off to four decimal places.

Table 2 in Appendix C gives a four place approximation to the common logarithm of any number between 1 and 10.

Example 1:   Find the common logarithm of 5.27.

First, look down the left column in Table 2 to locate 5.2. Second, look across the 5.2 row and locate the entry under the column labeled 7. This is the four place approximation of the log 5.27. Therefore, log 5.27 = 0.7218, that is, $10^{0.7218} = 5.27$.

If a calculator is used instead of Table 2,

Enter	Press	Display
5.27	$\boxed{\log}$	.72181061, which is .7218 when rounded to four decimal places.

Example 2:   Find the common logarithm of 9.01.

First, look down the left column to locate 9.0. Second, look across the 9.0 row to locate the entry under the column labeled 1. The entry on the row labeled 9.0 and the column labeled 1 is the logarithm of 9.01. Therefore, log 9.01 = 0.9547, that is, $10^{0.9547} = 9.01$.

Consider the logarithms of numbers other than those between 1 and 10, for example, the logarithm of 425. The log 425 will equal some number x such that $10^x = 425$. This number x will be between 2 and 3 because $10^2 = 100$ and $10^3 = 1000$, and 425 is between 100 and 1000. Expressing 425 in scientific notation:

$$
\begin{aligned}
425 &= (4.25)(10^2) \\
&= (10^{0.6284})(10^2) \quad \textbf{Using Table 2, log 4.25 = .6284} \\
&= 10^{0.6284+2} \quad\quad\;\; \textbf{Law of Exponents}
\end{aligned}
$$

Therefore,

$$425 = 10^{(0.6284+2)}$$

Expressed in logarithmic notation,

$$\log 425 = 0.6284 + 2$$

It has been shown that the logarithm of any number can be expressed approximately as the sum of a four place decimal and an integer.

It has been shown that the logarithm of any non-negative number may be expressed as the sum of a positive decimal less than 1 and an integer. The positive decimal less than 1 is referred to as the *mantissa* and the integer is referred to as the *characteristic*.

Rule: To find the common logarithm of any non-negative number N:

Step 1: Express N in scientific notation.

Step 2: The mantissa is the logarithm of the first factor when N is expressed in scientific notation.

Step 3: The characteristic is the exponent of the 10 when N is expressed in scientific notation.

Step 4: $\log N$ = mantissa + characteristic

Example 3: Find the log 425.

Step 1: $425 = (4.25)(10^2)$
Step 2: The mantissa is log 4.25 = 0.6284 **Using Table 2**
Step 3: The characteristic is $+2$
Step 4: $\log 425 = 0.6284 + 2 = 2.6284$

Example 4: Find the log 0.0081.

Step 1: $0.0081 = (8.1)(10^{-3})$
Step 2: The mantissa is log 8.1 = 0.9085 **Using Table 2**
Step 3: The characteristic is -3
Step 4: $\log 0.0081 = 0.9085 + (-3)$
$$= 0.9085 - 3$$

The result of Step 4 in Example 4 may be simplified by adding $(+0.9085)$ and (-3). Using the rule in Unit 3 for adding a positive and a negative number, subtract 0.9085 from 3 and make the result negative because the number with the largest absolute value is negative. Therefore,

$$0.9085 - 3 = -2.0915$$

However, by convention, logarithms are expressed as a sum of a decimal between 0 and 1 and an integer. That is the reason for leaving the answer in Example 4 as $0.9085 - 3$.

If a calculator is used,

Enter	Press	Display	Press	Enter	Press	Display
0.0081	log	-2.0915150	$+$	3	$=$	.9084850 $-$ 3

To obtain the form used in the example, add 3 and subtract 3 from the display.

EXERCISES 12.3

Using Table 2, find the common logarithm approximations of the following:

1. log 3.91

2. log 6.07

3. log 8.2   **4.** log 1.1

5. log 6   **6.** log 2

7. log 36.9   **8.** log 509

9. log 0.07   **10.** log 0.573

11. log 6040   **12.** log 47,000

13. log 0.0071   **14.** log 0.0511

15. log 560   **16.** log 29

17. An x-ray tube operates with a voltage of 180,000 V. Find the logarithm of the voltage.

18. The smallest number on the display of a pocket calculator is 10^{-12}.

Find the logarithm of the number.

12.4   Antilogarithms

The antilogarithm of a number is derived by reversing the process of obtaining the logarithm. If the log 4.73 = 0.6749, then the antilogarithm of 0.6749 = 4.73.

Definition: The antilogarithm of x equals N, provided that the logarithm of N equals x; that is,

antilog $x = N$ if and only if   log $N = x$

Example 1:   Find the antilog 0.3118.

Antilog 0.3118 = 2.05 because log 2.05 = 0.3118

Rule 1:   To find the antilogarithm of a positive decimal less than 1:

Step 1:   Look for the entry equal to the number on the inner part of the table.

Step 2:   Look up the row and column headings for that entry. The antilog of the number will be the decimal with the row heading as its units and tenths digit and the column heading as its hundredths digit.

Example 2:   Find antilog 0.1818.

Step 1:   Locate 0.1818 on the inner part of Table 2.

Step 2:   Row heading = 1.5. Column heading = 2

Therefore, antilog 0.1818 = 1.52

If a calculator is used instead of Table 2,

Enter	Press	Enter	Press	Display
10	y^x	.1818	$=$	1.5198475 which is equal to 1.52 when rounded off.

Consider the antilogarithm of a number other than a positive decimal less than 1. The object is to find a number whose log is the original number. If the original number is the log of a number, it can be divided into a positive decimal part, the mantissa, and an integral part, the characteristic. The antilog will be found by finding the antilog of the mantissa and moving the decimal point as many places to the right or the left as the characteristic indicates.

Example 3:   Find the antilog of 4.6042.

This means to find N, where log N = 4.6042. By definition of a logarithm,

$$
\begin{aligned}
N &= 10^{4.6042} \\
&= 10^{(4 + 0.6042)} \\
&= 10^4 \cdot 10^{0.6042} \quad \textbf{Law I of exponents} \\
&= 10^4 \cdot 4.02 \quad \textbf{Using Table 2 log 4.02 = .6042} \\
&= 4.02 \cdot 10^4 \quad \textbf{Commutative property}
\end{aligned}
$$

Rule 2:   To find the antilog of any number:

Step 1:   Divide the number into a positive decimal part, the mantissa, and an integral part, the characteristic.

Step 2:   Find the antilog of the mantissa using Rule 1.

Step 3:   Multiply the result of Step 2 times 10 raised to a power equal to the characteristic.

The result of Step 3 is the antilog of the number, expressed in scientific notation.

Example 4:   Find the antilog 4.6042.

Step 1:   The mantissa is 0.6042. The characteristic is 4.

Step 2:   Antilog 0.6042 = 4.02

Step 3: $4.02 \cdot 10^4$

Therefore, antilog 4.6042 = $4.02 \cdot 10^4$

Example 5:   Find the antilog (0.9605 − 2).

Step 1:   The mantissa is 0.9605. The characteristic is −2.

Step 2:   Antilog 0.9605 = 9.13

Step 3: $9.13 \cdot 10^{-2}$

Therefore, antilog (0.9605 − 2) = $9.13 \cdot 10^{-2}$

With practice, Steps 1, 2, and 3 may be performed simultaneously.

EXERCISES 12.4

Using the table, find the antilogarithmic approximation of the following:

1. antilog 0.5575

2. antilog 0.1038

3. antilog 0.9547

4. antilog 0.8102

5. antilog 0.017

6. antilog 0.705

7. antilog 2.6551

8. antilog 3.7796

9. antilog $(0.5172 - 1)$

10. antilog $(0.8663 - 3)$

11. antilog 3.42

12. antilog 6.711

13. antilog $(0.721 - 3)$

14. antilog $(0.847 - 4)$

15. antilog 1.017

16. antilog 9.909

12.5 Properties of Logarithms

By definition, logarithms are another way of writing an exponential expression. Thus, logarithms will inherit some properties of the Laws of Exponents.

Law I of Exponents states: $10^m \cdot 10^n = 10^{m+n}$
Let $M = 10^m$ and $N = 10^n$
Then $MN = 10^m \cdot 10^n = 10^{m+n}$.
Therefore, $MN = 10^{m+n}$.
Writing it in logarithmic form, $\log MN = m + n$.
But since $10^m = M$, then $m = \log M$
and since $10^n = N$, then $n = \log N$.
Therefore, $\log MN = m + n = \log M + \log N$.

Property 1:   The log of a product equals the sum of the logs of the factors.

$\log MN = \log M + \log N$

Example 1: $\log (100)(10) = \log 100 + \log 10$ **Property 1**
$= \quad 2 \quad + \quad 1$
$= \quad 3$

Law II of Exponents states: $\dfrac{10m}{10n} = 10^{m-n}$.

Let $\qquad\qquad 10^m = M$ and $10^n = N$

Then $\qquad\qquad \dfrac{M}{N} = \dfrac{10m}{10n} = 10^{m-n}$

Therefore, $\qquad\qquad \dfrac{M}{N} = 10^{m-1}.$

Writing it in logarithmic form, $\quad \log \dfrac{M}{N} = m - n.$

But since $\qquad\qquad 10^m = M,$ then $\log M = m$
and since $\qquad\qquad 10^n = N,$ then $\log N = n.$

Therefore, $\qquad\qquad \log \dfrac{M}{N} = \log M - \log N.$

Property 2: The log of a quotient equals the difference of the logs.

$$\log \dfrac{M}{N} = \log M - \log N$$

Example 2:

$\qquad \log \dfrac{100}{10} = \log 100 - \log 10$ **Property 2**

$\qquad\qquad\qquad = 2 - 1$
$\qquad\qquad\qquad = 1$

Law V of Exponents states: $\qquad (10^n)^x = 10^{nx} = 10^{xn}.$
Let $\qquad\qquad\qquad\qquad\qquad 10^n = N$
Then $\qquad\qquad\qquad\qquad\qquad (10^n)^x = N^x$
Therefore, $\qquad\qquad\qquad\qquad 10^{xn} = N^x$
Writing it in logarithmic form, $\qquad \log N^x = xn.$
Since $\qquad\qquad\qquad\qquad\qquad 10^n = N,$ then $\log N = n.$
Therefore, $\qquad\qquad\qquad\qquad \log N^x = x \log N$

Property 3: The log of a number raised to an exponent equals the exponent times the log of the number.

$\log N^x = x \log N$

Example 3: $\log (100^5) = 5(\log 100)$ **Property 3**
$\qquad\qquad\qquad = 5(2)$
$\qquad\qquad\qquad = 10$

Example 4: $\log \sqrt{10{,}000} = \log (10{,}000^{1/2})$ **Definition of fractional exponent**

$\qquad\qquad\qquad = \dfrac{1}{2} (\log 10{,}000)$ **Property 3**

$\qquad\qquad\qquad = \dfrac{1}{2} (4)$

$\qquad\qquad\qquad = 2$

Example 5: $\log (0.0034)(45.3) = \log 0.0034 + \log 45.3$ **Property 1**
$\qquad\qquad\qquad = (0.5315 - 3) + (0.6561 + 1)$
$\qquad\qquad\qquad = 1.1876 - 2$
$\qquad\qquad\qquad = 0.1876 + 1 - 2$
$\qquad\qquad\qquad = 0.1876 - 1$

Example 6: $\log \dfrac{639}{27} = \log 639 - \log 27$ **Property 2**

$\qquad\qquad = (0.8055 + 2) - (0.4314 + 1)$
$\qquad\qquad = 0.8055 + 2 - 0.4314 - 1$
$\qquad\qquad = 0.3741 + 1$

Example 7: $\log \dfrac{0.429}{0.0682} = \log 0.429 - 0.0682$ **Property 2**

$\log 0.429 = .6325 - 1$
$\underline{\log 0.0682 = .8338 - 2}$

In order to avoid a negative mantissa, add 1 and subtract 1 from the minuend.

$\log 0.429 = 1.6325 - 2$
$\underline{\log 0.0682 = 0.8338 - 2}$
$\qquad\qquad\qquad\quad 0.7987$

$\log \dfrac{0.429}{0.0682} = 0.7987$

Example 8: $\log (0.0588)^3 = 3(\log 0.0588)$ **Property 3**
$\qquad\qquad\qquad = 3(0.7694 - 2)$
$\qquad\qquad\qquad = 2.3082 - 6$
$\qquad\qquad\qquad = 0.3082 + 2 - 6$
$\qquad\qquad\qquad = 0.3082 - 4$

Example 9: $\log \sqrt[5]{731{,}000} = \log (731{,}000^{1/5})$
$\qquad\qquad\qquad\quad = \dfrac{1}{5}(\log 731{,}000)$ **Property 3**
$\qquad\qquad\qquad\quad = \dfrac{1}{5}(0.8639 + 5)$
$\qquad\qquad\qquad\quad = 0.1728 + 1$

On the calculator,

Enter	Press	Press	Enter	Press	Display
731,000	$\boxed{\log}$	$\div$	5	$=$	1.1727835

which is 1.1728 when rounded to four decimal places.

Example 10: $\log \sqrt[4]{0.62} = \log (0.62)^{1/4}$
$\qquad\qquad\qquad = \dfrac{1}{4}(\log 0.62)$ **Property 3**
$\qquad\qquad\qquad = \dfrac{1}{4}(.7924 - 1)$
$\qquad\qquad\qquad = \dfrac{1}{4}(.7924 - 1 + 3 - 3)$ **In order to avoid a fractional characteristic, add 3 and subtract 3.**
$\qquad\qquad\qquad = \dfrac{1}{4}(3.7924 - 4)$
$\qquad\qquad\qquad = 0.9481 - 1$

Therefore, $\log \sqrt[4]{0.62} = 0.9481$

Properties of logarithms are used to solve exponential equations that cannot be solved simply. The equation $2^{x+1} = 8$ was solved simply because 8 could be expressed as 2^3. To solve $2^{x+1} = 9$, take the logarithm of each side.

Example 11: Solve: $2^{x+1} = 9$

$\log 2^{x+1} = \log 9$ **Take the log of each side**
$(x + 1)\log 2 = \log 9$ **Property 3**
$x + 1 = \dfrac{\log 9}{\log 2}$

$$x = \frac{\log 9}{\log 2} - 1$$

$$x \doteq \frac{.9542}{.3010} - 1$$

$$x \doteq 2.1701$$

Example 12: Solve: $\log x + \log (x - 5) = \log 6$

$$\log x(x - 5) = \log 6 \quad \textbf{Property 1}$$
$$x(x - 5) = 6$$
$$x^2 - 5x - 6 = 0$$
$$(x - 6)(x + 1) = 0$$
$$x = 6 \quad \text{or} \quad x = -1$$

Check: Since $\log a$ is defined only if a is positive, -1 is not a solution.

$$\log x + \log(x - 5) = \log 6$$
$$\log 6 + \log(6 - 5) = \log 6$$
$$\log 6 + \log 1 = \log 6$$
$$\log 6(1) = \log 6 \quad \textbf{Property 1}$$

Example 13: Solve: $\log(5x + 6) - \log(x + 2) = \log 4$

$$\log \frac{5x + 6}{x + 2} = \log 4 \quad \textbf{Property 2}$$

$$\frac{5x + 6}{x + 2} = 4$$

$$5x + 6 = 4(x + 2)$$
$$5x + 6 = 4x + 8$$
$$x + 6 = 8$$
$$x = 2$$

Check: $\log(5x + 6) - \log(x + 2) = \log 4$
$$\log(5 \cdot 2 + 6) - \log(2 + 2) = \log 4$$
$$\log \frac{16}{4} = \log 4 \quad \textbf{Property 2}$$
$$\log 4 = \log 4$$

EXERCISES 12.5

Using the properties of logarithms, find the following logs:

1. $\log(0.031)(4680)$

2. $\log(0.00411)(0.00579)$

3. $\log(65,000)(423)$

4. $\log(0.633)(.481)$

5. $\log\dfrac{812}{905}$

6. $\log\dfrac{0.00412}{0.062}$

7. $\log\dfrac{71.4}{0.48}$

8. $\log\dfrac{0.311}{42,500}$

9. $\log(36.9)^4$

10. $\log(0.0083)^7$

11. $\log(0.00639)^5$

12. $\log(674,000)^9$

13. $\log\sqrt{0.0674}$

14. $\log\sqrt{73,400}$

15. $\log\sqrt[3]{854,000}$

16. $\log\sqrt[3]{0.0074}$

17. $\log\sqrt[4]{8.92}$

18. $\log\sqrt[6]{0.943}$

19. $\log(42\sqrt{5})$

20. $\log(.6\sqrt{3})$

21. $\log\dfrac{\sqrt{7}}{3.1}$

22. $\log\dfrac{0.55}{\sqrt{8}}$

Solve for x:

23. $3^x = 5$

24. $5^{x+1} = 7$

25. $4^{x-1} = 11$

26. $3^{2x-3} = 8$

27. $12^{2x-3} = 478$

28. $4^{5x} = 276$

29. $\log(9x + 2) - \log(x + 2) = \log 8$

30. $\log 4 = \log(3x - 2) - \log(x - 2)$

31. $\log(x + 3) - \log 3 = \log(x - 5)$

32. $\log x + \log 3 = \log(2x + 1)$

33. $\log(6x + 2) = \log(x + 3) + \log 4$

34. $2 \log x = \log 25$

35. $\log(x - 2) + \log(x + 1) = \log 1$

12.6   Computation Using Logarithms

Logarithms may be used to simplify arithmetic computations. They are used to multiply, divide, raise to a power, and extract a root.

Rule: To find the result of an arithmetic expression using logarithms:

Step 1:   Find the log of the arithmetic expression and express it as the sum of a decimal between 0 and 1 and an integer.

Step 2:   Find the antilog of the result of Step 1.

The result of Step 2 is the answer.

Example 1:   Find $(0.00451)(0.57)$.

Step 1: $\log (0.00451)(0.57) = \log 0.00451 + \log 0.57$
$= \log (4.51 \cdot 10^{-3}) + \log (5.7 \cdot 10^{-1})$
$= 0.6542 - 3 + 0.7559 - 1$
$= 1.4101 - 4$
$= 0.4101 - 3$

Step 2: $\text{antilog } (0.4101 - 3) = (2.57)(10^{-3})$
$= 0.00257$

The mantissa, .4101, cannot be found exactly from Table 2. Use the number 2.57, having a mantissa closest to .4101.
Therefore, $(0.00451)(0.57) \doteq 0.00257$
    On the calculator,

Enter	Press	Press	Enter	Press	Press	Display
0.00451	$\boxed{\log}$	$\boxed{+}$	0.57	$\boxed{\log}$	$\boxed{=}$	-2.589948602

Press	Enter	Press	Display
$\boxed{+}$	3	$\boxed{=}$	.4100513976 $- 3$, which is .4101 $- 3$ when rounded.

Press	Enter	Press	Enter	Press	Display
$\boxed{c}$	10	$\boxed{y^x}$	.4101	$\boxed{=}$	2.578987706, which is 2.57 when rounded.

Therefore, $2.57 \times 10^{-3} \doteq 0.00257$

Example 2: Find $\dfrac{0.683}{570}$.

Step 1: $\log \dfrac{0.683}{570} = \log 0.683 - \log 570$

$= \log(6.83 \cdot 10^{-1}) - \log(5.7 \cdot 10^2)$
$= (0.8344 - 1) - (0.7559 + 2)$
$= 0.8344 - 1 - 0.7559 - 2$
$= 0.0785 - 3$

Step 2: $\text{antilog}(0.0785 - 3) = (1.18)(10^{-3}) = 0.00118$

Therefore, $\dfrac{0.683}{570} \doteq 0.00118$

Example 3: Find $(58.3)^4$.

Step 1: $\log 58.3^4 = 4(\log 58.3)$
$= 4(0.7657 + 1)$
$= 3.0628 + 4$
$= 0.0628 + 3 + 4$
$= 0.0628 + 7$

Step 2: $\text{antilog}(0.0628 + 7) = (1.16)(10^7)$
$= 11,600,000$

Therefore, $(58.3)^4 \doteq 11,600,000$

Example 4: Find $\sqrt[5]{0.0023}$

Step 1: $\log \sqrt[5]{0.0023} = \log(0.0023)^{1/5}$

$= \dfrac{1}{5} \log 0.0023$

$= \dfrac{1}{5}(0.3617 - 3)$

$= \dfrac{1}{5}(2.3617 - 5)$ **Add 2 and subtract 2.**

$= .4723 - 1$

Step 2: $\text{antilog}(0.4723 - 1) = (2.97)(10^{-1})$
$= 0.297$

Therefore, $\sqrt[5]{0.0023} \doteq 0.297$

EXERCISES 12.6

Compute the following using logarithms:

1. $(95,700)(780)$　　　　　　**2.** $(854)(67,000)$

3. (0.00581)(0.0045)

4. (0.043)(0.99)

5. $\dfrac{7850}{11,200}$

6. $\dfrac{67}{4910}$

7. $\dfrac{0.00712}{0.77}$

8. $\dfrac{0.899}{0.56}$

9. $(783,000)^6$

10. $(56,000)^3$

11. $(0.0749)^3$

12. $(0.071)^5$

13. $\sqrt{3}$

14. $\sqrt{2}$

15. $\sqrt{677}$

16. $\sqrt{85}$

17. $\sqrt[3]{0.068}$

18. $\sqrt[5]{0.00439}$

19. $\sqrt[4]{57,000}$

20. $\sqrt[3]{86.3}$

21. $\dfrac{(8660)(0.047)}{(0.355)}$

22. $\dfrac{(63.1)(895)}{87.9}$

Solve the following problems using logarithms:

23. Find the amount of power needed by a car exerting a force of 192 kg at $30\dfrac{\text{km}}{\text{hr}}$. (Power = force × speed.)

24. Potential energy (*PE*) is the product of weight and height. Find the potential energy of a 3200-kg automobile at the top of a 100-meter hill.

25. Find the current, i, drawn by a 0.5 horsepower electric motor operated from a 120-volt source of electricity, given by

$$i = \sqrt{\dfrac{0.5\ \text{hp} \times 746\ \dfrac{\text{watts}}{\text{hp}}}{120\ \text{volts}}}$$

26. The maximum electric field intensity from the sun is

$$E = 2 \times 3 \times 10^8 \times 1.26 \times 10^{-6} \times 1400\ \dfrac{\text{volts}}{\text{m}}$$

Find E.

27. The wave velocity of a string weighing $0.2\ \dfrac{\text{kg}}{\text{m}}$ and under a tension of 10 kg is

$$v = \sqrt{\dfrac{10}{6.25 \times 10^{-3}}}\ \dfrac{\text{m}}{\text{sec}}$$

Find v.

12.7   Natural Logarithms

Common logarithms are logarithms to the base 10. In many of the sciences the base e is used. Logarithms to the base e are called *natural logarithms*. The base e is an irrational number with an approximate value of 2.718.

Logarithms to the base e are written $\ln N$ to distinguish them from logarithms to the base 10.

Rule: To change from base 10 to base e the following formula is used:

$$\ln N = \frac{\log N}{\log e}$$

Since $\qquad\qquad e \doteq 2.718$
then $\qquad\quad \log e \doteq \log 2.718$
and $\qquad\quad \log e \doteq 0.4343$ **From Table 2**

Example 1:   Find $\ln 364$

$$\ln 364 = \frac{\log 364}{\log 2.718} = \frac{2.5611}{0.4343} \qquad \textbf{Using change of base formula}$$

$$\log \frac{2.5611}{0.4343} = \log 2.5611 - \log 0.4343 \qquad \textbf{To divide 2.5611 by 0.4343, use}$$
$$= (1.4082 - 1) - (0.6375 - 1) \quad \textbf{logarithms.}$$
$$= 0.7707$$

The antilog of $0.7707 \doteq 5.90$ **Using Table 2**
Therefore, $\ln 364 \doteq 5.90$.

Check by using the calculator.

Enter | Press | Display
364 | $\boxed{\ln x}$ | 5.897153868, which rounded off is 5.90.

Example 2:   Find $\ln 0.362$

$$\ln 0.362 = \frac{\log 0.362}{\log e} = \frac{0.5587 - 1}{0.4343} = \frac{-0.4413}{0.4343} = -\frac{0.4413}{0.4343}$$

$$\log \frac{0.4413}{0.4343} = \log 0.4413 - \log 0.4343$$

$$= (0.6444 - 1) - (0.6375 - 1)$$

$$= 0.0069$$

The antilog of $0.0069 \doteq 1.02$.
Therefore, $\ln 0.362 \doteq -1.02$.

EXERCISES 12.7

Find the natural logarithms of the following:

1. $\ln 37$

2. $\ln 81$

3. ln 574

4. ln 768

5. ln 2.82

6. ln 4.33

7. ln 44.5

8. ln 92.1

9. ln 0.66

10. ln 0.82

11. The work done in expanding a gas is $W = k \ln \left(\dfrac{v_2}{v_1} \right)$. Find W if $k = 1.6$ ft-lb, $v_1 = 10$ sq. in., and $v_2 = 60$ sq. in.

12. Audio level in decibels is given by $d = 20 \ln \left(\dfrac{v_2}{v_1} \right)$. Find d if $v_1 = 20$ volts and $v_2 = 200$ volts.

Unit 12 Self-Evaluation

1. Graph: $y = \log_5 x$

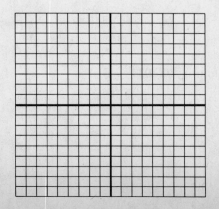

2. Solve: $\log_2 16 = x$

3. Solve: $\log_3 \dfrac{1}{81} = x$

4. Express in logarithmic form: $3^{-1} = 0.33$

5. Express in exponential form: $\log_{10} 0.01 = -2$

6. Evaluate: $\log_5 625$

7. Find: log 3520

8. Find: log 0.0378

9. Find: antilog 2.525

10. Find: antilog $(0.7033 - 3)$

11. Solve: $4^x = 11$

12. Solve: $2 \log x = \log 16$

13. Use logarithms to compute: $(756{,}000)(0.0443)$

14. Use logarithms to compute: $\sqrt[3]{0.784}$

15. Find: $\ln 721$

Trigonometry

Unit 13 Objectives
1. To determine the trigonometric ratios of an angle in a given triangle.
2. To find the value of trigonometric ratios for angles from 0° to 90° inclusive.
3. To solve right triangles.
4. To find the trigonometric ratios of angles larger than 90°.
5. To solve problems using vectors.
6. To express complex numbers in polar form, to multiply and divide complex numbers in polar form, and to express complex numbers in exponential form.
7. To solve oblique triangles using the Law of Sines.
8. To solve oblique triangles using the Law of Cosines.
9. To graph curves of the form $y = a \sin bx$, and $y = a \cos bx$.
10. To graph equations of the form $y = a \sin (bx + \theta)$, and $y = a \cos (bx + \theta)$.
11. To graph equations of the form $y = a \tan bx$.
12. To solve trigonometric equations.

unit 13

13.1 Trigonometric Ratios

A triangle that has one right angle is called a *right triangle*. The sum of the angles of any triangle is 180°. Therefore, a right triangle has one right angle and two acute angles. Since the sum of the two acute angles must be 90°, the acute angles of a right triangle are complementary.

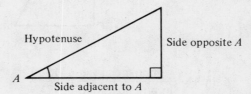

Figure 13.1

The six relations of an acute angle of a right triangle, defined in terms of the two sides and hypotenuse, are called *trigonometric ratios,* or *trigonometric functions*. They are:

Rule: The trigonometric ratios are:

$$\text{sine } A = \frac{\text{Side opposite } A}{\text{hypotenuse}} \qquad \text{cosecant } A = \frac{\text{hypotenuse}}{\text{Side opposite } A}$$

$$\text{cosine } A = \frac{\text{Side adjacent } A}{\text{hypotenuse}} \qquad \text{secant } A = \frac{\text{hypotenuse}}{\text{Side adjacent } A}$$

$$\text{tangent } A = \frac{\text{Side opposite } A}{\text{Side adjacent } A} \qquad \text{cotangent } A = \frac{\text{Side adjacent } A}{\text{Side opposite } A}$$

Note from the given definitions and from Example 1 below that the csc is the reciprocal of the sin; the sec is the reciprocal of the cos; and the cot is the reciprocal of the tan.

Example 1: Find the trigonometric ratios of angle A in the following triangle.

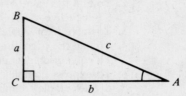

Figure 13.2

$$\sin A = \frac{a}{c} \qquad \csc A = \frac{c}{a}$$

$$\cos A = \frac{b}{c} \qquad \sec A = \frac{c}{b}$$

$$\tan A = \frac{a}{b} \qquad \cot A = \frac{b}{a}$$

Example 2: Given $\sin A = \dfrac{15}{17}$, determine $\sec A$.

To determine the secant of A, sketch the given triangle.

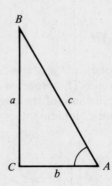

Figure 13.3

$$\sin A = \frac{\text{opposite}}{\text{hypotenuse}} = \frac{a}{c} = \frac{15}{17}$$

$$b^2 = c^2 - a^2 \quad \textbf{Using the Pythagorean Theorem}$$

$$b^2 = 17^2 - 15^2$$

$$b^2 = 289 - 225 = 64$$

$$b = 8$$

Therefore, $\sec A = \dfrac{\text{hypotenuse}}{\text{adjacent}} = \dfrac{c}{b} = \dfrac{17}{8}$

Example 3: Given $\sin A = \dfrac{15}{17}$, verify $\sin^2 A + \cos^2 A = 1$

$$\sin A = \frac{15}{17} \quad \textbf{Given}$$

$$\cos A = \frac{8}{17} \quad \textbf{Using results of Example 2}$$

Then, $\sin^2 A + \cos^2 A = 1$

$$\left(\frac{15}{17}\right)^2 + \left(\frac{8}{17}\right)^2 = 1$$

$$\frac{225}{289} + \frac{64}{289} = 1$$

$$\frac{289}{289} = 1$$

$$1 = 1$$

EXERCISES 13.1

Given: Find:

1.

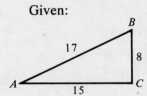

sin A =   csc A =
cos A =   sec A =
tan A =   cot A =

2.

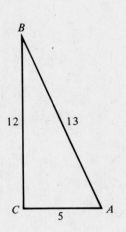

sin B =   csc B =
cos B =   sec B =
tan B =   cot B =

3.

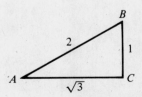

cos B =   cot A =
sin A =   sec A =
csc B =   tan B =

4.

sin A =   cot A =
tan B =   sec B =
csc A =   cos A =

Given triangle ABC, C = 90°.

Given:	Find:
5. $a = 3$	$\sin A =$
$b = 4$	$\cot B =$

6. $a = 12$	$\cos B =$
$c = 13$	$\tan A =$

7. $b = 8$	$\csc A =$
$c = 17$	$\sin B =$

8. $a = 3$	$\cos B =$
$b = 4$	$\sec A =$

Given:	Find:
9. $\sin A = \dfrac{3}{5}$	$\sec A =$

10. $\cos A = \dfrac{4}{5}$	$\csc A =$

11. $\cot A = \dfrac{8}{15}$	$\sin A =$

12. $\sec A = \dfrac{13}{12}$ $\sin A =$

13. $\sin B = \dfrac{7}{25}$ $\cot B =$

14. $\tan B = \dfrac{40}{9}$ $\cos B =$

15. $\cos B = \dfrac{9}{41}$ $\tan B =$

Given: Verify: $\sin^2 \theta + \cos^2 \theta = 1$

16. $\cot \theta = \dfrac{4}{3}$

17. $\sec \theta = \dfrac{2\sqrt{3}}{3}$

18. $\csc \theta = \dfrac{17}{8}$

19. $\sin \theta = \dfrac{40}{41}$

20. $\tan \theta = 1$

13.2   Values of the Trigonometric Ratios

The trigonometric ratios for 0°, 30°, 45°, 60°, and 90° can be easily determined. In geometry, a triangle that has three equal sides and three equal angles is called an *equilateral triangle*. Since the sum of the three angles of a triangle must equal 180°, each angle in an equilateral triangle must equal 60°. By letting each side equal two, we have the following triangle:

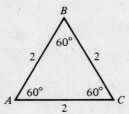

Figure 13.4

If angle *B* is divided into equal angles with a line perpendicular to side *AC*, it will divide the side *AC* into two equal parts and form two right triangles:

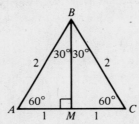

Figure 13.5

Using the Pythagorean Theorem, the unknown side, *MB*, is determined to be $\sqrt{3}$. Consider one of the right triangles formed:

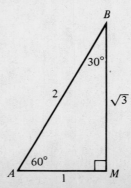

Figure 13.6

$$\sin 30° = \frac{1}{2} \qquad\qquad \csc 30° = 2$$

$$\cos 30° = \frac{\sqrt{3}}{2} \qquad\qquad \sec 30° = \frac{2}{\sqrt{3}}$$

$$\tan 30° = \frac{1}{\sqrt{3}} \qquad\qquad \cot 30° = \sqrt{3}$$

$$\sin 60° = \frac{\sqrt{3}}{2} \qquad\qquad \csc 60° = \frac{2}{\sqrt{3}}$$

$$\cos 60° = \frac{1}{2} \qquad\qquad \sec 60° = 2$$

$$\tan 60° = \sqrt{3} \qquad\qquad \cot 60° = \frac{1}{\sqrt{3}}$$

An *isosceles right triangle* is a triangle with two equal sides and two equal angles. Since the two acute angles of a right triangle must equal 90°, each angle in an isosceles right triangle is 45°. Letting each of the two equal sides be equal to one, we have the following triangle:

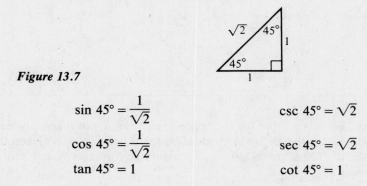

Figure 13.7

$$\sin 45° = \frac{1}{\sqrt{2}} \qquad\qquad \csc 45° = \sqrt{2}$$

$$\cos 45° = \frac{1}{\sqrt{2}} \qquad\qquad \sec 45° = \sqrt{2}$$

$$\tan 45° = 1 \qquad\qquad \cot 45° = 1$$

To determine the trigonometric ratios of 0°, picture an angle as it becomes smaller and smaller. At 0°, the side opposite, a, is zero. The hypotenuse is the same as the adjacent side. Therefore $c = b$. Using the definitions of the trigonometric ratios,

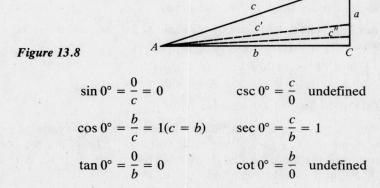

Figure 13.8

$$\sin 0° = \frac{0}{c} = 0 \qquad\qquad \csc 0° = \frac{c}{0} \text{ undefined}$$

$$\cos 0° = \frac{b}{c} = 1 (c = b) \qquad \sec 0° = \frac{c}{b} = 1$$

$$\tan 0° = \frac{0}{b} = 0 \qquad\qquad \cot 0° = \frac{b}{0} \text{ undefined}$$

To determine the trigonometric ratios of 90°, picture an angle as it gets larger and approaches 90°. At 90°, the side adjacent, b, is equal to zero. The hypotenuse is the same as the side opposite. Therefore, $c = a$. Using the definitions of the trignometric ratios,

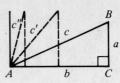

Figure 13.9

$$\sin 90° = \frac{a}{c} = 1(c = a) \qquad\qquad \csc 90° = \frac{c}{a} = 1$$

$$\cos 90° = \frac{b}{c} = \frac{0}{c} = 0 \qquad\qquad \sec 90° = \frac{c}{b} = \frac{c}{0} \text{ undefined}$$

$$\tan 90° = \frac{a}{b} = \frac{a}{0} \text{ undefined} \qquad\qquad \cot 90° = \frac{b}{a} = \frac{0}{a} = 0$$

Degree	Radians	Sine	Cosine	Tangent
0°	0	0	1	0
30°	$\frac{\pi}{6}$	$\frac{1}{2}$	$\frac{\sqrt{3}}{2}$	$\frac{1}{\sqrt{3}}$
45°	$\frac{\pi}{4}$	$\frac{1}{\sqrt{2}}$	$\frac{1}{\sqrt{2}}$	1
60°	$\frac{\pi}{3}$	$\frac{\sqrt{3}}{2}$	$\frac{1}{2}$	$\sqrt{3}$
90°	$\frac{\pi}{2}$	1	0	undefined

The above is a table for the trigonometric ratios of a few angles commonly used. Computing the trigonometric ratios for other angles between 0° and 90° is more difficult. These ratios have been calculated for us and can be found in Table 3, Appendix C.

Example 1: Use Table 3 to find the tangent of 28°30′.

Read down the left column until you find 28°30′. Find the column *across the top* of the table headed "tan θ." The tangent of 28°30′ is directly across from 28°30′ in the column headed tan θ.

tan 28°30′ ≐ .5430

On the calculator: $30′ = \frac{30}{60} = .5$ degrees.

To enter 28.5° check your calculator manual for the radian/degree switch. Slide the switch to degree measure.

Enter	Press	Display
28.5	tan	0.5429557 ≐ 0.5430

Example 2: Use Table 3 to find the cosine of 72°10′.

Angles between 45° and 90° are found in the columns at the right of the tables. *Read up* the right column until you find 72°10′. Find the column *across the bottom* of the table headed "cos θ." The cosine of 72°10′ is directly across from 72°10′ in the column marked "cos θ."

cos 72°10′ ≐ .3062

On the calculator:

Enter	Press	Enter	Press	Enter	Press	Press	Display
72	+	1	÷	6	=	cos	.30624918 ≐ .3062

Example 3:   Use Table 3 to find x if the tangent of x is .2247.

To find the angle whose tangent is .2247, look down the columns headed "tan θ" at the top of the table and up the columns headed "tan θ" at the bottom of the page. Since .2247 is found in the column marked "tan θ" at the top of the page, the angle in the left column of the page is read:

$$\tan x = .2247$$
$$x = 12°40'$$

On the calculator: Find the angle whose tan is .2247. Some calculators have a key marked $\boxed{TAN^{-1}}$. If your calculator does not have this key, it will have a key marked $\boxed{arc}$ or $\boxed{INV}$. Check your manual.

Enter	Press	Press	Display
.2247	$\boxed{arc}$	$\boxed{tan}$	12.664022

Round off .664022 to two decimal places, .66, and multiply by 60. (.66)(60) = 39.60, or 40′.

Example 4:   Find the csc 52° using a calculator.

The csc 52° is the reciprocal of the sin 52°.

Enter	Press	Press	Display
52	$\boxed{sin}$	$\boxed{\frac{1}{x}}$	1.26901821 ≐ 1.269

Example 5:   Given cot x = 1.376, find x in degree measure using a calculator.

The cot is the reciprocal of the tan.

Enter	Press	Press	Press	Display
1.376	$\boxed{\frac{1}{x}}$	$\boxed{arc}$	$\boxed{tan}$	36.007562 ≐ 36°.

EXERCISES 13.2

In problems 1 through 12, use Table 3 to determine the value of the indicated trigonometric ratio. If a calculator is used instead of Table 3, round answers to four decimal places.

1. sin 28°

2. cot 15°50′

3. sec 67°10′

4. tan 84°20′

5. cos 9°

6. csc 77°50′

7. sin 39°10′

8. sec 36°40′

9. cot 42°30′

10. csc 81°

11. $\sin 13°30'$

12. $\tan 51°40'$

In problems 13 through 24, determine the value of the angle between 0° and 90° from Table 3 or by using a calculator.

13. $\sin x = .2588$

14. $\csc x = 6.392$

15. $\cos x = .6947$

16. $\tan x = 2.246$

17. $\cos x = .8949$

18. $\cot x = .6128$

19. $\csc x = 1.386$

20. $\sec x = 1.159$

21. $\tan x = .7627$

22. $\sec x = 1.074$

23. $\sin x = .9528$

24. $\cos x = .3987$

Torque (T) depends on the magnitude of the force (F), the distance (r) from the force to the axis of rotation, and the direction (θ) of the force. $T = Fr \sin \theta$. Calculate the torque, given:

25. $F = 3600$ kg $r = 2.3$ m $\theta = 61°20'$

26. $F = 400$ kg $r = 3.0$ m $\theta = 55°10'$

27. $F = 1600$ kg $r = 2.5$ m $\theta = 86°$

28. $F = 900$ kg $r = 1.8$ m $\theta = 72°40'$

13.3 Elementary Right Triangle Applications

It is now possible to find the unknown parts of a right triangle using Table 3. When solving right triangle problems: first, draw the triangle; second, mark the known parts; and third, select the trigonometric ratio involving two known facts and one unknown.

Example 1: Given a right triangle with $A = 32°$ and $c = 7$, solve for the unknown parts.

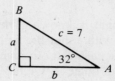

Figure 13.10

When one acute angle is given, the other may be obtained by subtracting from 90°. Recall that the two acute angles in a right triangle are complementary.

$$B = 90° - 32°$$
$$B = 58°$$

It is now possible to work with either angle A or angle B, since both are known. Selecting angle A and the hypotenuse (which is given to be 7), we may work with the sine and cosine of angle A.

$$\sin 32° = \frac{a}{7} \quad \textbf{Solve for } a$$
$$7 \cdot (\sin 32°) = a$$
$$7 \cdot (.5299) \doteq a$$
$$3.7093 \doteq a$$
$$a \doteq 4$$

$$\cos 32° = \frac{b}{7} \quad \textbf{Solve for } b$$
$$7 \cdot (\cos 32°) = b$$
$$7 \cdot (.8480) = b$$
$$5.9360 = b$$
$$b \doteq 6$$

Therefore, $B \doteq 32°$ $a \doteq 4$
            $C = 90°$ $b \doteq 6$

On the calculator: $a = 7 \sin 32°$

Enter	Press	Enter	Press	Press	Display
7	$\boxed{\times}$	32	$\boxed{\text{sin}}$	$\boxed{=}$	3.7094349

3.7094349 rounded to four decimal places = 3.7094, which is more accurate than 3.7093, since a closer approximation of the sin of 32° was used on the calculator. The original problem used only one significant figure, so both answers are rounded to one significant figure, 4.

On the calculator: $b = 7 \cos 32°$

Enter	Press	Enter	Press	Press	Display
7	$\boxed{\times}$	32	$\boxed{\text{cos}}$	$\boxed{=}$	5.936337 $\doteq$ 6

Example 2: Given a right triangle with $a = 3.10$ and $b = 4.28$, solve the triangle for the unknown parts.

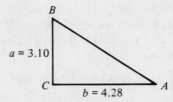

Figure 13.11

Neither acute angle is known. Therefore, either angle may be selected. It is possible to work with either the tangent or the cotangent since the side opposite and the side adjacent are both given. Selecting angle A and choosing to work with the tangent of angle A, we have the ratio,

$$\tan A = \frac{3.1}{4.28} = .7243$$

$$A = 36°$$

$$B = 90° - 36°$$

$$B = 54°$$

In order to find c, the hypotenuse, any ratio involving c may be used. Selecting angle A and the cosecant of angle A, we have,

$$\csc 36° = \frac{c}{3.1}$$

$$3.1 \csc 36° = c$$

$$3.1(1.701) = c$$

$$5.2731 = c$$

$$c \doteq 5.27$$

Therefore, $A \doteq 36°$ $\qquad c \doteq 5.27$

$$B \doteq 54°$$

$$C = 90°$$

Example 3: The angle of elevation to the top of a skyscraper is 53°10′ from a point 100 meters from the base. How high is the skyscraper?

First sketch the figure. The angle of elevation is the angle between the line of sight and an object above the horizontal plane. To find a, the height of the building, the tan or cot of A may be used. Choosing the tan, $\tan A = \dfrac{a}{b}$

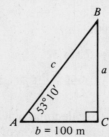

Figure 13.12

$$\tan 53°10′ = \frac{a}{100}$$

$$100 \cdot \tan 53°10′ = a$$

$$100(1.335) \doteq a$$

$$134 \text{ m} \doteq a$$

Therefore, the skyscraper is approximately 134 m high.

Example 4: An airplane pilot spots a hangar whose angle of depression is 21°. If the plane is at 2100 m, how far is the hangar from a point directly below the plane?

The angle of depression is the angle between the line of sight and an object below the horizontal plane.

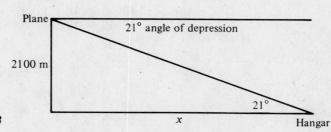

Figure 13.13

$$\cot 21° = \frac{x}{2100}$$
$$2100 \cdot \cot 21° = x$$
$$2100(2.605) \doteq x$$
$$5470.5 \doteq x$$
$$x \doteq 5500 \text{ m}$$

Therefore, the plane is approximately 5500 m from the hangar.

EXERCISES 13.3

Solve the following right triangles with given parts. The triangle in each case is the triangle *ACB* with angle *C* given as the right angle.

1. $a = 3$
    $b = 4$

2. $a = 12.5$
    $b = 9.1$

3. $a = 2$
    $A = 32°$

4. $a = 127$
    $c = 164$

5. $a = 101$
    $b = 120$

6. $a = 843$
    $A = 44°$

7. $c = 4217$
$B = 48°40'$

8. $B = 14°20'$
$c = 345$

9. $b = 86.7$
$c = 167$

10. $c = 18$
$B = 12°10'$

11. $B = 42°50'$
$a = 8.8$

12. $A = 32°30'$
$c = 56.8$

13. From the top of a cliff 164 meters high the angle of depression of a boat is 85°50′. How far away is the boat from the foot of the cliff?

14. The angle of elevation to the top of a radio tower is 34°20′ from a point 270 meters from the foot of the tower. How high is the tower?

15. A piece of sheet metal is 12 cm by 19 cm. Find the angle the diagonal makes with the longer side.

16. A ladder leaning against a building makes an angle of 27°30′ with the side of the building. If the ladder is 2.1 meters from the base of the building, how long is the ladder?

17. A surveyor, using a transit, notes that the angle of elevation to the top of a tree is 18°10′ as measured from a point 150 meters from its base. How tall is the tree?

18. A boy is flying a kite. If the string on the kite makes an angle of 26°30′ with the ground, how much string is needed to get the kite 100 meters above the ground?

19. The top of a hill is 300 meters above a level line. If the measure of an angle of elevation from a surveyor's position is 68°10′ to the top of the hill, how far is the surveyor from a point directly below the top of the hill?

20. The base angle of an isosceles triangle is 40°10′ and a side is 6.8 cm. Find the altitude and the base of the triangle.

21. A country road has a rise of 2 km for every 7 km along the road. What is the angle of rise?

22. From a seaplane the angle of depression to a buoy in the water is 36°20′. If the seaplane is at 10,000 meters, how far is the buoy from a point directly below the plane?

23. From the top of a 60-meter building, the angle of elevation to the top of a building next door is 20°40′. If the two buildings are 30 meters apart, how high is the building next door?

24. A rectangle is 14.2 cm by 8.4 cm. Find the length of the diagonal and the measure of the angles formed by a diagonal with the sides.

13.4   Angles Larger Than 90°

To study angles greater than 90°, place the angle in standard position. Take a point on the terminal side of the angle and drop a line perpendicular to the *x*-axis. The distance *r* is always taken to be positive.

Quadrant I

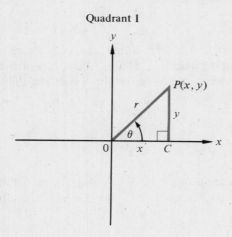

$$\sin \theta = \frac{y}{r} \qquad \csc \theta = \frac{r}{y}$$

$$\cos \theta = \frac{x}{r} \qquad \sec \theta = \frac{r}{x}$$

$$\tan \theta = \frac{y}{x} \qquad \cot \theta = \frac{x}{y}$$

Figure 13.14

Quadrant II

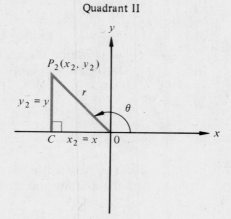

$$\sin \theta = \frac{y}{r} \qquad \csc \theta = \frac{r}{y}$$

$$\cos \theta = \frac{-x}{r} \qquad \sec \theta = \frac{r}{-x}$$

$$\tan \theta = \frac{y}{-x} \qquad \cot \theta = \frac{-x}{y}$$

Figure 13.15

Quadrant III

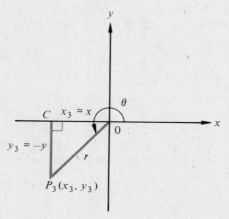

$$\sin \theta = \frac{-y}{r} \qquad \csc \theta = \frac{r}{-y}$$

$$\cos \theta = \frac{-x}{r} \qquad \sec \theta = \frac{r}{-x}$$

$$\tan \theta = \frac{-y}{-x} \qquad \cot \theta = \frac{-x}{-y}$$

Figure 13.16

Quadrant IV

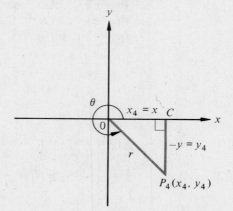

$$\sin \theta = \frac{-y}{r} \qquad \csc \theta = \frac{r}{-y}$$

$$\cos \theta = \frac{x}{r} \qquad \sec \theta = \frac{r}{x}$$

$$\tan \theta = \frac{-y}{x} \qquad \cot \theta = \frac{x}{-y}$$

Figure 13.17

Note that the sine and cosecant are the only positive ratios in the second quadrant. The tangent and cotangent are positive in the third quadrant, and the cosine and secant are positive in the fourth:

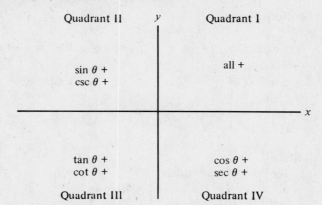

Figure 13.18

In Figures 13.15, 13.16, and 13.17, the acute angle formed between the terminal side of θ and the *x*-axis is the *reference angle* for θ. In each figure, this angle is ∠*POC*.

In Figure 13.15,   ∠POC = 180° − θ
If θ = 135°, then   ∠POC = 180° − 135° = 45°

In Figure 13.16,   ∠POC = θ − 180°
If θ = 240°, then   ∠POC = 240° − 180° = 60°

In Figure 13.17,   ∠POC = 360° − θ
If θ = 330°, then   ∠POC = 360° − 330° = 30°

Rule 1:   To find the reference angle, θ′, for an angle θ, where θ is greater than 90°:

Step 1:   Determine the quadrant of θ.

Step 2:   If θ is in Quadrant II, θ′ = 180° − θ.
If θ is in Quadrant III, θ′ = θ − 180°.
If θ is in Quadrant IV, θ′ = 360° − θ.

Rule 2:   To find trigonometric ratios of θ, when 90° < θ < 360°:

Step 1:   Determine the reference angle, θ′, of θ using Rule 1.

Step 2:   Find the trigonometric ratio of θ′, using Table 3 in Appendix C or finding the exact value.

Step 3:   Determine the sign of the trigonometric ratio of θ, using Figure 13.18.

The trigonometric ratio of θ is equal to the result of Steps 2 and 3.

Example 1:   Find the exact values of the trigonometric ratios of 300°, using Rule 2.

Step 1:   θ′ = 360° − 300°   **Using Rule 1; θ is in Quadrant IV.**
θ′ = 60°

Step 2: $\sin 60° = \dfrac{\sqrt{3}}{2}$ $\csc 60° = \dfrac{2}{\sqrt{3}}$ **Using Table T-1.**

$\cos 60° = \dfrac{1}{2}$ $\sec 60° = 2$

$\tan 60° = \sqrt{3}$ $\cot 60° = \dfrac{1}{\sqrt{3}}$

Step 3:   cos θ and sec θ are positive, and   **Using Figure 13.18.**
all other trigonometric ratios are
negative.

Therefore,

$$\sin 300° = -\frac{\sqrt{3}}{2} \qquad \csc 300° = -\frac{2}{\sqrt{3}}$$

$$\cos 300° = \frac{1}{2} \qquad \sec 300° = 2$$

$$\tan 300° = -\sqrt{3} \qquad \cot 300° = -\frac{1}{\sqrt{3}}$$

Example 2:   Find tan 168°.

168° is in Quadrant II
$\theta' = 180° - 168° = 12°$ **Rule 1**
$\tan 12° \doteq 0.2126$ **Table 3**
$\tan 168° \doteq -0.2126$ **Figure 13.18**

Rule 3:   To find θ, given trigonometric ratios of θ:

Step 1:   Determine the quadrant of θ from the given information, using Figure 13.18.

Step 2:   Find the reference angle, θ′, of θ using Table 3.

Step 3:   If θ is in Quadrant II, θ = 180° − θ′.
If θ is in Quadrant III, θ = 180° + θ′.
If θ is in Quadrant IV, θ = 360° − θ′.

Example 3:   Find θ where 0° ≤ θ < 360°, given that cos θ = −.7969 and tan θ > 0.

Step 1:   θ is in Quadrant III   **cos θ is negative**
tan θ is positive

Step 2:   If cos θ′ = 0.7969,   **Table 3**
then θ′ = 37°10′

Step 3:   θ = 180° + 37°10′

Therefore, θ = 217°10′.

Example 4:   Find θ where 0° ≤ θ < 360°, given that sec θ = 3.42 and sin θ < 0.

Step 1:   θ is in Quadrant IV   **sec θ is positive**
sin θ is negative

Step 2:   If sec θ′ = 3.42,   **Table 3**
then θ′ ≐ 73°

Step 3:   θ = 360° − 73°

Therefore, θ = 287°.

EXERCISES 13.4

Find the exact values of the six trigonometric ratios of the following angles.

1. 225°   **2.** 330°   **3.** 210°

4. 150°   **5.** 240°   **6.** 135°

7. 120°   **8.** 315°   **9.** 180°

10. 270°

Determine the value of the indicated trigonometric ratio.

11. tan 317°   **12.** cos 109°   **13.** sin 251°

14. sec 98°   **15.** cot 189°   **16.** sin 274°

17. csc 95°10′   **18.** cos 301°30′   **19.** cot 176°40′

20. sec 289°20′   **21.** cos 312°20′   **22.** tan 142°

Find θ where 0° ≤ θ < 360°, given:

23. tan θ = −.4734, cos θ > 0   **24.** sin θ = .5373, cot θ < 0

25. csc θ = 1.153, sec θ < 0   **26.** cot θ = .2962, cos θ < 0

27. $\cos \theta = .7808$, $\tan \theta < 0$

28. $\sec \theta = -14.96$, $\cot \theta > 0$

29. $\sin \theta = -.0349$, $\cot \theta < 0$

30. $\csc \theta = 9.567$, $\cos \theta < 0$

31. $\cot \theta = -.8591$, $\sec \theta > 0$

32. $\tan \theta = -.2278$, $\cos \theta < 0$

13.5   Vectors

In previous problems we have used such quantities as time, length, speed, and area. For example, the statement that a car travels 40 kilometers per hour mentions a quantity that includes size or magnitude but not direction. Such quantities are *scalar quantities* and the numbers that represent them are scalars.

A *vector quantity* is one that has both magnitude and direction, such as velocity, force, acceleration, or momentum. The rate of speed of a car traveling southwest at 40 kilometers per hour is an example of a vector quantity.

Symbolically, a vector can be denoted by two letters, the first indicating the initial point, and the second indicating the terminal point. An arrow is placed over the two letters. The vector from O to A is written $\overrightarrow{OA}$. Another representation for a vector quantity is a capital letter in bold type such as **V** or **A**. The magnitude of the vector is represented by the absolute value symbol, $|\overrightarrow{OA}|$ or $|\mathbf{A}|$.

Graphically, the direction of the vector is indicated by a line segment with an arrow on one end. Its magnitude is indicated by the length of the line segment.

Figure 13.19

To represent the vector, **A**, in the rectangular coordinate system, use the origin as the initial point. The vector is represented by its terminal point, (x,y).

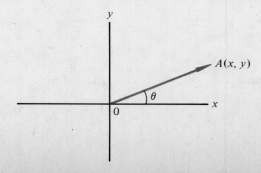

Figure 13.20

Drop a line from **A** perpendicular to the x-axis. This line segment is called the y-scalar component of the vector **A** and is represented by A_y. Draw a line from **A** perpendicular to the y-axis. This line is called the x-scalar component of the vector **A** and is represented by A_x. Note that $|\mathbf{A}|$ is the resultant of A_x and A_y.

The direction of this vector is the angle formed with the positive x-axis as the initial side and the vector $\overrightarrow{OA}$ as the terminal side.

The magnitude, r, and the direction, θ, of the vector may be expressed as an ordered pair, (r,θ). (r,θ) is called the *polar form* of the vector.

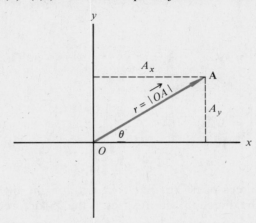

Figure 13.21

From Figure 13.21, the following relationships to a right triangle can be established:

$$\tan \theta = \frac{A_y}{A_x} \qquad \text{θ is the direction of A.}$$

Since $\sin \theta = \dfrac{A_y}{|r|}$ A_y is the y-scalar component of **A**.

then $A_y = |r| \sin \theta$ **Solving for A_y**

Since $\cos \theta = \dfrac{A_x}{|r|}$ A_x is the x-scalar component of **A**.

then $A_x = |r| \cos \theta$ **Solving for A_x.**

$$|r|^2 = A_x{}^2 + A_y{}^2$$

$$|r| = \sqrt{A_x{}^2 + A_y{}^2} \qquad \textbf{Pythagorean Theorem.}$$

Example 1:   Express the vector $(8,32°)$ in rectangular form.

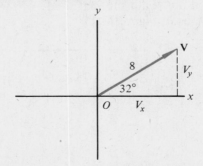

Figure 13.22

$$A_x = |r| \cos \theta \qquad\qquad A_y = |r| \sin \theta$$
$$A_x = 8 \cos 32° \qquad\qquad A_y = 8 \sin 32°$$
$$A_x \doteq 8(.8480) \qquad\qquad A_y \doteq 8(.5299)$$
$$A_x \doteq 7 \qquad\qquad\qquad A_y \doteq 4$$

Therefore, $(8,32°) \doteq (7,4)$.

Example 2:   Express the vector (15,8.0) in polar form.

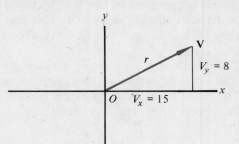

Figure 13.23

$$|r| = \sqrt{A_x^2 + A_y^2} \qquad\qquad \tan\theta = \frac{A_y}{A_x}$$

$$|r| = \sqrt{15^2 + 8^2} \qquad\qquad \tan\theta = \frac{8}{15}$$

$$|r| = \sqrt{225 + 64} \qquad\qquad \tan\theta \doteq .5333$$

$$r \doteq 17 \qquad\qquad\qquad\qquad \theta \doteq 28°$$

Therefore, $(15,8.0) = (17,28°)$.

Example 3:   Two forces of 60 kg and 75 kg act at right angles to each other on an object. Find the direction with respect to the 75-kg force and the resultant force.

The resultant force, **R,** is the vector sum of the two vectors representing the forces. The direction with respect to the 75 kg force is θ.

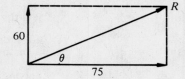

Figure 13.24

$$\tan\theta = \frac{60}{75} = .8$$

$$\theta = 38°40'$$

$$|R| = \sqrt{60^2 + 75°} = \sqrt{3600 + 5625} = \sqrt{9225}$$

$$|R| \doteq 96$$

Example 4:   Two cables support a weight of 500 lb. One cable is at 60° to the horizontal and the other at 30°. Find the magnitude of the force that each cable exerts at the junction D, as in Figure 13.25.

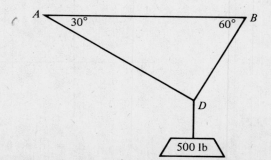

Figure 13.25

To support the weight, a force of 500 lb must be applied in an opposite direction to the pull of the weight. Constructing a vector diagram, force vectors **A** and **B** represent the forces on D exerted by **A** and **B**, as in Figure 13.26.

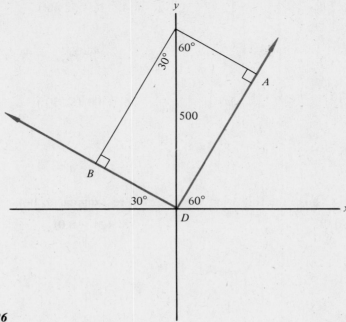

Figure 13.26

$$\overrightarrow{DA} = 500 \sin 60° \doteq 500(.8660) \doteq 433 \text{ lb}$$
$$\overrightarrow{DB} = 500 \sin 30° \doteq 500(.5000) \doteq 250 \text{ lb}$$

Example 5:   A ship is sailing on a course of 36° at a speed of 75 kilometers per hour. How many kilometers per hour is the ship moving in a direction due north? (Course is the angle measured from the north clockwise through the east.)

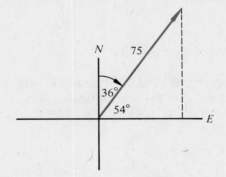

Figure 13.27

$$V_y = |r| \sin θ$$
$$V_y = 75 \sin 54°$$
$$V_y \doteq 75(.8090) \doteq 61 \text{ km per hour}$$

EXERCISES 13.5

Change the following vectors from polar to rectangular form:

1. $(2.00, 82°)$ **2.** $(34, 313°)$

3. $(200, 115°20')$

4. $(16.1, 72°40')$

5. $(102, 267°)$

6. $(500, 152°20')$

Change the following vectors from rectangular to polar form:

7. $(-18, -6.0)$

8. $(24, -8.0)$

9. $(13.2, 17.9)$

10. $(2.04, 3.25)$

11. $(-1.6, 6.4)$

12. $(24.6, 6.30)$

13. A rocket is fired at an angle of 76° with the horizontal. If its speed is 200 km per hour, find the horizontal and vertical components of the velocity.

14. A car is moving up an incline, making an angle of 33°40' with the horizontal. If the car is traveling at a rate of 853 centimeters per second, find its horizontal and vertical velocities.

15. An airplane is flying on a course of 47°10' at a speed of 260 km per hour. How many kilometers per hour is the plane moving in a direction due east?

16. The eastward component of the velocity of a ship is 6.7 km. Its northward component of velocity is 11.2 km. Find its direction.

17. Find the magnitude of a force if its horizontal component of force is 93.7 kg and its vertical component of force is 88.2 kg. What angle does its line of action make with the horizontal?

18. A resultant of 420 kg has a direction of 30°10′ with the horizontal. Find the horizontal force.

19. The resultant of two vectors acting at an angle of 90° with respect to each other is 300 kg. If one force is 110 kg, find the other force. Find the angle the unknown force makes with the resultant.

20. A force has a northward component of 40 kg and an eastward component of 60 kg. Find the direction and magnitude of the force.

21. A sled is pulled horizontally a distance of 200 m by a rope that is at an angle of 32° to the horizontal. If a constant force of 15 kg is used, find the work done. Work (W) is equal to the product of the force (F) and the distance (s) an object is moved. $W = Fs \cos \theta$. θ represents the angle between the direction of the force and the direction of motion.

22. A force of 30 kg at an angle of 20° to the horizontal is used to move a crate 10 m on a level floor. Find the work done.

23. A force of 9 kg makes an angle of 30° with the horizontal. Find the horizontal and vertical components of the force.

24. An object is held in position on an inclined plane by a force parallel to the plane and another perpendicular to it. Find the magnitudes of the forces holding the object in position if the object weighs 150 kg and the angle of inclination of the plane is 28° (see the figure below).

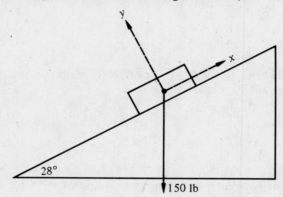

25. An 800-kg weight is held stationary on an inclined plane. The plane makes an angle of 40° with the horizontal. Determine the magnitudes of the forces holding the object in position.

26. Determine the forces exerted at point *D* on a rope from which the weight is suspended:

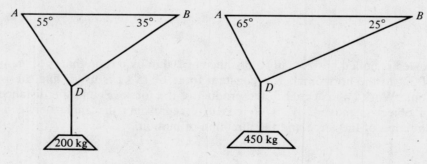

13.6   Complex Numbers in Rectangular, Polar, and Exponential Forms

In Unit 11, a complex number was defined to be a number of the form, $a + bj$, where a and b are members of the real numbers and $j = \sqrt{-1}$.

To illustrate a complex number graphically the horizontal axis is called the *axis of real numbers,* and the vertical axis is the *axis of imaginary numbers.* The rectangular system used in this way is called the *complex plane.* Every complex number, $a + bj$, can be represented in the complex plane by a vector where the initial point is the origin and the terminal point is (a,b).

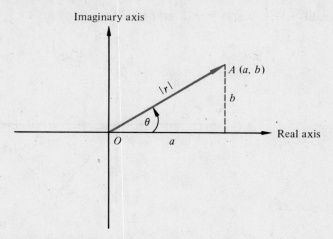

Figure 13.28

To illustrate the addition of two complex numbers, $a + bj$ and $c + dj$, graphically represent them as vectors. $a + bj = \overrightarrow{OA}$. $c + dj = \overrightarrow{OC}$. Draw a parallelogram with $\overrightarrow{OA}$ and $\overrightarrow{OC}$ as sides of the parallelogram. The diagonal of the parallelogram, $\overrightarrow{OE}$, is the sum, or resultant, of $\overrightarrow{OA}$ and $\overrightarrow{OC}$.

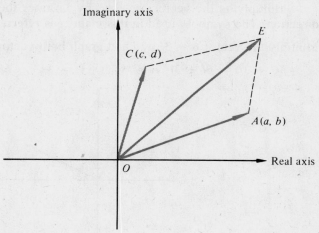

Figure 13.29

Example 1: Add $2 + 3j$ and $4 + 2j$ graphically.

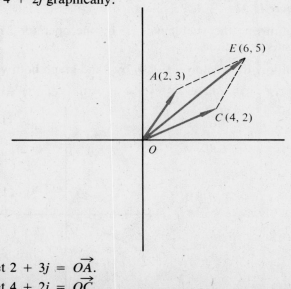

Figure 13.30

Let $2 + 3j = \overrightarrow{OA}$.

Let $4 + 2j = \overrightarrow{OC}$

$\overrightarrow{OA} + \overrightarrow{OC} = \overrightarrow{OE}$

$(2 + 3j) + (4 + 2j) = 6 + 5j$

Example 2:   Add $-3 - 4j$ and $8 - 2j$ graphically.

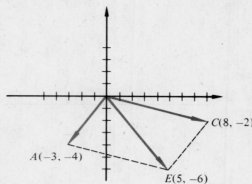

Figure 13.31

$$\text{Let } -3 - 4j = \overrightarrow{OA}.$$
$$\text{Let } 8 - 2j = \overrightarrow{OC}.$$
$$\overrightarrow{OA} + \overrightarrow{OC} = \overrightarrow{OE}$$
$$(-3 - 4j) + (8 - 2j) = 5 - 6j$$

Multiplying the vector, $a + bj$, by j rotates the vector 90° in the positive direction. The symbol j used in this manner is referred to as the *operator j*.

Example 3:   Multiply the vector $6 + 3j$ by j and graph both vectors in the complex plane.

$$j(6 + 3j) = 6j + 3j^2 = 6j + 3(-1) = -3 + 6j$$

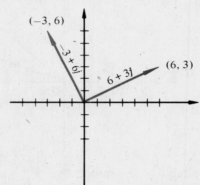

Figure 13.32

Multiplying the vector $6 + 3j$ by the operator j rotated the vector 90° in the positive direction.

Example 4:   Multiply the vector $-3 + 6j$ by j and graph both vectors.

$$j(-3 + 6j) = -3j + 6j^2 = -3j + 6(-1) = -6 - 3j$$

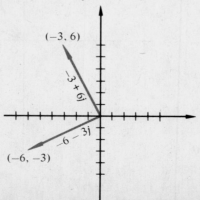

Figure 13.33

Multiplying the vector $-3 + 6j$ by the operator j rotated the vector 90° in the positive direction. Note from Example 3 that multiplying the vector, $6 + 3j$, by $j \cdot j$ or j^2 rotated the vector 180° in the positive direction.

$a + bj$ is the *algebraic* or *rectangular* form of a complex number.

a is the x-component of the vector $\overrightarrow{OA}$.

b is the y-component of the vector.

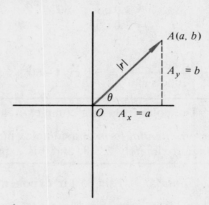

Figure 13.34

$$A_x = a = r \cos \theta$$
$$A_y = b = r \sin \theta$$

By substitution, $a + bj = r \cos \theta + r \sin \theta \, j$
$$= r(\cos \theta + j \sin \theta)$$

$r(\cos \theta + j \sin \theta)$ is called the *trigonometric* or *polar* form of a complex number, and is abbreviated, $r\angle\theta$ or r cis θ.

Rule 1: To find the product of two complex numbers in polar form, multiply the two magnitudes and add the two angles:

$$r_1(\cos \theta_1 + j \sin \theta_1)r_2(\cos \theta_2 + j \sin \theta_2) = r_1r_2[\cos(\theta_1 + \theta_2) + j \sin (\theta_1 + \theta_2)]$$
$$(r_1 \text{ cis } \theta)(r_2 \text{ cis } \theta_2) = r_1r_2 \text{ cis } (\theta_1 + \theta_2).$$

Rule 2: To find the quotient of two complex numbers in polar form, divide their magnitudes and subtract the two angles.

$$\frac{r_1(\cos \theta_1 + j \sin \theta_1)}{r_2 (\cos \theta_2 + j \sin \theta_2)} = \frac{r_1}{r_2}[\cos(\theta_1 - \theta_2) + j \sin (\theta_1 - \theta_2)]$$

or

$$\frac{r_1 \text{ cis } \theta_1}{r_2 \text{ cis } \theta_2} = \frac{r_1}{r_2} \text{ cis } (\theta_1 - \theta_2).$$

Example 5: Multiply (3.0 cis 16°) by (5.0 cis 47°) and express the answer in rectangular form.

$$\begin{aligned}(5.0 \text{ cis } 47°)(3.0 \text{ cis } 16°) &= (5.0)(3.0) \text{ cis } (47° + 16°) \\ &= 15 \text{ cis } 63°\end{aligned}$$

$$\begin{array}{ll} a = r \cos \theta & b = r \sin \theta \\ \quad = 15 \cos 63° & \quad = 15 \sin 63° \\ \quad = 15(.4540) & \quad = 15(.8910) \\ \quad \doteq 6.8 & \quad \doteq 13 \end{array}$$

Therefore, $(5.0 \text{ cis } 47°)(3.0 \text{ cis } 16°) = 15 \text{ cis } 63° \doteq (6.8, 13)$.

Example 6: Divide (30 cis 260°) by (10 cis 320°) and express the answer in rectangular form.

$$\frac{30 \text{ cis } 260°}{10 \text{ cis } 320°} = \frac{30}{10} \text{ cis } (260° - 320°)$$

$$= 3.0 \text{ cis } (-60°)$$

$$
\begin{array}{ll}
a = r \cos \theta & b = r \sin \theta \\
 = 3.0 \cos (-60°) & = 3.0 \sin (-60°) \\
 = 3.0 \,(.5) & = 3.0 \,(-.8660) \\
 \doteq 1.5 & \doteq -2.6
\end{array}
$$

Therefore, $\dfrac{30. \text{ cis } 260°}{10. \text{ cis } 320°} = 3.0 \text{ cis } (-60°) \doteq (1.5, -2.6)$.

Rule 3: The polar form of a complex number, $r (\cos \theta + j \sin \theta) = re^{j\theta}$.

$re^{j\theta}$ is the *exponential form* of a complex number, where e is an irrational number equal to approximately 2.718, and θ is expressed in radian measure.

Example 7: Express $4 (\cos 35° + j \sin 35°)$ in exponential form.

$$4 (\cos 35° + j \sin 35°) = 4e^{.611j} \quad \textbf{Using Table 3, } 35° \doteq \textbf{0.611 radian}$$

Example 8: Express $2 + 5j$ in exponential form.

$$
\begin{array}{ll}
2^2 + 5^2 = r^2 & \tan \theta = \dfrac{5}{2} = 2.5 \\
4 + 25 = r^2 & \theta = 1.19 \\
29 = r^2 & \\
\sqrt{29} = r & \\
5.38 \doteq r &
\end{array}
$$

Therefore, $2 + 5j = 5.38e^{1.19j}$

Example 9: Express the complex number $7e^{3.32j}$ in polar form and in rectangular form.

$$
\begin{array}{ll}
3.32 \text{ radian} = 190.24° \doteq 190°20' & \textbf{Using Table 3} \\
7e^{3.32j} = 7 (\cos 190°20' + j \sin 190°20') & \textbf{Rule 3} \\
\phantom{7e^{3.32j}} = 7 (-.9838 - .1794j) & \textbf{Table 3} \\
\phantom{7e^{3.32j}} = -6.89 - 1.26j & \\
\phantom{7e^{3.32j}} = (-6.89, -1.26) &
\end{array}
$$

EXERCISES 13.6

Add the following complex numbers graphically:

1. $(5 + 3j) + (2 - 4j)$

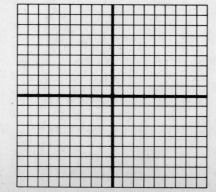

2. $(1 - j) + (1 + j)$

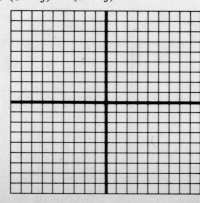

3. $(-3 - 2j) + (7 - j)$

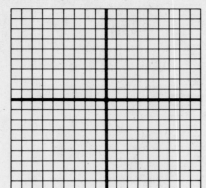

4. $(-6 + 5j) + (-1 + j)$

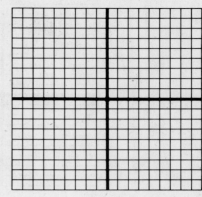

5. $(5 - 2j) + (3 - 2j)$

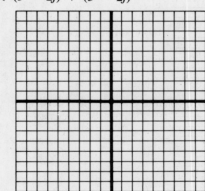

6. $(4 + 4j) + (-6 - 3j)$

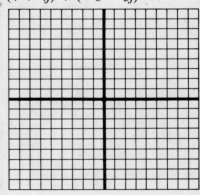

7. Multiply the vector $(4 - 3j)$ by j and graph both vectors.

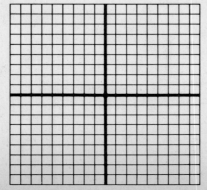

8. Multiply the vector $(-3 - 5j)$ by j^2 and graph both vectors.

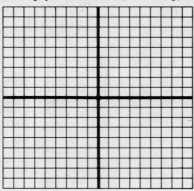

9. Multiply the vector $(-2 + 4j)$ by j^3 and graph both vectors.

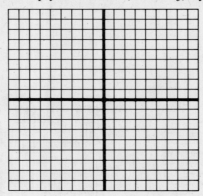

10. Multiply the vector $(5 + 3j)$ by j and graph both vectors.

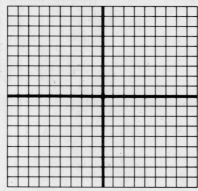

11. Multiply the vector $(6 - 5j)$ by j and graph both vectors.

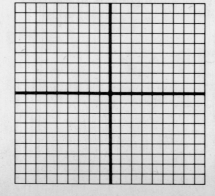

Multiply and divide the following vectors and express answers in rectangular form:

12. 6.1 cis 24° and 2.5 cis 55°

13. 2.04 cis 76° and 9.11 cis 72°

14. 14 cis 111° and 27 cis 89°

15. 3.2 cis 214° and 5.6 cis 189°

16. .600 cis 348° and .200 cis 291°

Express the following complex numbers in exponential form:

17. $3 + 7j$ **18.** $-2 + 3j$ **19.** $5 - 6j$ **20.** $-6 - 2j$

Express the following complex numbers in polar form and rectangular form:

21. $5e^{2.78j}$ **22.** $6e^{4.24j}$ **23.** $2.6e^{3.04j}$ **24.** $7.2e^{1.65j}$

13.7   The Law of Sines

An *oblique triangle* is a triangle that does not contain a right angle. To solve an oblique triangle, the *Law of Sines* may be used. This law states that in any triangle the sides are proportional to the sines of the opposite angles.

Construct a triangle *ABC*. The angle *A* may be acute (Fig. 13.35), or obtuse (Fig. 13.36).

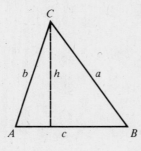

Figure 13.35   *Figure 13.36*

In either triangle, $\sin A = \dfrac{h}{b}$ and $\sin B = \dfrac{h}{a}$

then $b \cdot \sin A = h$ and $a \cdot \sin B = h$.

Therefore, $b \sin A = a \sin B$

Dividing both members by $\sin A \sin B$,

$$\frac{b}{\sin B} = \frac{a}{\sin A}$$

In the same way, construct an altitude from *B* to *AC*.

$$\sin A = \frac{h}{c} \text{ and } \sin C = \frac{h}{a}$$

$$c \cdot \sin A = h \text{ and } a \cdot \sin C = h$$

Therefore, $c \sin A = a \sin C$

Dividing both members by $\sin A \sin C$,

$$\frac{c}{\sin C} = \frac{a}{\sin A}$$

Rule: The Law of Sines states that

$$\frac{a}{\sin A} = \frac{b}{\sin B} = \frac{c}{\sin C}$$

Triangles may be classified according to the information given in the problem. *Case 1* is the case in which *two angles* and *one side* will be given.

Example 1: Solve the oblique triangle using the Law of Sines, given: $A = 141°$

$$B = 32°$$
$$a = 20.$$

$$C = 180° - (A + B) = 180° - (141° + 32°)$$
$$C = 180° - 173°$$
$$C = 7°$$

$$\frac{a}{\sin A} = \frac{b}{\sin B}$$

$$\frac{20}{\sin 141°} = \frac{b}{\sin 32°}$$

$$\frac{20}{.6293} = \frac{b}{.5299}$$

$$\frac{20(.5299)}{.6293} = b \qquad \textbf{To compute } b, \textbf{ use a calculator or logs.}$$

$$b \doteq 17$$

$$\frac{a}{\sin A} = \frac{c}{\sin C}$$

$$\frac{20}{\sin 141°} = \frac{c}{\sin 7°}$$

$$\frac{20 \sin 7°}{\sin 141°} = c$$

$$\frac{20(.1219)}{.6293} = c$$

$$c \doteq 3.8$$

Therefore, $C = 7°$ $b \doteq 17$

$$c \doteq 3.8$$

Case 2 is the case in which *two sides* and the *angle opposite one* of them are given. This case is called the *ambiguous case* because there may be two solutions, one solution, or no solution.

Example 2: Given: $A = 29°$

$\qquad\qquad a = 4.2$

$\qquad\qquad b = 6.2$

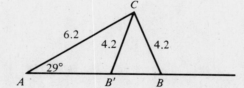

Figure 13.37

$$\frac{a}{\sin A} = \frac{b}{\sin B}$$

$$\frac{4.2}{\sin 29°} = \frac{6.2}{\sin B}$$

$$\sin B = \frac{6.2(.4848)}{4.2}$$

$$\sin B = .7157$$

B is a first or second quadrant angle.

If B is in Quadrant I:	If B is in Quadrant II:

If B is in Quadrant I:
$\quad B = 45°40'\qquad 0° < B < 90°$

$C = 180° - (A + B)$
$\quad = 180° - (29° + 45°40')$
$\quad = 180° - 74°40'$
$\quad = 105°20'$

$$\frac{a}{\sin A} = \frac{c}{\sin C}$$

$$\frac{4.2}{\sin 29°} = \frac{c}{\sin 105°20'}$$

$$\frac{4.2}{.4848} = \frac{c}{.9644}$$

$$c = \frac{4.2(.9644)}{.4848}$$

$$c \doteq 8.4$$

If B is in Quadrant II:
$\quad B' = 180° - 45°40' = 134°20',$
$\qquad\quad 90° < B < 180°$

$C' = 180° - (A + B')$
$\quad = 180° - (29° + 134°20')$
$\quad = 180° - 163°20'$
$\quad = 16°40'$

$$\frac{a}{\sin A} = \frac{c'}{\sin C'}$$

$$\frac{4.2}{\sin 29°} = \frac{c'}{\sin 16°40'}$$

$$\frac{4.2}{.4848} = \frac{c'}{.2868}$$

$$c' = \frac{4.2(.2868)}{.4848}$$

$$c' = 2.5$$

Therefore, the two solutions are :

$$B \doteq 45°40' \qquad B' \doteq 134°20'$$
$$C \doteq 105°20' \qquad C' \doteq 16°40'$$
$$c \doteq 8.4 \qquad c' \doteq 2.5$$

EXERCISES 13.7

Solve triangle ABC using the Law of Sines, given the following:

1. $A = 58°20'$
$B = 47°30'$
$a = 40$

2. $A = 123°$
$C = 35°$
$c = 101$

3. $B = 33°40'$
$C = 132°$
$b = 3.12$

4. $A = 118°$
$a = 2.6$
$B = 22°40'$

5. $B = 69°$
$C = 64°$
$b = 300$

6. $A = 115°$
$C = 51°$
$a = 41$

7. $a = 9.0$
   $b = 7.0$
   $A = 31°$

8. $b = 4.5$
   $a = 7.2$
   $A = 34°20'$

9. $a = 4.3$
   $b = 7.4$
   $A = 21°20'$

10. $a = 15.2$
    $b = 19.1$
    $A = 120°40'$

11. $a = 24$
    $b = 20$
    $A = 137°10'$

12. $b = 3.8$
    $a = 4.9$
    $B = 43°30'$

13. A surveyor measured a triangular lot. One side is 300 ft and another side is 120 ft. If the angle opposite the 300-ft side is 80°, find the length of the third side.

14. Find the length of the equal legs of a frame shaped like an isosceles triangle if the third side measures 80 ft and makes angles of 25° with each of the other sides.

13.8   The Law of Cosines

Not all oblique triangles may be solved by using the Law of Sines. For example, given three sides, a, b, and c, not enough information is given to use the law, $\frac{a}{\sin A} = \frac{b}{\sin B} = \frac{c}{\sin C}$. Also given two sides, a and b, and the angle C, there again is not enough information given to use the Law of Sines. These cases may be solved by using the *Law of Cosines*.

The Law of Cosines states that in any triangle, the square of any side of a triangle is equal to the sum of the squares of the other two sides minus twice the product of those two sides and the cosine of the angle between them:

Rule: The Law of Cosines states that

(1) $a^2 = b^2 + c^2 - 2bc \cos A$
(2) $b^2 = a^2 + c^2 - 2ac \cos B$
(3) $c^2 = a^2 + b^2 - 2ab \cos C$

To verify the Law of Cosines, consider Figure 13.38.

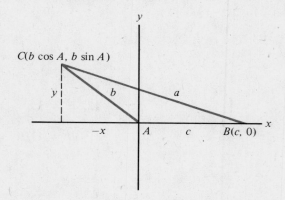

Figure 13.38

Point at $C = (-x, y) = (b \cos A, b \sin A)$

$$\cos A = \frac{-x}{b} \qquad\qquad \sin A = \frac{y}{b}$$

$$b \cos A = -x \qquad\qquad b \sin A = y$$

Using the distance formula, find side a, which is the distance between B and C.

$a^2 = (b \cos A - c)^2 + (b \sin A - 0)^2$
$a^2 = b^2 \cos^2 A - 2bc \cos A + c^2 + b^2 \sin^2 A$
$a^2 = b^2 \cos^2 A + b^2 \sin^2 A + c^2 - 2bc \cos A$ $(\cos^2 A + \sin^2 A = 1)$
$a^2 = b^2 (\cos^2 A + \sin^2 A) + c^2 - 2bc \cos A$
$a^2 = b^2 + c^2 - 2bc \cos A$

In this same manner, the validity of equations (2) and (3) can be shown.

Example 1:   Solve triangle ABC, given: $a = 50$, $b = 40$, and $C = 48°$.

Since angle C is given and sides a and b are given, choose equation (3).

$c^2 = a^2 + b^2 - 2ab \cos C$
$c^2 = 50^2 + 40^2 - 2(50)(40)(\cos 48°)$
$c^2 = 2500 + 1600 - 4000(.6691)$
$c^2 = 1424$
$c \doteq 38$

Now that side c and angle C are known, the Law of Sines may be used to find angle A or angle B.

$$\frac{c}{\sin C} = \frac{a}{\sin A}$$

$$\frac{38}{\sin 48°} = \frac{50}{\sin A}$$

$$\sin A = \frac{50(.7431)}{38}$$

$$\sin A = .9778$$

$$A \doteq 78°$$

$$B \doteq 180° - (78° + 48°)$$

$$B \doteq 54°$$

Therefore, $A \doteq 78°$ $B \doteq 54°$ $c \doteq 38$

Example 2: Solve triangle ABC, given: $a = 3.4$, $b = 2.8$, and $c = 1.4$. Since all three sides are given, any of the equations for the Law of Cosines may be used. Choosing (1),

$$a^2 = b^2 + c^2 - 2bc \cos A$$

$$3.4^2 = 2.8^2 + 1.4^2 - 2(2.8)(1.4) \cos A$$

$$11.56 = 7.84 + 1.96 - 7.84 \cos A$$

$$11.56 = 9.80 - 7.84 \cos A$$

$$1.76 = -7.84 \cos A$$

$$\frac{1.76}{-7.84} = \cos A$$

$$-.2245 \doteq \cos A$$

From Table 3, the angle whose cosine is .2245 is found to be 77°. However, the cosine of the angle is negative and, therefore, must be an angle in the second quadrant. Therefore, $A = 180° - 77° = 103°$. The Law of Sines may now be used to find either angle B or angle C.

$$\frac{a}{\sin A} = \frac{b}{\sin B}$$

$$\frac{3.4}{\sin 103°} = \frac{2.8}{\sin B}$$

$$\sin B = \frac{2.8 \sin 103°}{3.4} = \frac{2.8(.9744)}{3.4}$$

$$\sin B = .8024$$

$$B = 53°20'$$

$$C = 180° - (103° + 53°20')$$

$$C = 23°40'$$

Therefore, $A \doteq 103°$

$$B \doteq 53°20'$$

$$C \doteq 23°40'$$

EXERCISES 13.8

Solve the triangle ABC, given the following information:

1. $a = 9$
   $b = 7$
   $c = 4$

2. $a = 20$
   $b = 70$
   $C = 39°$

3. $b = 3.6$
   $a = 2.2$
   $C = 121°40'$

4. $b = 52$
   $c = 54$
   $A = 70°40'$

5. $a = 5.42$
   $c = 3.14$
   $B = 126°$

6. $b = 4.1$
   $c = 3.5$
   $A = 70°50'$

7. $a = 4.4$
   $b = 7.6$
   $c = 5.5$

8. $a = 43$
   $b = 64$
   $C = 31°50'$

9. $a = 18$
   $b = 25$
   $c = 38$

10. $a = 11.2$
    $c = 11.8$
    $B = 28°$

11. $b = 6.2$
    $c = 1.3$
    $A = 130°10'$

12. $b = 72$
    $c = 84$
    $A = 51°$

13.9 Sine Curves; Cosine Curves

In the equation $y = \sin x$, the variable x represents the measure of an angle the variable y, a real number. For example, if $x = \frac{\pi}{6}$, then $y = \sin \frac{\pi}{6} = \frac{1}{2}$.

Assign x values between 0 and 2π and note the increase or decrease in y. Although degree measure could be chosen for x, radian measure is preferred in order to express x and y in terms of a common unit of measurement.

Table 13.1

x	0	$\frac{\pi}{4}$	$\frac{\pi}{2}$	$\frac{3\pi}{4}$	π	$\frac{5\pi}{4}$	$\frac{3\pi}{2}$	$\frac{7\pi}{4}$	2π
$y = \sin x$	0	0.7	1	0.7	0	-0.7	-1	-0.7	0

It is shown in calculus that the sine curve is continuous, and that:

as x increases from 0 to $\frac{\pi}{2}$, y increases from 0 to 1;

as x increases from $\frac{\pi}{2}$ to π, y decreases from 1 to 0;

as x increases from π to $\frac{3\pi}{2}$, y decreases from 0 to -1;

as x increases from $\frac{3\pi}{2}$ to 2π, y increases from -1 to 0.

To picture this more clearly, plot the points from Table 13.2 in the plane, marking the x-axis off in units of $\frac{\pi}{4}$, and the y-axis in units of tenths. Connect these points with a smooth curve (Fig. 13.39).

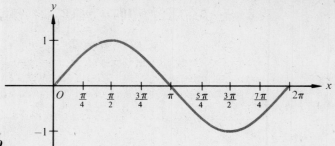

Figure 13.39

The sine curve is periodic. The sine of $\frac{\pi}{4} + 2\pi$ is equal to the sine of $\frac{\pi}{4}$.

The sine of $\frac{\pi}{2} + 2\pi$ is equal to the sine of $\frac{\pi}{2}$, and so on. In general, $\sin(x + 2\pi)$ $= \sin x$. Therefore, the pattern observed in Table 13.2 can be repeated in both directions every 2π (Fig. 13.40).

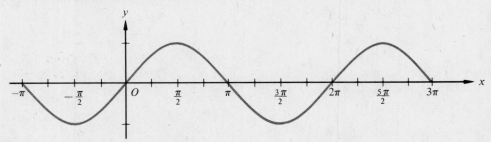

Figure 13.40

The smallest fixed interval after which the curve repeats itself is called the *period* of the curve. The period of the sine curve is 2π.

Note that the maximum value of y is $+1$, and the minimum value of y is -1. The absolute value of either the maximum or minimum value of y is called the *amplitude*. Therefore, the amplitude of the sine curve is 1.

To sketch the curve $y = 3 \sin x$, use the information from Table 13.1 to form a similar table.

Table 13.2

x	0	$\frac{\pi}{4}$	$\frac{\pi}{2}$	$\frac{3\pi}{4}$	π	$\frac{5\pi}{4}$	$\frac{3\pi}{2}$	$\frac{7\pi}{4}$	2π
$y = 3 \sin x$	0	2.1	3	2.1	0	-2.1	-3	-2.1	0

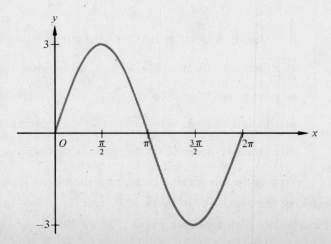

Figure 13.41

Note that the period of $y = 3 \sin x$ is 2π, the same as the period of $y = \sin x$. However, the amplitude of $y = 3 \sin x$ is 3.

Rule 1:   In general, the equation $y = a \sin x$ has an amplitude of $|a|$.

To sketch $y = \sin 3x$, the angle will be multiplied by 3 to find each value of y. To form a table, let x increase $\dfrac{\pi}{18}$.

Table 13.3

x	0	$\dfrac{\pi}{18}$	$\dfrac{\pi}{9}$	$\dfrac{\pi}{6}$	$\dfrac{2\pi}{9}$	$\dfrac{5\pi}{18}$	$\dfrac{\pi}{3}$	$\dfrac{7\pi}{18}$	$\dfrac{4\pi}{9}$	$\dfrac{\pi}{2}$	$\dfrac{5\pi}{9}$	$\dfrac{11\pi}{18}$	$\dfrac{2\pi}{3}$
$y = \sin 3x$	0	0.5	0.9	1	0.9	0.5	0	−0.5	−0.9	−1	−0.9	−0.5	0

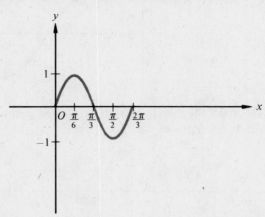

Figure 13.42

Note that the amplitude of $y = \sin 3x$ is 1, the same as the amplitude of $y = \sin x$. The period is $\frac{1}{3}$ of the period of the sine curve, $\frac{1}{3}$ of 2π or $\dfrac{2\pi}{3}$.

Rule 2:   In general, the equation $y = \sin bx$ has a period of $\dfrac{2\pi}{|b|}$.

In summary, then, the equation $y = a \sin bx$ has a period of $\dfrac{2\pi}{|b|}$ and an amplitude of $|a|$.

Rule 3:   To sketch one period of $y = a \sin bx$ beginning at 0.

Step 1:   Determine the amplitude, $|a|$.

Step 2:   Determine the period, $\dfrac{2\pi}{|b|}$.

Step 3:   Mark off the x-axis in four equal segments such that each is $\frac{1}{4}$ of the period.

Step 4:   Sketch the curve using the pattern of the sine curve:

as x increases from	$y = a \sin bx$
0 to $\frac{1}{4}$ of the period,	increases from 0 to its maximum value.
$\frac{1}{4}$ to $\frac{1}{2}$ of the period,	decreases from its maximum value to 0.
$\frac{1}{2}$ to $\frac{3}{4}$ of the period,	decreases from 0 to its minimum value.
$\frac{3}{4}$ to $\frac{4}{4}$ of the period,	increases from its minimum value to 0.

Example 1:   Sketch one period of the curve $y = 2 \sin \frac{1}{2}x$, beginning at $x = 0$.

Step 1:   The amplitude is $|a| = |2| = 2$.

Step 2:   The period is $\dfrac{2\pi}{|b|} = \dfrac{2\pi}{\left|\frac{1}{2}\right|} = 4\pi$.

Steps 3 and 4:

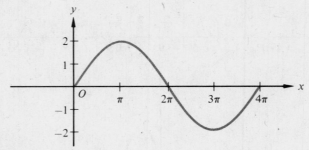

Figure 13.43

To graph the curve $y = \cos x$, assign x values from 0 to 2π, and note the increase or decrease in y.

Table 13.4

x	0	$\frac{\pi}{4}$	$\frac{\pi}{2}$	$\frac{3\pi}{4}$	π	$\frac{5\pi}{4}$	$\frac{3\pi}{2}$	$\frac{7\pi}{4}$	2π
$y = \cos x$	1	0.7	0	−0.7	−1	−0.7	0	0.7	1

Again, from calculus, it can be shown that the cosine curve is continuous and that, as shown in Figure 13.46:

as x increases from 0 to $\frac{\pi}{2}$, y decreases from 1 to 0;

$\frac{\pi}{2}$ to π, y decreases from 0 to -1;

π to $\frac{3\pi}{2}$, y increases from -1 to 0;

$\frac{3\pi}{2}$ to 2π, y increases from 0 to 1.

The period of the cosine curve is 2π. The amplitude is 1. In the equation $y = a \cos bx$, the amplitude is $|a|$ and the period is $\dfrac{2\pi}{|b|}$.

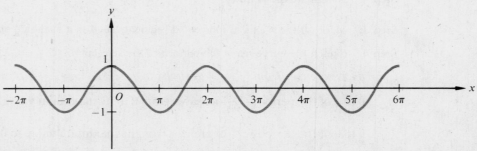

Figure 13.44

Rule 4:   To sketch one period of $y = a \cos bx$ beginning at 0.

Step 1:   State the amplitude, $|a|$.

Step 2:   State the period, $\dfrac{2\pi}{|b|}$.

Step 3:   Mark off the x-axis in four equal segments such that each is $\frac{1}{4}$ of the period.

Step 4:   Sketch the curve using the pattern of the cosine curve:

as x increases from	$y = a \cos bx$
0 to $\frac{1}{4}$ of the period,	decreases from its maximum value to 0.
$\frac{1}{4}$ to $\frac{1}{2}$ of the period,	decreases from 0 to its minimum value.
$\frac{1}{2}$ to $\frac{3}{4}$ of the period,	increases from its minimum value to 0.
$\frac{3}{4}$ to $\frac{4}{4}$ of the period,	increases from 0 to its maximum value.

Example 2:   Sketch the curve of $y = 3 \cos 2x$.

Step 1:   The amplitude is $|a| = |3| = 3$.

Step 2:   The period is $\dfrac{2\pi}{|b|} = \dfrac{2\pi}{|2|} = \pi$.

Steps 3 and 4:

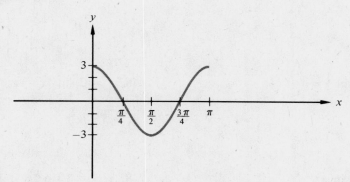

Figure 13.45

Example 3:   Sketch the curve of $y = -2 \sin \frac{1}{3}x$.

Step 1:   The amplitude is $|a| = |-2| = 2$.

Step 2:   The period is $\dfrac{2\pi}{|b|} = \dfrac{2\pi}{\left|\dfrac{1}{3}\right|} = 6\pi$.

Step 3:   Mark off the x-axis in units of $\frac{1}{4}(6\pi)$ or $\frac{3}{2}\pi$.

Step 4:   Note that $a = -2$. As x increases from 0 to $\frac{3}{2}\pi$, the value of y will not increase from 0 to the maximum value of 2, but y will decrease from 0 to the minimum value of -2. The pattern will be reversed.

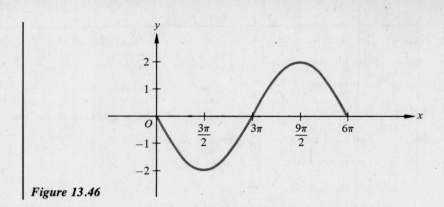

Figure 13.46

EXERCISES 13.9

Sketch the following. Begin at 0 and show one complete period.

1. $y = 4 \sin x$

2. $y = \sin 2x$

3. $y = \sin \frac{1}{4}x$

4. $y = \frac{1}{3} \sin x$

5. $y = 5 \sin \frac{1}{3} x$

6. $y = 2 \sin 4x$

7. $y = -\sin x$

8. $y = -3 \sin 2x$

9. $y = 3 \cos x$

10. $y = \cos \frac{1}{3}x$

11. $y = \cos 2x$

12. $y = 4 \cos x$

13. $y = 5 \cos \frac{1}{2}x$

14. $y = 2 \cos 3x$

15. $y = -\cos x$

16. $y = \frac{2}{3} \cos x$

17. $y = -2 \cos \frac{1}{2}x$

18. $y = 4 \sin \frac{1}{5}x$

19. $y = \frac{7}{5} \sin \frac{1}{2}x$

20. $y = \frac{9}{4} \cos \frac{9}{4}x$

13.10   The Graphs of $y = a \sin(bx + \theta)$ and $y = a \cos(bx + \theta)$

In the equations $y = a \sin (bx + \theta)$ and $y = a \cos (bx + \theta)$, θ is called the *phase angle*. The curve is shifted to the left if θ is positive and to the right if θ is negative. The amount of the displacement from the origin is called the *phase shift*, and the amount of shift is determined by $-\dfrac{\theta}{b}$.

Compare the graphs of $y = \sin x$ and $y_1 = \sin (x + \theta)$ in Figure 13.47:

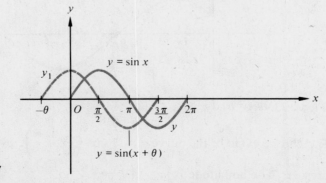

Figure 13.47

The graph over one period, regardless of its starting point, is called a *cycle* of the curve. The graph of $y_1 = \sin(x + \theta)$ from $x = -\theta$ to $x = 2\pi - \theta$ illustrates one cycle of the curve.

Rule: To sketch the graph of a sine or cosine curve involving a phase angle:

Step 1:   Determine the amplitude, $|a|$.

Step 2:   Determine the period, $\dfrac{2\pi}{|b|}$.

Step 3:   Determine the amount of shift, $-\dfrac{\theta}{b}$.

Step 4:   Mark off the x-axis into four equal segments from $\left(-\dfrac{\theta}{b}\right)$ to $\left(-\dfrac{\theta}{b} + \dfrac{2\pi}{b}\right)$.

Step 5:   Sketch the curve using the pattern of the sine or the cosine curve.

Example 1:   Sketch one cycle of the curve, $y = 2 \sin\left(\dfrac{1}{2}x + \dfrac{\pi}{6}\right)$.

Step 1:   The amplitude is $|a| = |2| = 2$.

Step 2:   The period is $\dfrac{2\pi}{|b|} = \dfrac{2\pi}{\left|\dfrac{1}{2}\right|} = 4\pi$.

Step 3:   The phase shift is $-\dfrac{\theta}{b} = -\dfrac{\dfrac{\pi}{6}}{\dfrac{1}{2}} = -\dfrac{\pi}{3}$.

Step 4: $-\dfrac{\pi}{3} + 4\pi = \dfrac{11}{3}\pi$

$-\dfrac{\pi}{3} + \left(\dfrac{1}{4} \text{ of } 4\pi\right) = \dfrac{2}{3}\pi$

$-\dfrac{\pi}{3} + \left(\dfrac{1}{2} \text{ of } 4\pi\right) = \dfrac{5}{3}\pi$

$-\dfrac{\pi}{3} + \left(\dfrac{3}{4} \text{ of } 4\pi\right) = \dfrac{8}{3}\pi$

Step 5:

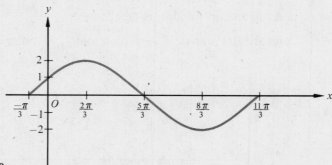

Figure 13.48

Example 2:   Sketch one cycle of the curve $y = 3 \cos\left(\dfrac{1}{3}x - \dfrac{\pi}{3}\right)$.

Step 1:   The amplitude is $|a| = |3| = 3$.

Step 2:   The period is $\dfrac{2\pi}{|b|} = \dfrac{2\pi}{\left|\dfrac{1}{3}\right|} = 6\pi$.

Step 3:   The phase shift is $-\dfrac{\theta}{b} = -\dfrac{-\dfrac{\pi}{3}}{\dfrac{1}{3}} = \pi.$

Step 4: $\pi + 6\pi = 7\pi$

$\pi + \left(\dfrac{1}{4} \text{ of } 6\pi\right) = \dfrac{5}{2}\pi$

$\pi + \left(\dfrac{1}{2} \text{ of } 6\pi\right) = 4\pi$

$\pi + \left(\dfrac{3}{4} \text{ of } 6\pi\right) = \dfrac{11}{2}\pi$

Step 5:

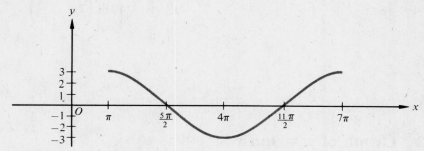

Figure 13.49

EXERCISES 13.10

Sketch one cycle of the following curves.

1. $y = \sin\left(x + \dfrac{\pi}{2}\right)$

2. $y = \sin\left(x - \dfrac{\pi}{4}\right)$

3. $y = \cos\left(x - \dfrac{\pi}{6}\right)$

4. $y = \cos(x + \pi)$

5. $y = \cos\left(2x - \dfrac{\pi}{3}\right)$

6. $y = \sin\left(\dfrac{1}{2}x + \dfrac{\pi}{6}\right)$

7. $y = 4 \sin \left(\dfrac{1}{4}x - \dfrac{2\pi}{3} \right)$ **8.** $y = 3 \cos \left(2x - \dfrac{3}{4}\pi \right)$

9. $y = 5 \sin \left(\dfrac{1}{5}x - \pi \right)$ **10.** $y = 2 \cos \left(\dfrac{1}{3}x - \dfrac{2\pi}{3} \right)$

11. $y = 2 \cos \left(\dfrac{1}{2}x + \dfrac{\pi}{2} \right)$ **12.** $y = 4 \sin \left(2x + \dfrac{5\pi}{6} \right)$

13.11   Graph of $y = \tan x$

To graph the equation $y = \tan x$, assign x values from 0 to 2π.

Table 13.5

x	0	$\frac{\pi}{6}$	$\frac{\pi}{4}$	$\frac{\pi}{3}$	$\frac{\pi}{2}$	$\frac{2\pi}{3}$	$\frac{3\pi}{4}$	$\frac{5\pi}{6}$	π	$\frac{7\pi}{6}$	$\frac{5\pi}{4}$	$\frac{4\pi}{3}$	$\frac{3\pi}{2}$	$\frac{5\pi}{3}$	$\frac{7\pi}{4}$	$\frac{11\pi}{6}$	2π
$y = \tan x$	0	.6	1	1.7	*	-1.7	-1	$-.6$	0	.6	1	1.7	*	-1.7	-1	$-.6$	0

Before plotting the points from Table 13.5, note from Table 3 in the Appendix C that the value of y becomes very large between $x = \dfrac{\pi}{3}$ and $x = \dfrac{\pi}{2}$. However, the value of y does not exist at $x = \dfrac{\pi}{2}$, or at $x = \dfrac{3\pi}{2}$ (Fig. 13.50).

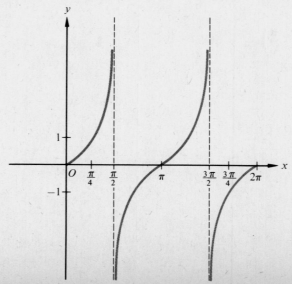

Figure 13.50

Rule 1:   In general, the period of $y = \tan bx$ is $\dfrac{\pi}{|b|}$.

The dotted lines in the figure are called *asymptotes*. These lines appear at values of x for which the tangent is undefined. The curve approaches, but never touches, the asymptotes.

Note that the tangent curve differs from the sine and cosine curves in that there is no maximum or minimum point for the tangent curve, the tangent curve is not continuous, and the period of the tangent curve is half the period of the sine or cosine curve.

Rule 2:   To sketch a period of $y = a \tan bx$:

Step 1:   Determine the period, $\dfrac{\pi}{|b|}$.

Step 2:   The curve will be undefined at $x = \frac{1}{2}$ of the period. Draw an asymptote through this point.

Step 3:   Sketch the curve following the tangent pattern:

as x increases from:	*y increases from:*
0 to $\frac{1}{2}$ of the period,	0 to infinity
$\frac{1}{2}$ of the period to the completion of a period	minus infinity to 0.

Example 1:   Sketch two periods of the curve $y = 2 \tan \frac{1}{2} x$.

Step 1:   The period is $\dfrac{\pi}{|b|} = \dfrac{\pi}{\left|\dfrac{1}{2}\right|} = 2\pi$.

Step 2:   The curve is undefined at $\frac{1}{2}(2\pi)$ or π. Draw an asymptote through $x = \pi$.

Step 3:

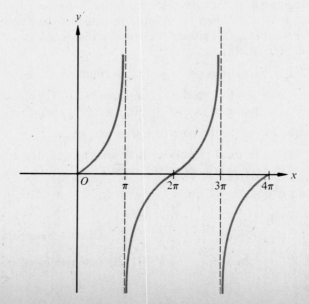

Figure 13.51

EXERCISES 13.11

Sketch two periods of the following curves:

1. $y = \tan \frac{1}{4}x$ **2.** $y = 3 \tan x$

3. $y = \tan 2x$ **4.** $y = 4 \tan \frac{1}{3}x$

5. $y = 2 \tan 3x$ **6.** $y = 5 \tan \frac{1}{5}x$

7. $y = 2 \tan \frac{1}{2}x$ **8.** $y = 3 \tan 4x$

13.12   Trigonometric Equations

Conditional trigonometric equations are equations that are not true for all replacements of the variable. To solve a trigonometric equation is to find all angles that will, when substituted in place of the variable in the equation, yield a true statement. Solutions are usually restricted to values of the angle between 0° and 360°, including 0°.

In the equation $\sin x = \frac{1}{2}$, x is equal to 30° since the sine of 30° $= \frac{1}{2}$. The sine is also positive in the second quadrant. The angle corresponding to 30° in the second quadrant is 180° − 30°, or 150°. Therefore, 150° is also a solution of the equation since the sine of 150° $= \frac{1}{2}$.

Methods used in solving first and second degree equations in algebra are applied in solving trigonometric equations.

Example 1:   Solve the equation $2 \sin \theta + 1 = 0$, for $0 \le \theta < 2\pi$.

$$2 \sin \theta + 1 = 0$$
$$2 \sin \theta = -1 \qquad \text{Subtract 1 from both members of the equation.}$$
$$\sin \theta = -\frac{1}{2} \qquad \text{Divide both members by 2.}$$

$$\theta = \frac{7\pi}{6} \quad \text{or} \quad \theta = \frac{11\pi}{6}$$

The sine is negative in the third and fourth quadrants. The sine of $\frac{\pi}{6}$ is $\frac{1}{2}$. The angle corresponding to $\frac{\pi}{6}$ in the third quadrant is $\pi + \frac{\pi}{6} = \frac{7\pi}{6}$. The angle corresponding to $\frac{\pi}{6}$ in the fourth quadrant is $2\pi - \frac{\pi}{6} = \frac{11\pi}{6}$.

Therefore, the solutions are $\frac{7\pi}{6}$ and $\frac{11\pi}{6}$.

Example 2: Solve the equation $\sin^2 u - \sin u = 0$ for $0 \le u < 2\pi$.

$$\sin^2 u - \sin u = 0$$ The sin u is a factor common to both terms.

$$\sin u (\sin u - 1) = 0$$ Factor.

$$\sin u = 0 \quad \text{or} \quad \sin u - 1 = 0$$ Set each factor equal to 0 and solve. The sin of both 0 and π is 0.

$$\sin u = 0 \quad \text{or} \quad \sin u = 1$$

$$u = 0 \quad \text{or} \quad u = \pi \quad \text{or} \quad u = \frac{\pi}{2}$$

Therefore, the solutions are 0, π, and $\frac{\pi}{2}$.

Example 3: Solve the equation $2\cos^2 x - \cos x - 1 = 0$ for $0° \le x < 360°$.

$$2 \cos^2 x - \cos x - 1 = 0$$
$$(2 \cos x + 1)(\cos x - 1) = 0$$ Factor.
$$2 \cos x + 1 = 0 \quad \text{or} \quad \cos x - 1 = 0$$ Set each factor equal to 0 and solve.

$$2 \cos x + 1 = 0 \quad \text{or} \quad \cos x - 1 = 0$$ Set each factor equal to 0 and solve.

$$2 \cos x = -1 \quad \text{or} \quad \cos x = 1$$ Solve for x.

$$\cos x = -\frac{1}{2} \quad \text{or} \quad \cos x = 1$$

$$x = 120° \text{ or } x = 240° \quad \text{or} \quad x = 0°$$

Therefore, the solutions are $120°$, $240°$, $0°$.

Example 4: Solve the equation $3 \sin^2 x + 4 \sin x - 1 = 0$ for $0° \le x < 360°$.

This equation may not be factored simply. The quadratic formula is used with $\sin x$ as the variable, $a = 3$, $b = 4$, and $c = -1$.

$$\sin x = \frac{-4 \pm \sqrt{4^2 - 4(3)(-1)}}{2(3)}$$

$$\sin x = \frac{-4 \pm \sqrt{28}}{6} = \frac{-4 \pm 2\sqrt{7}}{6} = \frac{-2 \pm \sqrt{7}}{3}$$

$$\sin x = \frac{-2 \pm 2.6458}{3}$$

$\sin x = -1.5486$ No solution since the sine x cannot be less than -1.

$\sin x = .2153$

$x \doteq 12°30'$ Use Table 3 or a calculator.

$x \doteq 167°30'$

Solutions are $12°30'$, $167°30'$.

EXERCISES 13.12

Solve the following equations for $0° \leq x < 360°$, or for $0 \leq \theta < 2\pi$.

1. $\sqrt{2} \sin \theta = 1$

2. $\sqrt{3} \tan x = 1$

3. $\sec^2 \theta = 4$

4. $2 \sin \theta = -\sqrt{3}$

5. $4 \sin^2 \theta = 3$

6. $\cot^2 \theta + \sqrt{3} \cot \theta = 0$

7. $\cos x - \cos x \cot x = 0$

8. $\cos x \sin^2 x - \cos x = 0$

9. $2 \sin^2 \theta - \sin \theta - 1 = 0$

10. $2 \sec^2 x + 2 \sec x - 4 = 0$

11. $\cot^2 x - 5 \cot x + 6 = 0$

12. $2 \sec^2 x + 11 \sec x + 15 = 0$

13. $2 \sin^2 \theta + 3 \sin \theta - 2 = 0$

14. $2 \sin^2 x + \sin x - 5 = 0$

15. $9 \sin^2 x - 21 \sin x = 8$

16. $15 \cot^2 x - \cot x - 2 = 0$

17. $3 \sin^2 x + \sin x - 5 = 0$

18. $\tan^2 x + 5 \tan x - 3 = 0$

19. $\sin^2 x + \sin x - 1 = 0$

20. $2 \cos^2 x + 2 \cos x - 3 = 0$

Unit 13 Self-Evaluation

Give the exact value of:

1. $\tan 300°$

2. $\sec 240°$

3. $\sin 210°$

Find θ where $0° \leq \theta < 360°$, given:

4. $\tan \theta = -.3640, \cos \theta > 0$

5. $\sec \theta = -3.236, \cot \theta > 0$

6. Given $\tan A = \frac{3}{4}$, $\csc A < 0$, determine the other five trigonometric ratios.

Use Table 3 or a calculator to determine the indicated trigonometric ratios.

7a. $\sin 140°$

7b. $\tan 212°40'$

7c. $\csc 340°10'$

8. $N = W \cos \theta$ is a formula for finding the normal force produced by any weight (W) resting on a surface inclined at an angle θ. Find N, given $W = 168$ kg and $\theta = 24°10'$.

9. Solve the right triangle ABC, given $C = 90°$; $c = 39$; $b = 7$.

10. The angle of elevation to the top of a tree is $27°20'$ from a point 12 meters from the foot of the tree. Find the height of the tree.

11. Sketch the graphs of the given curves:

    a. $y = 4 \cos \frac{1}{2}x$

    b. $y = \sin \left(x + \frac{\pi}{2} \right)$

    c. $y = \tan \frac{1}{3}x$

12. Express the vector $(5, 37°)$ in rectangular form.

13. Express the vector $(12, 5)$ in polar form.

14. Solve the oblique triangle ABC, given $a = 3.8$, $b = 6.1$, and $c = 4.2$.

15. A force of 148 kilograms has a direction of $36°50'$ with the horizontal. Find the horizontal force.

16. Add $(8 - 2j)$ and $(4 + 6j)$ graphically.

17. Graph $j(-6 + 8j)$.

18. Solve the oblique triangle ABC, given: $A = 37°$, $a = 18.2$, and $b = 20.1$.

19. Multiply 2.5 cis 116° by 3.1 cis 98°, and express the answer in rectangular form.

20. Solve the equation: $6 \tan^2 \theta + 11 \tan \theta - 10 = 0$ for $0° \leq \theta < 360°$.

appendix A English and Metric Systems of Measures

		English System	Metric Equivalent	
L I N E A R		1 inch (in.)	= 2.54 cm.	
	12 in.	= 1 foot (ft.)	= 30.48 cm.	
	3 ft.	= 1 yard (yd.)	= .9144 m.	
	5½ yd.	= 1 rod (rd.)	= 5.0292 m.	
	5280 ft.	= 1 mile (mi.)	= 1.6093 km.	
S Q U A R E		1 square inch (sq. in.)	= 6.4516 cm²	
	144 sq. in.	= 1 square foot (sq. ft.)	= .0929 m²	
	9 sq. ft.	= 1 square yard (sq. yd.)	= .8361 m²	
	30¼ sq. yd.	= 1 square rod (sq. rd.)	= 25.293 m²	
	160 sq. rd.	= 1 acre (A.)	= 4047.9 m²	
	640 A.	= 1 square mile (sq. mi.)	= 2.5899 km²	
C U B I C		1 cubic inch (cu. in.)	= 16.387 cm³	
	1.728 cu. in.	= 1 cubic foot (cu. ft.)	= .0283 m³	
	27 cu. ft.	= 1 cubic yard (cu. yd.)	= .7646 m³	
L I Q U I D		1 pint (pt.)	= .4732 l.	= 473.176 cc.
	2 pt.	= 1 quart (qt.)	= .9463 l.	
	4 qt.	= 1 gallon (gal.)	= 3.7853 l.	= .0038 cu. m.
W E I G H T		1 ounce (oz.)	= 28.35 g	
	16 oz.	= 1 pound (lb.)	= 453.59 g	
	2000 lb.	= 1 short ton (s.t.)		

		Metric System	English Equivalent
LINEAR		1 millimeter (mm.)	= .0394 in.
	10 mm.	= 1 centimeter (cm.)	= .3937 in.
	10 cm.	= 1 decimeter (dm.)	= 3.937 in.
	10 dm.	= 1 meter (m.)	= 39.37 in.
	10 m.	= 1 dekameter (dkm.)	= 393.7 in.
	10 dkm.	= 1 hectometer (hm.)	= 328.083 ft.
	10 hm.	= 1 kilometer (km.)	= .6214 mi.
SQUARE		1 sq. millimeter (mm²)	= .002 sq. in.
	100 mm²	= 1 sq. centimeter (cm²)	= .1549 sq. in.
	100 cm²	= 1 sq. decimeter (dm²)	= 15.49 sq. in.
	100 dm²	= 1 sq. meter (m²)	= 1549 sq. in.
	100 m²	= 1 sq. dekameter (dkm²)	= 119.6 sq. yd.
	100 dkm²	= 1 sq. hectometer (hm²)	= 2.471 A.
	100 hm²	= 1 sq. kilometer (km²)	= 247.104 A.
CUBIC		1 cu. millimeter (cu. mm.)	= .000061 cu. in.
	1000 cu. mm.	= 1 cu. centimeter (cc.)	= .061 cu. in.
	1000 cc.	= 1 cu. decimeter (cu. dm.)	= 61 cu. in.
	1000 cu. dm.	= 1 cu. meter (cu. m.)	= 35.315 cu. ft.
	1000 cu. m.	= 1 cu. dekameter (cu. dkm.)	= 1631.225 cu. yd.
	1000 cu. dkm.	= 1 cu. hectometer (cu. hm.)	= 1,631,225 cu. yd.
	1000 cu. hm.	= 1 cu. kilometer (cu. km.)	= .2395 cu. mi.
LIQUID		1 milliliter (ml.)	= .061 cu. in.
	10 ml.	= 1 centiliter (cl.)	= .6102 cu. in.
	10 cl.	= 1 deciliter (dl.)	= 6.102 cu. in.
	10 dl.	= 1 liter (l.)	= 1.057 qt.
	10 l.	= 1 dekaliter (dkl.)	= 610.2 cu. in.
	10 dkl.	= 1 hectoliter (hl.)	= 6102.5 cu. in.
	10 hl.	= 1 kiloliter (kl.)	= 264.178 gal.
WEIGHT		1 milligram (mg.)	= .0000353 oz.
	10 mg.	= 1 centigram (cg.)	= .000353 oz.
	10 cg.	= 1 decigram (dg.)	= .00353 oz.
	10 dg.	= 1 gram (g.)	= .03527 oz.
	10 g.	= 1 decagram (dkg.)	= .3527 oz.
	10 dkg.	= 1 hectogram (hg.)	= 3.527 oz.
	10 hg.	= 1 kilogram (kg.)	= 2.2046 lb.

appendix B Tables

Table 1 Powers and Roots

No.	Sq.	Sq. Root	Cube	Cube Root	No.	Sq.	Sq. Root	Cube	Cube Root
1	1	1.000	1	1.000	51	2,601	7.141	132,651	3.708
2	4	1.414	8	1.260	52	2,704	7.211	140,608	3.733
3	9	1.732	27	1.442	53	2,809	7.280	148,877	3.756
4	16	2.000	64	1.587	54	2,916	7.348	157,464	3.780
5	25	2.236	125	1.710	55	3,025	7.416	166,375	3.803
6	36	2.449	216	1.817	56	3,136	7.483	175,616	3.826
7	49	2.646	343	1.913	57	3,249	7.550	185,193	3.849
8	64	2.828	512	2.000	58	3,364	7.616	195,112	3.871
9	81	3.000	729	2.080	59	3,481	7.681	205,379	3.893
10	100	3.162	1,000	2.154	60	3,600	7.746	216,000	3.915
11	121	3.317	1,331	2.224	61	3,721	7.810	226,981	3.936
12	144	3.464	1,728	2.289	62	3,844	7.874	238,328	3.958
13	169	3.606	2,197	2.351	63	3,969	7.937	250,047	3.979
14	196	3.742	2,744	2.410	64	4,096	8.000	262,144	4.000
15	225	3.873	3,375	2.466	65	4,225	8.062	274,625	4.021
16	256	4.000	4,096	2.520	66	4,356	8.124	287,496	4.041
17	289	4.123	4,913	2.571	67	4,489	8.185	300,763	4.062
18	324	4.243	5,832	2.621	68	4,624	8.246	314,432	4.082
19	361	4.359	6,859	2.668	69	4,761	8.307	328,509	4.102
20	400	4.472	8,000	2.714	70	4,900	8.367	343,000	4.121
21	441	4.583	9,261	2.759	71	5,041	8.426	357,911	4.141
22	484	4.690	10,648	2.802	72	5,184	8.485	373,248	4.160
23	529	4.796	12,167	2.844	73	5,329	8.544	389,017	4.179
24	576	4.899	13,824	2.884	74	5,476	8.602	405,224	4.198
25	625	5.000	15,625	2.924	75	5,625	8.660	421,875	4.217
26	676	5.099	17,576	2.962	76	5,776	8.718	438,976	4.236
27	729	5.196	19,683	3.000	77	5,929	8.775	456,533	4.254
28	784	5.292	21,952	3.037	78	6,084	8.832	474,552	4.273
29	841	5.385	24,389	3.072	79	6,241	8.888	493,039	4.291
30	900	5.477	27,000	3.107	80	6,400	8.944	512,000	4.309
31	961	5.568	29,791	3.141	81	6,561	9.000	531,441	4.327
32	1,024	5.657	32,768	3.175	82	6,724	9.055	551,368	4.344
33	1,089	5.745	35,937	3.208	83	6,889	9.110	571,787	4.362
34	1,156	5.831	39,304	3.240	84	7,056	9.165	592,704	4.380
35	1,225	5.916	42,875	3.271	85	7,225	9.220	614,125	4.397
36	1,296	6.000	46,656	3.302	86	7,396	9.274	636,056	4.414
37	1,369	6.083	50,653	3.332	87	7,569	9.327	658,503	4.431
38	1,444	6.164	54,872	3.362	88	7,744	9.381	681,472	4.448
39	1,521	6.245	59,319	3.391	89	7,921	9.434	704,969	4.465
40	1,600	6.325	64,000	3.420	90	8,100	9.487	729,000	4.481
41	1,681	6.403	68,921	3.448	91	8,281	9.539	753,571	4.498
42	1,764	6.481	74,088	3.476	92	8,464	9.592	778,688	4.514
43	1,849	6.557	79,507	3.503	93	8,649	9.644	804,357	4.531
44	1,936	6.633	85,184	3.530	94	8,836	9.695	830,584	4.547
45	2,025	6.708	91,125	3.557	95	9,025	9.747	857,375	4.563
46	2,116	6.782	97,336	3.583	96	9,216	9.798	884,736	4.579
47	2,209	6.856	103,823	3.609	97	9,409	9.849	912,673	4.595
48	2,304	6.928	110,592	3.634	98	9,604	9.899	941,192	4.610
49	2,401	7.000	117,649	3.659	99	9,801	9.950	970,299	4.626
50	2,500	7.071	125,000	3.684	100	10,000	10.000	1,000,000	4.642

Table 2   Four-Place Logarithms of Numbers

n	0	1	2	3	4	5	6	7	8	9
10	0000	0043	0086	0128	0170	0212	0253	0294	0334	0374
11	0414	0453	0492	0531	0569	0607	0645	0682	0719	0755
12	0792	0828	0864	0899	0934	0969	1004	1038	1072	1106
13	1139	1173	1206	1239	1271	1303	1335	1367	1399	1430
14	1461	1492	1523	1553	1584	1614	1644	1673	1703	1732
15	1761	1790	1818	1847	1875	1903	1931	1959	1987	2014
16	2041	2068	2095	2122	2148	2175	2201	2227	2253	2279
17	2304	2330	2355	2380	2405	2430	2455	2480	2504	2529
18	2553	2577	2601	2625	2648	2672	2695	2718	2742	2765
19	2788	2810	2833	2856	2878	2900	2923	2945	2967	2989
20	3010	3032	3054	3075	3096	3118	3139	3160	3181	3201
21	3222	3243	3263	3284	3304	3324	3345	3365	3385	3404
22	3424	3444	3464	3483	3502	3522	3541	3560	3579	3598
23	3617	3636	3655	3674	3692	3711	3729	3747	3766	3784
24	3802	3820	3838	3856	3874	3892	3909	3927	3945	3962
25	3979	3997	4014	4031	4048	4065	4082	4099	4116	4133
26	4150	4166	4183	4200	4216	4232	4249	4265	4281	4298
27	4314	4330	4346	4362	4378	4393	4409	4425	4440	4456
28	4472	4487	4502	4518	4533	4548	4564	4579	4594	4609
29	4624	4639	4654	4669	4683	4698	4713	4728	4742	4757
30	4771	4786	4800	4814	4829	4843	4857	4871	4886	4900
31	4914	4928	4942	4955	4969	4983	4997	5011	5024	5038
32	5051	5065	5079	5092	5105	5119	5132	5145	5159	5172
33	5185	5198	5211	5224	5237	5250	5263	5276	5289	5302
34	5315	5328	5340	5353	5366	5378	5391	5403	5416	5428
35	5441	5453	5465	5478	5490	5502	5514	5527	5539	5551
36	5563	5575	5587	5599	5611	5623	5635	5647	5658	5670
37	5682	5694	5705	5717	5729	5740	5752	5763	5775	5786
38	5798	5809	5821	5832	5843	5855	5866	5877	5888	5899
39	5911	5922	5933	5944	5955	5966	5977	5988	5999	6010
40	6021	6031	6042	6053	6064	6075	6085	6096	6107	6117
41	6128	6138	6149	6160	6170	6180	6191	6201	6212	6222
42	6232	6243	6253	6263	6274	6284	6294	6304	6314	6325
43	6335	6345	6355	6365	6375	6385	6395	6405	6415	6425
44	6435	6444	6454	6464	6474	6484	6493	6503	6513	6522
45	6532	6542	6551	6561	6571	6580	6590	6599	6609	6618
46	6628	6637	6646	6656	6665	6675	6684	6693	6702	6712
47	6721	6730	6739	6749	6758	6767	6776	6785	6794	6803
48	6812	6821	6830	6839	6848	6857	6866	6875	6884	6893
49	6902	6911	6920	6928	6937	6946	6955	6964	6972	6981
50	6990	6998	7007	7016	7024	7033	7042	7050	7059	7067
51	7076	7084	7093	7101	7110	7118	7126	7135	7143	7152
52	7160	7168	7177	7185	7193	7202	7210	7218	7226	7235
53	7243	7251	7259	7267	7275	7284	7292	7300	7308	7316
54	7324	7332	7340	7348	7356	7364	7372	7380	7388	7396

Table 2   Four-Place Logarithms of Numbers (Continued)

n	0	1	2	3	4	5	6	7	8	9
55	7404	7412	7419	7427	7435	7443	7451	7459	7466	7474
56	7482	7490	7497	7505	7513	7520	7528	7536	7543	7551
57	7559	7566	7574	7582	7589	7597	7604	7612	7619	7627
58	7634	7642	7649	7657	7664	7672	7679	7686	7694	7701
59	7709	7716	7723	7731	7738	7745	7752	7760	7767	7774
60	7782	7789	7796	7803	7810	7818	7825	7832	7839	7846
61	7853	7860	7868	7875	7882	7889	7896	7903	7910	7917
62	7924	7931	7938	7945	7952	7959	7966	7973	7980	7987
63	7993	8000	8007	8014	8021	8028	8035	8041	8048	8055
64	8062	8069	8075	8082	8089	8096	8102	8109	8116	8122
65	8129	8136	8142	8149	8156	8162	8169	8176	8182	8189
66	8195	8202	8209	8215	8222	8228	8235	8241	8248	8254
67	8261	8267	8274	8280	8287	8293	8299	8306	8312	8319
68	8325	8331	8338	8344	8351	8357	8363	8370	8376	8382
69	8388	8395	8401	8407	8414	8420	8426	8432	8439	8445
70	8451	8457	8463	8470	8476	8482	8488	8494	8500	8506
71	8513	8519	8525	8531	8537	8543	8549	8555	8561	8567
72	8573	8579	8585	8591	8597	8603	8609	8615	8621	8627
73	8633	8639	8645	8651	8657	8663	8669	8675	8681	8686
74	8692	8698	8704	8710	8716	8722	8727	8733	8739	8745
75	8751	8756	8762	8768	8774	8779	8785	8791	8797	8802
76	8808	8814	8820	8825	8831	8837	8842	8848	8854	8859
77	8865	8871	8876	8882	8887	8893	8899	8904	8910	8915
78	8921	8927	8932	8938	8943	8949	8954	8960	8965	8971
79	8976	8982	8987	8993	8998	9004	9009	9015	9020	9025
80	9031	9036	9042	9047	9053	9058	9063	9069	9074	9079
81	9085	9090	9096	9101	9106	9112	9117	9122	9128	9133
82	9138	9143	9149	9154	9159	9165	9170	9175	9180	9186
83	9191	9196	9201	9206	9212	9217	9222	9227	9232	9238
84	9243	9248	9253	9258	9263	9269	9274	9279	9284	9289
85	9294	9299	9304	9309	9315	9320	9325	9330	9335	9340
86	9345	9350	9355	9360	9365	9370	9375	9380	9385	9390
87	9395	9400	9405	9410	9415	9420	9425	9430	9435	9440
88	9445	9450	9455	9460	9465	9469	9474	9479	9484	9489
89	9494	9499	9504	9509	9513	9518	9523	9528	9533	9538
90	9542	9547	9552	9557	9562	9566	9571	9576	9581	9586
91	9590	9595	9600	9605	9609	9614	9619	9624	9628	9633
92	9638	9643	9647	9652	9657	9661	9666	9671	9675	9680
93	9685	9689	9694	9699	9703	9708	9713	9717	9722	9727
94	9731	9736	9741	9745	9750	9754	9759	9763	9768	9773
95	9777	9782	9786	9791	9795	9800	9805	9809	9814	9818
96	9823	9827	9832	9836	9841	9845	9850	9854	9859	9863
97	9868	9872	9877	9881	9886	9890	9894	9899	9903	9908
98	9912	9917	9921	9926	9930	9934	9939	9943	9948	9952
99	9956	9961	9965	9969	9974	9978	9983	9987	9991	9996

Table 3   Four-Place Values of Functions and Radians

Degrees	Radians	Sin θ	Cos θ	Tan θ	Cot θ	Sec θ	Csc θ		
0° 00′	.0000	.0000	1.0000	.0000		1.000		1.5708	90° 00′
10	.0029	.0029	1.0000	.0029	343.8	1.000	343.8	1.5679	50
20	.0058	.0058	1.0000	.0058	171.9	1.000	171.9	1.5650	40
30	.0087	.0087	1.0000	.0087	114.6	1.000	114.6	1.5621	30
40	.0116	.0116	.9999	.0116	85.94	1.000	85.95	1.5592	20
50	.0145	.0145	.9999	.0145	68.75	1.000	68.76	1.5563	10
1° 00′	.0175	.0175	.9998	.0175	57.29	1.000	57.30	1.5533	89° 00′
10	.0204	.0204	.9998	.0204	49.10	1.000	49.11	1.5504	50
20	.0233	.0233	.9997	.0233	42.96	1.000	42.98	1.5475	40
30	.0262	.0262	.9997	.0262	38.19	1.000	38.20	1.5446	30
40	.0291	.0291	.9996	.0291	34.37	1.000	34.38	1.5417	20
50	.0320	.0320	.9995	.0320	31.24	1.001	31.26	1.5388	10
2° 00′	.0349	.0349	.9994	.0349	28.64	1.001	28.65	1.5359	88° 00′
10	.0378	.0378	.9993	.0378	26.43	1.001	26.45	1.5330	50
20	.0407	.0407	.9992	.0407	24.54	1.001	24.56	1.5301	40
30	.0436	.0436	.9990	.0437	22.90	1.001	22.93	1.5272	30
40	.0465	.0465	.9989	.0466	21.47	1.001	21.49	1.5243	20
50	.0495	.0494	.9988	.0495	20.21	1.001	20.23	1.5213	10
3° 00′	.0524	.0523	.9986	.0524	19.08	1.001	19.11	1.5184	87° 00′
10	.0553	.0552	.9985	.0553	18.07	1.002	18.10	1.5155	50
20	.0582	.0581	.9983	.0582	17.17	1.002	17.20	1.5126	40
30	.0611	.0610	.9981	.0612	16.35	1.002	16.38	1.5097	30
40	.0640	.0640	.9980	.0641	15.60	1.002	15.64	1.5068	20
50	.0669	.0669	.9978	.0670	14.92	1.002	14.96	1.5039	10
4° 00′	.0698	.0698	.9976	.0699	14.30	1.002	14.34	1.5010	86° 00′
10	.0727	.0727	.9974	.0729	13.73	1.003	13.76	1.4981	50
20	.0756	.0756	.9971	.0758	13.20	1.003	13.23	1.4952	40
30	.0785	.0785	.9969	.0787	12.71	1.003	12.75	1.4923	30
40	.0814	.0814	.9967	.0816	12.25	1.003	12.29	1.4893	20
50	.0844	.0843	.9964	.0846	11.83	1.004	11.87	1.4864	10
5° 00′	.0873	.0872	.9962	.0875	11.43	1.004	11.47	1.4835	85° 00′
10	.0902	.0901	.9959	.0904	11.06	1.004	11.10	1.4806	50
20	.0931	.0929	.9957	.0934	10.71	1.004	10.76	1.4777	40
30	.0960	.0958	.9954	.0963	10.39	1.005	10.43	1.4748	30
40	.0989	.0987	.9951	.0992	10.08	1.005	10.13	1.4719	20
50	.1018	.1016	.9948	.1022	9.788	1.005	9.839	1.4690	10
6° 00′	.1047	.1045	.9945	.1051	9.514	1.006	9.567	1.4661	84° 00′
		Cos θ	Sin θ	Cot θ	Tan θ	Csc θ	Sec θ	Radians	Degrees

Table 3 Four-Place Values of Functions and Radians (Continued)

Degrees	Radians	Sin θ	Cos θ	Tan θ	Cot θ	Sec θ	Csc θ		
6°00'	.1047	.1045	.9945	.1051	9.514	1.006	9.567	1.4661	84°00'
10	.1076	.1074	.9942	.1080	9.255	1.006	9.309	1.4632	50
20	.1105	.1103	.9939	.1110	9.010	1.006	9.065	1.4603	40
30	.1134	.1132	.9936	.1139	8.777	1.006	8.834	1.4573	30
40	.1164	.1161	.9932	.1169	8.556	1.007	8.614	1.4544	20
50	.1193	.1190	.9929	.1198	8.345	1.007	8.405	1.4515	10
7°00'	.1222	.1219	.9925	.1228	8.144	1.008	8.206	1.4486	83°00'
10	.1251	.1248	.9922	.1257	7.953	1.008	8.016	1.4457	50
20	.1280	.1276	.9918	.1287	7.770	1.008	7.834	1.4428	40
30	.1309	.1305	.9914	.1317	7.596	1.009	7.661	1.4399	30
40	.1338	.1334	.9911	.1346	7.429	1.009	7.496	1.4370	20
50	.1367	.1363	.9907	.1376	7.269	1.009	7.337	1.4341	10
8°00'	.1396	.1392	.9903	.1405	7.115	1.010	7.185	1.4312	82°00'
10	.1425	.1421	.9899	.1435	6.968	1.010	7.040	1.4283	50
20	.1454	.1449	.9894	.1465	6.827	1.011	6.900	1.4254	40
30	.1484	.1478	.9890	.1495	6.691	1.011	6.765	1.4224	30
40	.1513	.1507	.9886	.1524	6.561	1.012	6.636	1.4195	20
50	.1542	.1536	.9881	.1554	6.435	1.012	6.512	1.4166	10
9°00'	.1571	.1564	.9877	.1584	6.314	1.012	6.392	1.4137	81°00'
10	.1600	.1593	.9872	.1614	6.197	1.013	6.277	1.4108	50
20	.1629	.1622	.9868	.1644	6.084	1.013	6.166	1.4079	40
30	.1658	.1650	.9863	.1673	5.976	1.014	6.059	1.4050	30
40	.1687	.1679	.9858	.1703	5.871	1.014	5.955	1.4021	20
50	.1716	.1708	.9853	.1733	5.769	1.015	6.855	1.3992	10
10°00'	.1745	.1736	.9848	.1763	5.761	1.015	5.759	1.3963	80°00'
10	.1774	.1765	.9843	.1793	5.576	1.016	5.665	1.3934	50
20	.1804	.1794	.9838	.1823	5.485	1.016	5.575	1.3904	40
30	.1833	.1822	.9833	.1853	5.396	1.017	5.487	1.3875	30
40	.1862	.1851	.9827	.1883	5.309	1.018	5.403	1.3846	20
50	.1891	.1880	.9822	.1914	5.226	1.018	5.320	1.3817	10
11°00'	.1920	.1908	.9816	.1944	5.145	1.019	5.241	1.3788	79°00'
10	.1949	.1937	.9811	.1974	5.066	1.019	5.164	1.3759	50
20	.1978	.1965	.9805	.2004	4.989	1.020	5.089	1.3730	40
30	.2007	.1994	.9799	.2035	4.915	1.020	5.016	1.3701	30
40	.2036	.2022	.9793	.2065	4.843	1.021	4.945	1.3672	20
50	.2065	.2051	.9787	.2095	4.773	1.022	4.876	1.3643	10
12°00'	.2094	.2079	.9781	.2126	4.705	1.022	4.810	1.3614	78°00'
		Cos θ	Sin θ	Cot θ	Tan θ	Csc θ	Sec θ	Radians	Degrees

Table 3   Four-Place Values of Functions and Radians (Continued)

Degrees	Radians	Sin θ	Cos θ	Tan θ	Cot θ	Sec θ	Csc θ		
12°00′	.2094	.2079	.9781	.2126	4.705	1.022	4.810	1.3614	78°00′
10	.2123	.2108	.9775	.2156	4.638	1.023	4.745	1.3584	50
20	.2153	.2136	.9769	.2186	4.574	1.024	4.682	1.3555	40
30	.2182	.2164	.9763	.2217	4.511	1.024	4.620	1.3526	30
40	.2211	.2193	.9757	.2247	4.449	1.025	4.560	1.3497	20
50	.2240	.2221	.9750	.2278	4.390	1.026	4.502	1.3468	10
13°00′	.2269	.2250	.9744	.2309	4.331	1.026	4.445	1.3439	77°00′
10	.2298	.2278	.9737	.2339	4.275	1.027	4.390	1.3410	50
20	.2327	.2306	.9730	.2370	4.219	1.028	4.336	1.3381	40
30	.2356	.2334	.9724	.2401	4.165	1.028	4.284	1.3352	30
40	.2385	.2363	.9717	.2432	4.113	1.029	4.232	1.3323	20
50	.2414	.2391	.9710	.2462	4.061	1.030	4.182	1.3294	10
14°00′	.2443	.2419	.9703	.2493	4.011	1.031	4.134	1.3265	76°00′
10	.2473	.2447	.9696	.2524	3.962	1.031	4.086	1.3235	50
20	.2502	.2476	.9689	.2555	3.914	1.032	4.039	1.3206	40
30	.2531	.2504	.9681	.2586	3.867	1.033	3.994	1.3177	30
40	.2560	.2532	.9674	.2617	3.821	1.034	3.950	1.3148	20
50	.2589	.2560	.9667	.2648	3.776	1.034	3.906	1.3119	10
15°00′	.2618	.2588	.9659	.2679	3.732	1.035	3.864	1.3090	75°00′
10	.2647	.2616	.9652	.2711	3.689	1.036	3.822	1.3061	50
20	.2676	.2644	.9644	.2742	3.647	1.037	3.782	1.3032	40
30	.2705	.2672	.9636	.2773	3.606	1.038	3.742	1.3003	30
40	.2734	.2700	.9628	.2805	3.566	1.039	3.703	1.3974	20
50	.2763	.2728	.9621	.2836	3.526	1.039	3.665	1.3945	10
16°00′	.2793	.2756	.9613	.2867	3.487	1.040	3.628	1.2915	74°00′
10	.2822	.2784	.9605	.2899	3.450	1.041	3.592	1.2886	50
20	.2851	.2812	.9596	.2931	3.412	1.042	3.556	1.2857	40
30	.2880	.2840	.9588	.2962	3.376	1.043	3.521	1.2828	30
40	.2909	.2868	.9580	.2994	3.340	1.044	3.487	1.2799	20
50	.2938	.2896	.9572	.3026	3.305	1.045	3.453	1.2770	10
17°00′	.2967	.2924	.9563	.3057	3.271	1.046	3.420	1.2741	73°00′
10	.2996	.2952	.9555	.3089	3.237	1.047	3.388	1.2712	50
20	.3025	.2979	.9546	.3121	3.204	1.048	3.356	1.2683	40
30	.3054	.3007	.9537	.3153	3.172	1.049	3.326	1.2654	30
40	.3083	.3035	.9528	.3185	3.140	1.049	3.295	1.2625	20
50	.3113	.3062	.9520	.3217	3.108	1.050	3.265	1.2595	10
18°00′	.3142	.3090	.9511	.3249	3.078	1.051	3.236	1.2566	72°00′
		Cos θ	Sin θ	Cot θ	Tan θ	Csc θ	Sec θ	Radians	Degrees

Table 3   Four-Place Values of Functions and Radians (Continued)

Degrees	Radians	Sin θ	Cos θ	Tan θ	Cot θ	Sec θ	Csc θ		
18°00′	.3142	.3090	.9511	.3249	3.078	1.051	3.236	1.2566	72°00′
10	.3171	.3118	.9502	.3281	3.047	1.052	3.207	1.2537	50
20	.3200	.3145	.9492	.3314	3.018	1.053	3.179	1.2508	40
30	.3229	.3173	.9483	.3346	2.989	1.054	3.152	1.2479	30
40	.3258	.3201	.9474	.3378	2.960	1.056	3.124	1.2450	20
50	.3287	.3228	.9465	.3411	2.932	1.057	3.098	1.2421	10
19°00′	.3316	.3256	.9455	.3443	2.904	1.058	3.072	1.2392	71°00′
10	.3345	.3283	.9446	.3476	2.877	1.059	3.046	1.2363	50
20	.3374	.3311	.9436	.3508	2.850	1.060	3.021	1.2334	40
30	.3403	.3338	.9426	.3541	2.824	1.061	2.996	1.2305	30
40	.3432	.3365	.9417	.3574	2.798	1.062	2.971	1.2275	20
50	.3462	.3393	.9407	.3607	2.773	1.063	2.947	1.2246	10
20°00′	.3491	.3420	.9397	.3640	2.747	1.064	2.924	1.2217	70°00′
10	.3520	.3448	.9387	.3673	2.723	1.065	2.901	1.2188	50
20	.3549	.3475	.9377	.3706	2.699	1.066	2.878	1.2159	40
30	.3578	.3502	.9367	.3739	2.675	1.068	2.855	1.2130	30
40	.3607	.3529	.9356	.3772	2.651	1.069	2.833	1.2101	20
50	.3636	.3557	.9346	.3805	2.628	1.070	2.812	1.2072	10
21°00′	.3665	.3584	.9336	.3839	2.605	1.071	2.790	1.2043	69°00′
10	.3694	.3611	.9325	.3872	2.583	1.072	2.769	1.2014	50
20	.3723	.3638	.9315	.3906	2.560	1.074	2.749	1.1985	40
30	.3752	.3665	.9304	.3939	2.539	1.075	2.729	1.1956	30
40	.3782	.3692	.9293	.3973	2.517	1.076	2.709	1.1926	20
50	.3811	.3719	.9283	.4006	2.496	1.077	2.689	1.1897	10
22°00′	.3840	.3746	.9272	.4040	2.475	1.079	2.669	1.1868	68°00′
10	.3869	.3773	.9261	.4074	2.455	1.080	2.650	1.1839	50
20	.3898	.3800	.9250	.4108	2.434	1.081	2.632	1.1810	40
30	.3927	.3827	.9239	.4142	2.414	1.082	2.613	1.1781	30
40	.3956	.3854	.9228	.4176	2.394	1.084	2.595	1.1752	20
50	.3985	.3881	.9216	.4210	2.375	1.085	2.577	1.1723	10
23°00′	.4014	.3907	.9205	.4245	2.356	1.086	2.559	1.1694	67°00′
10	.4043	.3934	.9194	.4279	2.337	1.088	2.542	1.1665	50
20	.4072	.3961	.9182	.4314	2.318	1.089	2.525	1.1636	40
30	.4102	.3987	.9171	.4348	2.300	1.090	2.508	1.1606	30
40	.4131	.4014	.9159	.4383	2.282	1.092	2.491	1.1577	20
50	.4160	.4041	.9147	.4417	2.264	1.093	2.475	1.1548	10
24°00′	.4189	.4067	.9135	.4452	2.246	1.095	2.459	1.1519	66°00′
		Cos θ	Sin θ	Cot θ	Tan θ	Csc θ	Sec θ	Radians	Degrees

Table 3 Four-Place Values of Functions and Radians (Continued)

Degrees	Radians	Sin θ	Cos θ	Tan θ	Cot θ	Sec θ	Csc θ		
24°00′	.4189	.4067	.9135	.4452	2.246	1.095	2.459	1.1519	66°00′
10	.4218	.4094	.9124	.4487	2.229	1.096	2.443	1.1490	50
20	.4247	.4120	.9112	.4522	2.211	1.097	2.427	1.1461	40
30	.4276	.4147	.9100	.4557	2.194	1.099	2.411	1.1432	30
40	.4305	.4173	.9088	.4592	2.177	1.100	2.396	1.1403	20
50	.4334	.4200	.9075	.4628	2.161	1.102	2.381	1.1374	10
25°00′	.4363	.4226	.9063	.4663	2.145	1.103	2.366	1.1345	65°00′
10	.4392	.4253	.9051	.4699	2.128	1.105	2.352	1.1316	50
20	.4422	.4279	.9038	.4734	2.112	1.106	2.337	1.1286	40
30	.4451	.4305	.9026	.4770	2.097	1.108	2.323	1.1257	30
40	.4480	.4331	.9013	.4806	2.081	1.109	2.309	1.1228	20
50	.4509	.4358	.9001	.4841	2.066	1.111	2.295	1.1199	10
26°00′	.4538	.4384	.8988	.4877	2.050	1.113	2.281	1.1170	64°00′
10	.4567	.4410	.8975	.4913	2.035	1.114	2.268	1.1141	50
20	.4596	.4436	.8962	.4950	2.020	1.116	2.254	1.1112	40
30	.4625	.4462	.8949	.4986	2.006	1.117	2.241	1.1083	30
40	.4654	.4488	.8936	.5022	1.991	1.119	2.228	1.1054	20
50	.4683	.4514	.8923	.5059	1.977	1.121	2.215	1.1025	10
27°00′	.4712	.4540	.8910	.5095	1.963	1.122	2.203	1.0966	63°00′
10	.4741	.4566	.8897	.5132	1.949	1.124	2.190	1.0966	50
20	.4771	.4592	.8884	.5169	1.935	1.126	2.178	1.0937	40
30	.4800	.4617	.8870	.5206	1.921	1.127	2.166	1.0908	30
40	.4829	.4643	.8857	.5243	1.907	1.129	2.154	1.0879	20
50	.4858	.4669	.8843	.5280	1.894	1.131	2.142	1.0850	10
28°00′	.4887	.4695	.8829	.5317	1.881	1.133	2.130	1.0821	62°00′
10	.4916	.4720	.8816	.5354	1.868	1.134	2.118	1.0792	50
20	.4945	.4746	.8802	.5392	1.855	1.136	2.107	1.0763	40
30	.4974	.4772	.8788	.5430	1.842	1.138	2.096	1.0734	30
40	.5003	.4797	.8774	.5467	1.829	1.140	2.085	1.0705	20
50	.5032	.4823	.8760	.5505	1.816	1.142	2.074	1.0676	10
29°00′	.5061	.4848	.8746	.5543	1.804	1.143	2.063	1.0647	61°00′
10	.5091	.4874	.8732	.5581	1.792	1.145	2.052	1.0617	50
20	.5120	.4899	.8718	.5619	1.780	1.147	2.041	1.0588	40
30	.5149	.4924	.8704	.5658	1.767	1.149	2.031	1.0559	30
40	.5178	.4950	.8689	.5696	1.756	1.151	2.020	1.0530	20
50	.5207	.4975	.8675	.5735	1.744	1.153	2.010	1.0501	10
30°00′	.5236	.5000	.8660	.5774	1.732	1.155	2.000	1.0472	60°00′
		Cos θ	Sin θ	Cot θ	Tan θ	Csc θ	Sec θ	Radians	Degrees

Table 3   Four-Place Values of Functions and Radians (Continued)

Degrees	Radians	Sin θ	Cos θ	Tan θ	Cot θ	Sec θ	Csc θ		
30°00′	.5236	.5000	.8660	.5774	1.732	1.155	2.000	1.0472	60°00′
10	.5265	.5025	.8646	.5812	1.720	1.157	1.990	1.0443	50
20	.5294	.5050	.8631	.5851	1.709	1.159	1.980	1.0414	40
30	.5323	.5075	.8616	.5890	1.698	1.161	1.970	1.0385	30
40	.5352	.5100	.8601	.5930	1.686	1.163	1.961	1.0356	20
50	.5381	.5125	.8587	.5969	1.675	1.165	1.951	1.0327	10
31°00′	.5411	.5150	.8572	.6009	1.664	1.167	1.942	1.0297	59°00′
10	.5440	.5175	.8557	.6048	1.653	1.169	1.932	1.0268	50
20	.5469	.5200	.8542	.6088	1.643	1.171	1.923	1.0239	40
30	.5498	.5225	.8526	.6128	1.632	1.173	1.914	1.0210	30
40	.5527	.5250	.8511	.6168	1.621	1.175	1.905	1.0181	20
50	.5556	.5275	.8496	.6208	1.611	1.177	1.896	1.0152	10
32°00′	.5585	.5299	.8480	.6249	1.600	1.179	1.887	1.0123	58°00′
10	.5614	.5324	.8465	.6289	1.590	1.181	1.878	1.0094	50
20	.5643	.5348	.8450	.6330	1.580	1.184	1.870	1.0065	40
30	.5672	.5373	.8434	.6371	1.570	1.186	1.861	1.0036	30
40	.5701	.5398	.8418	.6412	1.560	1.188	1.853	1.0007	20
50	.5730	.5422	.8403	.6453	1.550	1.190	1.844	.9977	10
33°00′	.5760	.5446	.8387	.6494	1.540	1.192	1.836	.9948	57°00′
10	.5789	.5471	.8371	.6536	1.530	1.195	1.828	.9919	50
20	.5818	.5495	.8355	.6577	1.520	1.197	1.820	.9890	40
30	.5847	.5519	.8339	.6619	1.511	1.199	1.812	.9861	30
40	.5876	.5544	.8323	.6661	1.501	1.202	1.804	.9832	20
50	.5905	.5568	.8307	.6703	1.492	1.204	1.796	.9803	10
34°00′	.5934	.5592	.8290	.6745	1.483	1.206	1.788	.9774	56°00′
10	.5963	.5616	.8274	.6787	1.473	1.209	1.781	.9745	50
20	.5992	.5640	.8258	.6830	1.464	1.211	1.773	.9716	40
30	.6021	.5664	.8241	.6873	1.455	1.213	1.766	.9687	30
40	.6050	.5688	.8225	.6916	1.446	1.216	1.758	.9657	20
50	.6080	.5712	.8208	.6959	1.437	1.218	1.751	.9628	10
35°00′	.6109	.5736	.8192	.7002	1.428	1.221	1.743	.9599	55°00′
10	.6138	.5760	.8175	.7046	1.419	1.223	1.736	.9570	50
20	.6167	.5783	.8158	.7089	1.411	1.226	1.729	.9541	40
30	.6196	.5807	.8141	.7133	1.402	1.228	1.722	.9512	30
40	.6225	.5831	.8124	.7177	1.393	1.231	1.715	.9483	20
50	.6254	.5854	.8107	.7221	1.385	1.233	1.708	.9454	10
36°00′	.6283	.5878	.8900	.7265	1.376	1.236	1.701	.9425	54°00′
		Cos θ	Sin θ	Cot θ	Tan θ	Csc θ	Sec θ	Radians	Degrees

Table 3   Four-Place Values of Functions and Radians (Continued)

Degrees	Radians	Sin θ	Cos θ	Tan θ	Cot θ	Sec θ	Csc θ		
36°00'	.6283	.5878	.8090	.7265	1.376	1.236	1.701	.9425	54°00'
10	.6312	.5901	.8073	.7310	1.368	1.239	1.695	.9396	50
20	.6341	.5925	.8056	.7355	1.360	1.241	1.688	.9367	40
30	.6370	.5948	.8039	.7400	1.351	1.244	1.681	.9338	30
40	.6400	.5972	.8021	.7445	1.343	1.247	1.675	.9308	20
50	.6429	.5995	.8004	.7490	1.335	1.249	1.668	.9279	10
37°00'	.6458	.6018	.7986	.7536	1.327	1.252	1.662	.9250	53°00'
10	.6487	.6041	.7969	.7581	1.319	1.255	1.655	.9221	50
20	.6516	.6065	.7951	.7627	1.311	1.258	1.649	.9192	40
30	.6545	.6088	.7934	.7673	1.303	1.260	1.643	.9163	30
40	.6574	.6111	.7916	.7720	1.295	1.263	1.636	.9134	20
50	.6603	.6134	.7898	.7766	1.288	1.266	1.630	.9105	10
38°00'	.6632	.6157	.7880	.7813	1.280	1.269	1.624	.9076	52°00'
10	.6661	.6180	.7862	.7860	1.272	1.272	1.618	.9047	50
20	.6690	.6202	.7844	.7907	1.265	1.275	1.612	.9018	40
30	.6720	.6225	.7826	.7954	1.257	1.278	1.606	.8988	30
40	.6749	.6248	.7808	.8002	1.250	1.281	1.601	.8959	20
50	.6778	.6271	.7790	.8050	1.242	1.284	1.595	.8930	10
39°00'	.6807	.6293	.7771	.8098	1.235	1.287	1.589	.8901	51°00'
10	.6836	.6316	.7753	.8146	1.228	1.290	1.583	.8872	50
20	.6865	.6338	.7735	.8195	1.220	1.293	1.578	.8843	40
30	.6894	.6361	.7716	.8243	1.213	1.296	1.572	.8814	30
40	.6923	.6383	.7698	.8292	1.206	1.299	1.567	.8785	20
50	.6952	.6406	.7679	.8342	1.199	1.302	1.561	.8756	10
40°00'	.6981	.6428	.7660	.8391	1.192	1.305	1.556	.8727	50°00'
10	.7010	.6450	.7642	.8441	1.185	1.309	1.550	.8698	50
20	.7039	.6472	.7623	.8491	1.178	1.312	1.545	.8668	40
30	.7069	.6494	.7604	.8541	1.171	1.315	1.540	.8639	30
40	.7098	.6517	.7585	.8591	1.164	1.318	1.535	.8610	20
50	.7127	.6539	.7566	.8642	1.157	1.322	1.529	.8581	10
41°00'	.7156	.6561	.7547	.8693	1.150	1.325	1.524	.8552	49°00'
10	.7185	.6583	.7528	.8744	1.144	1.328	1.519	.8523	50
20	.7214	.6604	.7509	.8796	1.137	1.332	1.514	.8494	40
30	.7243	.6626	.7490	.8847	1.130	1.335	1.509	.8465	30
40	.7272	.6648	.7470	.8899	1.124	1.339	1.504	.8436	20
50	.7301	.6670	.7451	.8952	1.117	1.342	1.499	.8407	10
42°00'	.7330	.6691	.7431	.9004	1.111	1.346	1.494	.8378	48°00'
		Cos θ	Sin θ	Cot θ	Tan θ	Csc θ	Sec θ	Radians	Degrees

Table 3   Four-Place Values of Functions and Radians (Continued)

Degrees	Radians	Sin θ	Cos θ	Tan θ	Cot θ	Sec θ	Csc θ		
42°00′	.7330	.6691	.7431	.9004	1.111	1.346	1.494	.8378	48°00′
10	.7359	.6713	.7412	.9057	1.104	1.349	1.490	.8348	50
20	.7389	.6734	.7392	.9110	1.098	1.353	1.485	.8319	40
30	.7418	.6756	.7373	.9163	1.091	1.356	1.480	.8290	30
40	.7447	.6777	.7353	.9217	1.085	1.360	1.476	.8261	20
50	.7476	.6799	.7333	.9271	1.079	1.364	1.471	.8232	10
43°00′	.7505	.6820	.7314	.9325	1.072	1.367	1.466	.8203	47°00′
10	.7534	.6841	.7294	.9380	1.066	1.371	1.462	.8174	50
20	.7563	.6862	.7274	.9435	1.060	1.375	1.457	.8145	40
30	.7592	.6884	.7254	.9490	1.054	1.379	1.453	.8116	30
40	.7621	.6905	.7234	.9545	1.048	1.382	1.448	.8087	20
50	.7650	.6926	.7214	.9601	1.042	1.386	1.444	.8058	10
44°00′	.7679	.6947	.7193	.9657	1.036	1.390	1.440	.8029	46°00′
10	.7709	.6967	.7173	.9713	1.030	1.394	1.435	.7999	50
20	.7738	.6988	.7153	.9770	1.024	1.398	1.431	.7970	40
30	.7767	.7009	.7133	.9827	1.018	1.402	1.427	.7941	30
40	.7796	.7030	.7112	.9884	1.012	1.406	1.423	.7912	20
50	.7825	.7050	.7092	.9942	1.006	1.410	1.418	.7883	10
45°00′	.7854	.7071	.7071	1.000	1.000	1.414	1.414	.7854	45°00′
		Cos θ	Sin θ	Cot θ	Tan θ	Csc θ	Sec θ	Radians	Degrees

Answers to Selected Problems

Unit 1

EXERCISES 1.1

1. 389	**9.** 1384	**17.** 38,040 km	**25.** 22,751
3. 591	**11.** 6742	**19.** 159	**27.** 86 = 86
5. 935	**13.** 10247	**21.** $41,590	**29.** 140 = 140
7. 1805	**15.** 129 m	**23.** 836 km	

EXERCISES 1.2

1. 322	**11.** 1767	**21.** $31,296	**31.** 96
3. 519	**13.** 1565	**23.** 21 kg	**33.** 12
5. 273	**15.** 2001	**25.** 906 ohms	**35.** 14
7. 359	**17.** 2191	**27.** 128	
9. 506	**19.** 6 in.	**29.** 59	

EXERCISES 1.3

1. 148	**11.** 231,693	**21.** 288 cm^2	**31.** 4272
3. 1248	**13.** 510,816	**23.** 260 cm^2	**33.** 2020
5. 11928	**15.** 30,622,397	**25.** 189 cu kg	**35.** 616
7. 3933	**17.** 64 yr 8 mo	**27.** $810	**37.** 1239
9. 110,814	**19.** 162 kl	**29.** 615 lb	**39.** 18,704

EXERCISES 1.4

1. 3, $R = 1$	**15.** 383	**29.** 9 cm
3. 16, $R = 3$	**17.** 2, $R = 242$	**31.** 41 cm
5. 9, $R = 2$	**19.** 608 g	**33.** 6 kg
7. 41, $R = 2$	**21.** 616 ℓ	**35.** 632
9. 68	**23.** 168	**37.** 32
11. 2, $R = 1$	**25.** 17	**39.** 843
13. 32, $R = 3$	**27.** $13\frac{\text{kl}}{\text{min.}}$	

EXERCISES 1.5

1. $1\frac{1}{2}$

3. $2\frac{4}{9}$

5. $2\frac{5}{6}$

7. $6\frac{5}{6}$

9. $2\frac{1}{34}$

11. $\frac{11}{5}$

13. $\frac{23}{8}$

15. $\frac{32}{3}$

17. $\frac{638}{25}$

19. $\frac{20301}{100}$

21. $3\frac{7}{32}$

23. $\frac{43}{8}$ ft by $\frac{103}{8}$ ft

EXERCISES 1.6

1. 2

3. 12

5. 12

7. 124

9. 909

11. $\frac{12}{16}$

EXERCISES 1.7

1. $\frac{3}{4}$

3. $\frac{1}{5}$

5. $\frac{21}{25}$

7. $\frac{76}{51}$

9. $\frac{121}{211}$

11. $\frac{25}{64}$

EXERCISES 1.8

1. 8^2

3. 1^5

5. 5^4

7. 64

9. 1

11. 216

13. 64

15. 72

17. 124,416

19. $3 \times 10^2 + 2 \times 10^1 + 7 \times 10^0$

21. $7 \times 10^1 + 9 \times 10^0$

23. $7 \times 10^2 + 0 \times 10^1 + 5 \times 10^0$

EXERCISES 1.9

1. 2, 3, 5, 7, 11, 13, 17, 19, 23, 29, 31, 37, 41, 43, 47

3. 2×3^2

5. $2 \times 5 \times 7$

7. $2^3 \times 5^3$

9. 3×5^3

11. $\frac{1}{3}$

13. $\frac{1}{2}$

15. $\frac{3}{11}$

17. $\frac{16}{9}$

19. $\frac{220}{43}$

21. $\frac{27}{64}$

EXERCISES 1.10

1. $4, \frac{1}{4}, \frac{2}{4}$

3. $15, \frac{9}{15}, \frac{5}{15}$

5. $12, \frac{3}{12}, \frac{10}{12}, \frac{6}{12}$

7. $8, \frac{4}{8}, \frac{6}{8}, \frac{5}{8}$

9. $36, \frac{9}{36}, \frac{10}{36}, \frac{9}{36}$

11. $105, \frac{66}{105}, \frac{56}{105}$

13. $\frac{3}{16}$ and $\frac{14}{16}$

EXERCISES 1.11

1. $\frac{4}{5}$

3. $\frac{4}{5}$

5. $\frac{7}{4}$

7. $\frac{5}{3}$

9. 1

11. $\frac{3}{4}$

13. $\frac{14}{15}$

15. $\frac{19}{12}$

17. $\frac{7}{9}$

19. $\frac{1}{2}$

21. 0

23. $3\frac{3}{5}$

25. 16

27. $7\frac{3}{4}$

29. $4\frac{7}{12}$

31. $2\frac{5}{12}$ cm

33. $10\frac{5}{6}$ kg

35. $\frac{7}{64}$ cm

37. $3\frac{3}{4}$

39. $62\frac{11}{16}$ in

EXERCISES 1.12

1. $\frac{1}{2}$

3. $\frac{15}{32}$

5. $\frac{3}{8}$

7. 10

9. $\frac{9}{2}$

11. 2

13. $9\frac{3}{8}$

15. $\frac{5}{3}$

17. $\frac{9}{11}$

19. 45

21. $\frac{9}{20}$

23. $2\frac{2}{9}$

25. 30

27. $48

29. 15 ft

31. $2\frac{5}{64}$ kg

33. $4\frac{3}{8}$ ℓ

35. 46 km

37. 16

39. $\frac{1}{2}$

41. $\frac{13}{15}$

43. $1\frac{1}{2}$

EXERCISES 1.13

1. fifty-two and seventy-four hundredths
3. eight thousand four hundred twenty-three ten thousandths
5. seventy-eight and eight hundredths
7. four thousand five hundred sixty-eight and eighty-six hundredths
9. six thousand eight hundred ninety-four and two hundred seventy-five thousandths
11. 61.3
13. 5.172
15. 4,000,000.0003
17. 400.5
19. 7,000.1286
21. 4
23. 2
25. 3
27. 3
29. 4
31. 2
33. 0.0
35. 77.30
37. 200
39. 7.2
41. 5.6

EXERCISES 1.14

1. 1.1
3. 1.62
5. 394.4
7. 23
9. 19.65

11. 9.5
13. 1.5
15. 0.03
17. 7.2
19. 53.6 cm

21. $1.40
23. 11.21 kg
25. 76.4
27. 1072 m/sec

EXERCISES 1.15

1. 0.4
3. 100
5. 4.5
7. 39
9. 40
11. 4000

13. 3
15. 10
17. 2
19. 1.3
21. 1100
23. 0.2

25. 0.00002
27. $24.86
29. 3,100 kg
31. $96.39
33. $.47
35. $1.26

37. 9.98×10^2
39. 5.486×10^6
41. 2.44×10^2
43. 6.2526
45. .638
47. 4468.64

EXERCISES 1.16

1. 183.07
3. 6876
5. 9108

7. 6594
9. 26.714
11. 21381376

13. 108.92578
15. 61479.25
17. 4

19. 20.56696
21. 1176
23. 12

EXERCISES 1.17

1. $\frac{3}{5}$

9. $\frac{28,583}{1000}$

17. 0.5

3. $\frac{9}{25}$

11. $\frac{21}{500}$

19. 0.67

21. 0.6

5. $\frac{3}{50}$

13. $\frac{18,001}{200}$

23. $\frac{625}{10,000}$

7. $\frac{51}{10}$

15. $\frac{841}{2}$

25. .28125

EXERCISES 1.18

1. 75%

11. 25%

21. 0.02

31. $\frac{3}{10}$

3. 47.8%

13. 83%

23. 0.025

33. $\frac{31}{200}$

5. 1.3%

15. 87.5%

25. 0.843

35. $\frac{321}{10,000}$

7. 0.8%

17. 237.5%

27. 0.0325

37. $\frac{3}{4}$

9. 340%

19. 0.23

29. 1.23

39. $\frac{33}{4}$

EXERCISES 1.19

1. 480 mm

13. 4,540 dm²

25. .00030 oz

3. .04538 g

15. .222 m³

27. 113 ft

5. 16,830 dg

17. .004083 m³

29. 2.69 ft

7. 30,000 ℓ

19. 2369 g

31. .031 ft

9. .0000031 ℓ

21. 3 in.

33. 7.97 sq ft

11. 9.32 m²

23. 11 pt

35. 7650 cu ft

UNIT 1 SELF-EVALUATION

1. 68

14. 7660%

2. 713,5

15. 65.28

3. 3.1

16. 108

4. 14

17. 800 km

5. 6.88

18. 68 kg

6. 27.7

19. $3^2 \times 5^2 \times 11$

7. $1\frac{5}{24}$

20. 604.373

8. $2\frac{7}{10}$

21. six and seven ten-thousandths

9. $\frac{1}{4}$

22. 260.

10. $1\frac{11}{53}$

23. 471

11. $5\frac{5}{6}$

24. $330.05

12. $.3\overline{18}$

25. $336\frac{5}{6}$ cm

13. $\frac{4}{125}$

Unit 2   **EXERCISES 2.1**

1. 0

5. $\overline{XZ}, \overline{ZU}, \overline{XU}$

9. m

13. AB, AC, BC

3. 2

7. m

11. 0

15. $\overrightarrow{AB}, \overrightarrow{BC}, \overrightarrow{CB}, \overrightarrow{BA}$

EXERCISES 2.2

1.

3.
315°

5.
620°

7. 58°

9. 37°

11. 45°

13. acute

15. obtuse

17. obtuse

19. obtuse

21. 393°

23. 366°

25. 485°

27. 20°

29. a = c = 35°
b = 145°

31. p = 48°
q = t = 80°

33. α = 107°
γ = β = 73°

35. 115°

37. 50°

39. 90°

41. 105°

43. ∠PZS

45. ∠PZT, ∠PZU, ∠VZR, ∠VZQ

47. ∠FAB, ∠ABE, ∠EBC, ∠BCD

49. ∠DEB, ∠EFA

51. ∠b and ∠c, ∠e and ∠f, ∠i and ∠j, ∠m and ∠n, ∠f and ∠g, ∠j and ∠k

53. ∠UTV and ∠QSR, ∠PTU and ∠RSW

55. ∠ADE and ∠BDC, ∠DAB and ∠ECD, ∠ABD and ∠DEC, ∠BCD and ∠FCG

EXERCISES 2.3

1. ∠BAC

3. ∠B

5. △BDF and △CDE

7. 18.03 m, 15.6 m²

9. 15.8 cm, 9.1 cm²

11. 9 m

13. 24

15. 9

17. 13

19. 26 cm

21. 15 m

23. 6 ft 9 in.

25. 96.6 ft

27. 15.8 in.

29. 70.8 sq ft

EXERCISES 2.4

1. square, 32 m, 64 m²

3. polygon, 42 m, 72 m²

5. trapezoid, 24 km, 28 sq km

7. polygon, 20 m, 22 m²

9. hexagon, 24 cm, 36 cm²

11. polygon, 19.2 m, 24 m²

13. 4 meters, 1 sq. m.

15. 3.4 ft, .5 ft²

17. 13.9 ft, 6.6 sq ft

19. 47 ft

21. 270.7 sq ft

EXERCISES 2.5

1. 13.8 cm, 15.2 cm²

3. 38.9 m, 121 m²

5. 22.9 cm²

7. 7.84 cm²

9. 2.9 cm²

11. π

13. $\frac{11\pi}{6}$

15. $\frac{4\pi}{3}$

17. $\frac{2\pi}{9}$

19. $\frac{7\pi}{3}$

21. 135°

23. 120°

25. 50°

27. 240°

29. 100°

31. 50.24 sq ft

33. 1200 sq in.

EXERCISES 2.6

1. 214 cm², 210 cm³

3. 54 cm², 27 cm³

5. 60.7 m², 32.3 m³

7. 86,400 m², 960,000 m³

9. 17,800 m², 131,000 m³

11. 5460 cm², 37,900 cm³

13. 15.9 ft³

15. 288

UNIT 2 SELF-EVALUATION

1. $\overleftrightarrow{MN}$; $\overrightarrow{MN}$, $\overrightarrow{NM}$; NM

2. parallel lines: *PR* and *ST*, *SQ* and *TR*
   intersecting lines: *PS* and *PR*, *PS* and *QS*, *PS* and *ST*, *QS* and *PR*, *QS* and *ST*, *RT* and *PR*, *RT* and *ST*
   coincident lines: *PQ*, *PR* and *QR*
   acute angles: ∠*PQS*, ∠*SPQ*, ∠*QST*, ∠*QRT*
   right angle: ∠*PSQ*
   obtuse angles: ∠*SQR*, ∠*STR*

3. vertical angles: γ and φ, δ and μ, *a* and *d*, *b* and *c*, β and ε, α and θ, *e* and *h*, *g* and *f*
   corresponding angles: β and δ, α and γ, φ and θ, μ and ε, *a* and *e*, *c* and *g*, *b* and *f*, *d* and *h*
   alternate interior angles: β and μ, γ and θ, *c* and *f*, *d* and *e*
   alternate exterior angles: α and φ, δ and ε, *b* and *g*, *a* and *h*

4.

175°

5. 42°

6. obtuse

7. 50.5 m, 98.05 m²

8. 17

9. 56 cm, 196 cm²

10. 32 m, 60 m²

11. 210 m, 1800 m²

12. 7.85 cm, 4.91 cm²

13. 150°

14. $\frac{11\pi}{30}$

15. prism, 456 cm², 540 cm³

16. cube, 150 cm², 125 cm³

17. cylinder, 378 cm², 565 cm³

18. pyramid, 107 cm², 41.8 cm³

19. cone, 188 cm², 171 cm³

20. sphere, 3018 cm², 15,600 cm³

Unit 3   **EXERCISES 3.1**

1. 4

3. −4

5. >

7. >

9. 8

11. 0

13. $\frac{7}{8}$

15. 11

17. −10

19. $-\frac{3}{4}$

21. −12

23. +1530

EXERCISES 3.2

1. 11
3. 4
5. 2
7. -6

9. -10
11. 5
13. $-4,999$
15. -4.75

17. 8
19. -14
21. 2
23. $\frac{1}{10}$

25. $\frac{1}{14}$
27. 20 kg
29. profit of \$13,390

EXERCISES 3.3

1. 2
3. 6
5. -11
7. -4
9. 8
11. 28

13. 28
15. -28
17. \$3.45
19. 4
21. -48%
23. -9

25. $-\frac{1}{3}$
27. $\frac{31}{24}$
29. $27°$
31. $-85.9°C$

EXERCISES 3.4

1. $10R - 5$
3. $-4m - 3$
5. $-6m - 3mn$
7. $11C - 8C^2$
9. $2n + 7m$
11. $-R - R^2$
13. $2T_1 - 2T$
15. $-2P_1 - 8P_0 + 2P$

17. $xy^2 + 5xy - 4x^2y$
19. $21Z_0 - 8Z_1 - 18$
21. $.356a - 9.86b + 3.71ab$
23. $5 + 6p + 11p^2 + 3p^3$
25. $10 + 6a - 15a^2$
27. $-gt^2 - 19t - 3$
29. $16s - 4$

EXERCISES 3.5

1. -18
3. -28
5. 27
7. 64
9. 0
11. -3854
13. 10962

15. -138.446
17. 2280
19. -8
21. 18
23. $\frac{7}{40}$
25. -2
27. 2

29. 9
31. -8
33. $-\frac{4}{3}$
35. $\frac{2}{7}$
37. 2
39. 18
41. -24

43. 22
45. -4
47. 11
49. 13
51. -339
53. 26.36 sq in.

EXERCISES 3.6

1. x^{11}
3. y^{15}
5. $-15T^5$
7. 5^5
9. $-12a^2b^4c$
11. $-10m^6n^2$
13. $\frac{3}{7}D^3$
15. $-48a^2b^4c^2$
17. $-\frac{3}{4}R^3S^3$
19. $21x + 6$
21. $12R_0 - 24$
23. $m^2 - m$
25. $-14r^2s^2t + 6rs^2t^3$

27. $2a^3b - 9a^2b^4c$
29. $4m^2n - 2mn^2$
31. $4x^2 - 6x + 12$
33. $6t^3 + 2t^2 - 2t$
35. $6jk - 24k^2$
37. $25x^3y + 10x^2y$
39. $-8d^3e^7f^2 + 124de^8f^5 - 36de^4f^6$
41. $-4h^5i^2 - h^2i^3 - 72h^2i$
43. $x^2 - 6xy + 2x + 7y$
45. $-5a^2 - 49a + 11ab$
47. $\frac{2b^2 - 5b}{2}$
49. $I = \frac{110}{R}$

EXERCISES 3.7

1. x^6

3. y^2

5. $2z^3$

7. $12D$

9. $\frac{2}{3}j^2$

11. $5S_0{}^4$

13. $\dfrac{-35r^2t^4}{s^2}$

15. $2R^2 - 4$

17. $12T - \dfrac{3}{T}$

19. $-9R + \dfrac{4}{R} - \dfrac{3}{R^2}$

21. $2m^6 + 4m^5 - 8m^3 - 7m$

23. $-5a + 2 + \dfrac{3}{a} - \dfrac{2}{a^2} + \dfrac{1}{a^3}$

25. $BF = \dfrac{WtL}{12}$

27. $T = \dfrac{WL}{NV}$

EXERCISES 3.8

1. $7m - 9$

3. $T + 12$

5. $2m + 8$

7. $4H - 36$

9. $F - 10$

11. $2r - 30s + 49$

13. $-129p - 105$

15. $-16x - 12$

17. $V = 75{,}000\,E_0 - 75{,}000E_f$

19. $V = 9.4a(2a + 3)^2$

EXERCISES 3.9

1. 14

3. -9

5. 10

7. -5

9. -14

11. 36

13. 40

15. 8

17. -14

19. -2

21. 0

23. 48

25. -54

27. $\frac{3}{2}$

29a. 8.93 m²

29b. 57,800 m², 1020 m

31a. 20

31b. 100

31c. -20

33a. 10

33b. 9.8

35a. 746 m²

35b. 1,891 cm²

37a. 38%

37b. 59%

39. 633,750 π^2

UNIT 3 SELF-EVALUATION

1. 13

2. -13

3. $\frac{19}{40}$

4. 34

5. -12

6. -99

7. -9

8. $\frac{3a}{b^2}$

9. 12

10. 0

11. $-80a^3b^3c$

12. $-8m^2n + 6mn^2$

13. $-2p - q - 4$

14. $4R_1 + 18$

15. $-13H - HK - 3K$

16. $-15d^4 + 20d^2 - 30$

17. $-15e + 16$

18. -42

19. 3289 sq. m

20. $\frac{40}{17}$

Unit 4

EXERCISES 4.1

1. $x = 4$

3. $y = 14$

5. $y = 0$

7. $b = 6$

9. $p = 9$

11. $K = -.59$

13. $d_2 = -10$

15. $r = 6$

17. $W = 5$

19. $s = -1$

21. $R = -24$

23. $g = .4$

25. $y = 6$

27. $W = 9$

29. $y = 4$

31. $a = 45$

33. $x = -4$

35. $p = -20$

37. $D = \dfrac{28}{5}$

39. $a = \dfrac{27}{8}$

41. $y = \frac{20}{33}$

43. $e = -2$

45. $x = 2$

47. $a = 4$

49. $y = -2$

51. $Z = 0$

53. $Q_0 = 3$

55. $r = -2$

57. $T_1 = \frac{3}{5}$

59. $e = \frac{98}{3}$

61. $H = \frac{209}{3}$

63. $p = -45$

EXERCISES 4.2

1. $r = 3$

3. $b = 3$

5. $a = 3$

7. $d = 1$

9. $z = 2$

11. $Z_0 = -12$

13. $W = -4$

15. $x = 2$

17. $p = 0$

19. $c = 3$

21. $a = 1$

23. $Q = \frac{1}{13}$

25. $a = \frac{7}{11}$

27. $D = -\frac{10}{9}$

29. $p = -\frac{3}{2}$

31. $t = 6$

33. $r = .4$

35. $m = -1$

37. $x = -.01$

39. $x = -1$

41. $f = -3$

43. $L_0 = -\frac{1}{14}$

EXERCISES 4.3

1. $x > -6$

3. $x < -9$

5. $x \geq 4$

7. $x > 0$

9. $x < -3$

11. $x < 3$

13. $x > -\frac{3}{2}$

15. $t > -1$

17. $f_1 > 3\frac{1}{3}$

19. $x \geq \$39,625$

EXERCISES 4.4

1. $R = \frac{E}{I}$

3. $p = W$

5. $E = \frac{7I}{n}$

7. $Q_1 = \frac{PQ_2 - Q}{P}$

9. $g = \frac{2S}{t^2}$

11. $m = \frac{yd}{R}$

13. $h = \frac{V}{lw}$

15. $g = Mt - Lt$

17. $V_0 = 2V - V_t$

19. $h = \frac{2A}{a + b} = 8$

21. $t = \frac{V - V_0}{a} = 4$

23. $P_1 = \frac{P_2 T_1}{T_2} = 30$

25a. $m = \frac{2k}{v^2} = 40$

25b. 8

EXERCISES 4.5

1. $n = 2$
3. $n = 12$
5. $n = 26$
7. $n = 2$

9. $n = 18$
11. $n = 6$
13. $n = 43$
15. $n = 10$

17. $n = 1$
19. $n = 80$
21. $\frac{G}{2} + \frac{W}{20} + \frac{C}{200}$
23. $A = \frac{2}{3} \pi r^3$

EXERCISES 4.6

1. 11 m, 25 m
3. 6 cm, 8 cm, 24 cm
5. 10 cm by 14 cm by 17 cm
7. 15 cm
9. 7 cm
11. 61.5 cm

13. 60 liters
15. 210 kg
17. 32 lb
19. 13
21. 54.4 sq in.

EXERCISES 4.7

1. $\frac{6}{7}$
3. $\frac{3}{10}$
5. $\frac{7}{100}$
7. $\frac{2}{3}$
9. $\frac{1}{96,000}$
11. $\frac{31}{12}$
13. .46 ft
15. $\frac{7}{18}$

17. $E = 10$
19. $f = 22$
21. $m = 77$
23. $t = 17.6$
25. $x = \frac{24}{11}$
27. 3000 bricks
29. 27.36 kg
31. 0.56

33. 218
35. 6667 ℓ
37. 411 cm^3
39. 4 in.
41. 268.8
43. 300 rpm
45. \$3,760

EXERCISES 4.8

1. $V = kP$
3. $I = kR^2$
5. $A = klw$
7. $V = khr^2$
9. $F = kma$
11. $k = 5$
13. $k = \frac{4}{23}$
15. $k = 36$

17. $k = 2$
19. $k = \frac{1}{8}$
21. $w = 120$
23. $R_1 = \frac{64}{3}$
25. $F = 32$
27. $43\frac{1}{3}$ mm
29. 23
31. .09

33. 52.6
35. 32
37. 3.4 min
39. $a = 3$
41. 420 km
43. 445.6 ohms

UNIT 4 SELF-EVALUATION

1. $R_2 = 6$
2. $m = -21$

3. $x = 18$
4. $p = -7$

5. $t = 10$

6. $K_{ts} = \dfrac{K + q - 1}{q}$

7. $n = 36$

8. $7\dfrac{7}{8}$ cm, $39\dfrac{3}{8}$ cm, $23\dfrac{3}{4}$ cm

9. $\dfrac{11}{15}$

10. $R = \dfrac{40}{3}$

11. $48\dfrac{8}{9}$ lb and $61\dfrac{1}{9}$ lb

12. $\dfrac{5}{16}$ km

13. $y = \dfrac{3}{2}$

14. 3,000 kg

15. $a = \dfrac{2A - hb}{h}$

16. $a = \dfrac{V - V_0}{t}$

17. $x < -1$

18. $x < 2$

19. $x > -\dfrac{9}{2}$

20. $t < -\dfrac{25}{16}$

Unit 5   **EXERCISES 5.1**

1.

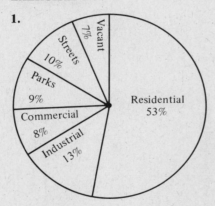

3.

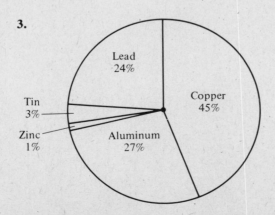

5.

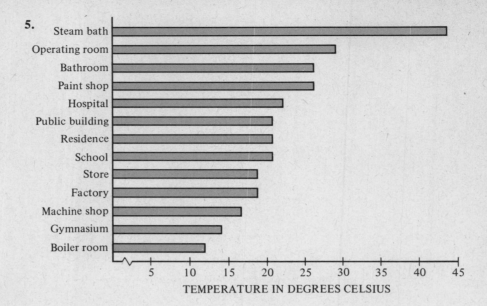

7.

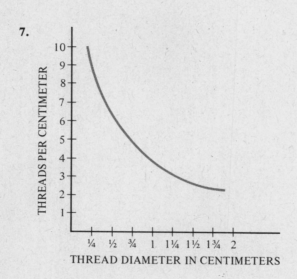

9.

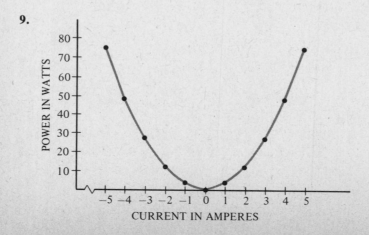

EXERCISES 5.2

1–9.

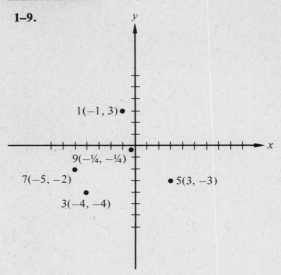

11. quadrilateral
13. rectangle
15. quadrilateral
17. triangle
19.

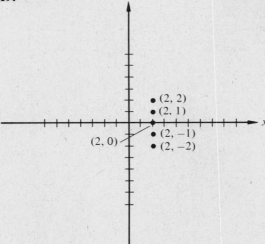

21.

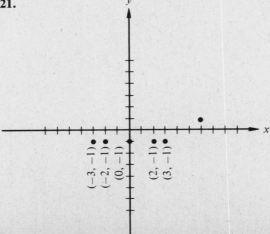

23. II and III

25. III and IV

27. $D = \{2,3,4\}$
$R = \{7,9,11\}$
Function

29. $D = \{5,7,9,11\}$
$R = \{10,14,18,22\}$
Function

31. $D = \{2,4,6\}$
$R = \{1,3,5,7\}$
Not a function

33. 4.1

35. 3.2

37. 8.1

39. 6.4, 9.2, 8.6

EXERCISES 5.3

1.

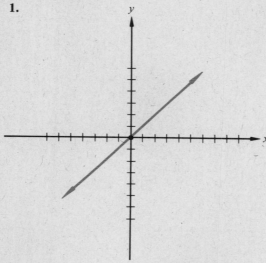

3.

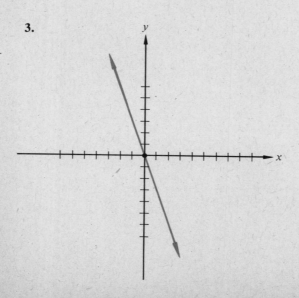

5.

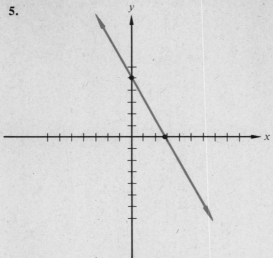

7.

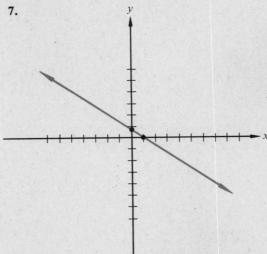

9.

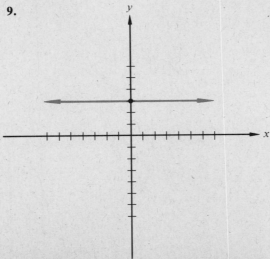

11.

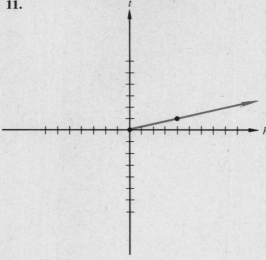

13. $y = 2x + 1$

15. $y = \dfrac{3}{4}x + \dfrac{5}{2}$

17. $x = 2$

19. $4x - y - 27 = 0$

21. $6x - 3y = 0$

UNIT 5 SELF-EVALUATION

1.

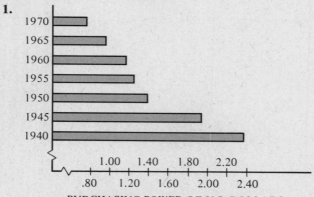

PURCHASING POWER OF U.S. DOLLARS

2.

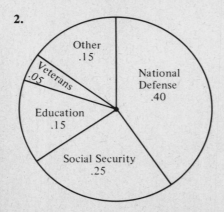

3.

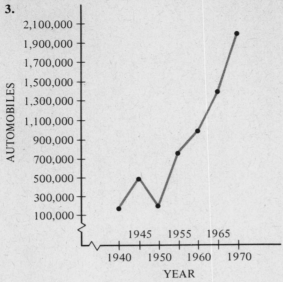

4.

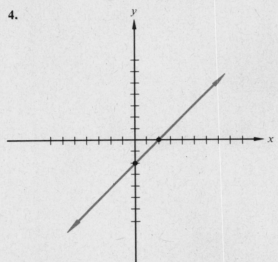

5.

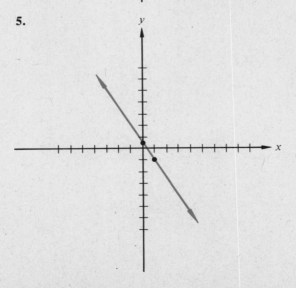

6.

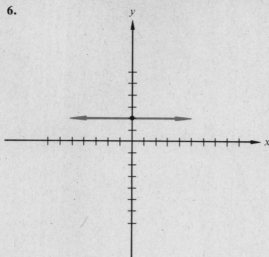

7.

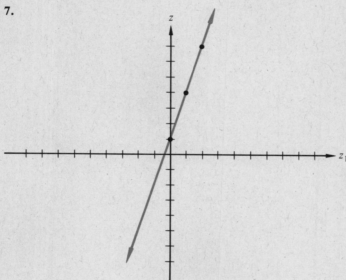

8.

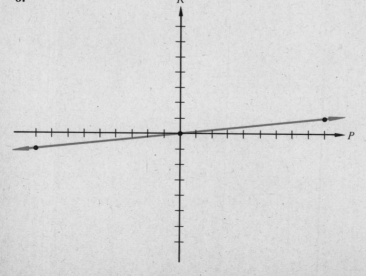

9. Function
$D = \{-8, -6, -2\}$
$R = \{-8, -6, -2\}$
10. Not a function
11. Not a function
12. Function
$D = R =$ set of real numbers
13. $y = -3x - 20$
14. $y = -2x + 4$
15. $5x + y = 0$

Unit 6

EXERCISES 6.1

1. $6(2x - y)$
3. $2y(4y + 1)$
5. $6a(1 - 7a)$
7. $g^2(4 - 5g + 6g^2)$
9. $2(2mn + 3m - n)$
11. $a(t - v + w)$
13. $3r^2(\pi - 3\pi h - 1)$
15. $2Q^2(3Q^3 - 2Q^2 + 5Q - 4)$
17. $.32R(R - 5)$
19. $\frac{1}{9} a^2 b^2(3ab + 5)$
21. $2mn(m + 2mn - 5n^2)$
23. $17bc(3a + 2cd)$
25. $2.7t^3(3t^3 - 10t + 2)$

27. $8(7V_u W^2 + 2V_u W - 2V_s)$
29. $-rs(r^2 + rs^2 + s^4)$
31. $EI(1 - I)$
33. $7(V_1 + 2V_2 - 4V)$
35. $P_1(V_1 + V_2 - 2)$
37. $at(L + L_0 - Lt)$
39. $-2(3f + 2f_s k)$
41. $(c + d)(e + 3)$
43. $(2n - 5)(m - 3)$
45. $(g + h)(f - 1)$
47. $13(2 + 3)$
49. $T = \frac{12}{l}(D - d)$

EXERCISES 6.2

1. $s^2 - 9$
3. $a^2 - 36$
5. $x^2 - 1$
7. $4a^2 - 25$
9. $25b^2 - 36$
11. $4R^2 - 25S^2$
13. $81d^2 - 49e^2$
15. $.09p^2 - q^2$
17. $\frac{9}{25} D^2 - E^2$
19. $\frac{1}{9} r^2 - \frac{1}{25} t^2$
21. $(20 + 1)(20 - 1)$

EXERCISES 6.3

1. $(B - 6)(B + 6)$
3. $(a - 1)(a + 1)$
5. $(8 - z)(8 + z)$
7. $(5p - 2)(5p + 2)$
9. $(11E + 7)(11E - 7)$
11. $(2a - 3y)(2a + 3y)$
13. $\left(\frac{1}{2} R - T\right)\left(\frac{1}{2} R + T\right)$
15. $\left(\frac{1}{3} x + y\right)\left(\frac{1}{3} x - y\right)$
17. $(.2 + t)(.2 - t)$
19. $(8f^2 g + 3h)(8f^2 g - 3h)$
21. $(20 + 5)(20 - 5)$

EXERCISES 6.4

1. $x^2 + 7x + 12$
3. $6a^2 + 17a + 5$
5. $d^2 - 9d + 20$
7. $18r^2 - 27r + 10$
9. $R^2 + 4R - 21$
11. $b^2 - 7b - 44$
13. $40P^2 + P - 6$
15. $9R_0^2 + 12R_0 R + 4R^2$
17. $\frac{2}{9} f^2 + \frac{1}{2} f + \frac{1}{4}$
19. $56p^2 - 13pq - 3q^2$
21. $3.68s^2 + 4.65s - .92$
23. $6r^2 - 9rs + 10rt - 15st$
25. $256t^2 + 256t - 3072$

EXERCISES 6.5

1. $(x + 2)(x + 3)$
3. $(p + 11)(p + 7)$
5. $(m - 2)(m - 4)$
7. $(b - 3)(b - 4)$
9. $(c + 3)(c - 2)$
11. $(n - 4)(n + 2)$
13. $(t - 2)(t + 9)$
15. $(f - 5)(f - 5)$
17. $(T + 3)(T + 4)$
19. $(n + 5)(n - 4)$

21. $(a + 2b)(a - b)$
23. $(5m + 1)(m + 2)$
25. $(c - 1)(3c + 5)$
27. $(6r + 5)(r - 1)$
29. $(V - 1)(16V + 5)$
31. $(8h + 1)(2h - 5)$
33. $(2r + 1)(r - 3)$
35. $(3X + 2Y)(3X - Y)$
37. $(3b - c)(b + 5c)$
39. $(A - L)(A - 2L)$

EXERCISES 6.6

1. $x^2 + 10x + 25$
3. $z^2 - 22z + 121$
5. $4a^2 + 12a + 9$
7. $25g^2 - 20g + 4$

9. $9D^2 + 30DE + 25E^2$
11. $36 - 12T + T^2$
13. $x^2 + 2xy + y^2$
15. $100 + 50 + 50 + 25$

EXERCISES 6.7

1. $(d + 9)^2$

3. $(a - 1)^2$
5. $(5 - c)^2$
7. $(2a + 7)^2$
9. $(5x - 2)^2$
11. $(12p - 1)^2$

13. $\left(\dfrac{2}{3}y - \dfrac{3}{5}\right)^2$
15. $(m + n)^2$
17. $(2T + 3R)^2$
19. $(3x + 5y)^2$
21. $5x + 1$

EXERCISES 6.8

1. $(x - 3)(x^2 + 3x + 9)$
3. $(5 + m)(25 - 5m + m^2)$
5. $(R - 1)(R^2 + R + 1)$
7. $(4 - s)(16 + 4s + s^2)$

9. $(2p - q^2)(4p^2 + 2pq^2 + q^4)$
11. $(b^3 + 2)(b^6 - 2b^3 + 4)$
13. $(3 + 2F)(9 - 6F + 4F^2)$

EXERCISES 6.9

1. $(x + y)(x + 7)$
3. $(D + 5)(3 + E)$
5. $(x^2 + 1)(3 + y)$

7. $(3m - 5)(17 - 2n)$
9. $(a + b)(c + 9)$
11. $(r^3 + s^2)(5 - 3q)$

EXERCISES 6.10

1. $6a(2x - 3a)$
3. $2ay(y - 1)^2$
5. $(2h - 3k)^2$
7. $3(2m + 25)(m - 12)$
9. $3(3r + 2)(r - 3)$
11. $4(t^2 + t - 8)$
13. $(16x + 9)(x - 1)$
15. $-4dy(5y - 1)$

17. $2(f - 9)(f + 9)$
19. $c(z - 3)(4z - 7)$
21. $(v - 15)(v + 9)$
23. $(4y - 7z)(4y + 7z)$
25. $(11m - 4)(m + 3)$
27. $(3a - 8b)^2$
29. $r = x + 7$

UNIT 6 SELF-EVALUATION

1. $4m^2 - n^2$
2. $35d^2 + 4d - 63$
3. $65r^2 - 81rs - 14s^2$
4. $15 - 82Z + 99Z^2$
5. $36T^2 + 60T + 25$
6. $(8 - x)(8 + x)$
7. $(z + 11)(z - 6)$

8. $5r^2t(3r - 4t^2)$
9. $(2B + 5)^2$
10. $(3x - y)(9x^2 + 3xy + y^2)$
11. $7(f - 3)(f + 3)$
12. $(P + R)(T - 3)$
13. $3a^2b(b - 2ab + 3a)$
14. $2(3s - 2)(5s + 2)$

15. $(W - 9)^2$

16. $3(x^2 + 5xy + 3y^2)$

17. $4(2A - 1)(2A + 1)$

18. $(5g + 3)(2g + 1)$

19. $(7y - 3)(6y - 5)$

20. $(2 + 11M)(3 + 4M)$

Unit 7

EXERCISES 7.1

1. $\dfrac{3x}{x - 5}$

3. $\dfrac{1}{p}$

5. $r + s$

7. $\dfrac{i}{i - r}$

9. $\dfrac{x + 2}{3}$

11. $m + 5$

13. $\dfrac{a}{A - 2}$

15. $\dfrac{d + 9f}{d - 6f}$

17. $\dfrac{Q + 2}{Q - 2}$

19. $\dfrac{x + 2}{x - 1}$

21. $\dfrac{2(w + 2)}{2w + 3}$

23. $\dfrac{5}{X + 2Y}$

25. $\dfrac{\theta}{\theta + 1}$

27. $\dfrac{2r}{r - R}$

EXERCISES 7.2

1. $\dfrac{4a}{3b}$

3. $\dfrac{4ms}{nt}$

5. $\dfrac{3XV}{8YW}$

7. $\dfrac{6p^2}{q}$

9. $\dfrac{3r^2}{r + s}$

11. $\dfrac{4R(P - 2)}{P + 3}$

13. $\dfrac{u - v}{(u - 3)(u + v)}$

15. $\dfrac{5e - 3}{e + 1}$

17. $\dfrac{(x - 1)(x - 4)}{3x - 1}$

19. $\dfrac{3(A - 1)}{5}$

21. 2

23. $\dfrac{s^2}{t}$

25. $\dfrac{2}{3B(B - 1)}$

27. $\dfrac{4y(x + 2)}{7}$

29. $\dfrac{p - 3q}{2(p + 2q)}$

31. $(2x - 3)(x - 1)$

33. $\dfrac{X(X + 3)}{(X + 1)(2X + 1)}$

35. $\dfrac{(3t - 2)(t + 1)}{(t - 1)^2}$

37. $\dfrac{S^2 - 8S + 5}{(S + 2)^2}$

39. $\dfrac{z + 4}{2(z - 3)}$

41. $\dfrac{wv^2}{gr}$

EXERCISES 7.3

1. $\dfrac{16b}{60a^3b^2}, \dfrac{25a^2}{60a^3b^2}$

3. $\dfrac{3r}{e^2i^2r}, \dfrac{-2ei}{e^2i^2r}$

5. $\dfrac{X^2 + X - 2}{(X + 3)(X - 3)(X + 2)}, \dfrac{3X^2 + 11X + 6}{(X + 3)(X - 3)(X + 2)}$

7. $\dfrac{E - I}{(E - I)^2(E + I)}, \dfrac{E + I}{(E - I)^2(E + I)}$

9. $\dfrac{2e^2}{Ee(E - e)}, \dfrac{E^2}{Ee(E - e)}$

11. $\dfrac{3R^2 - 3}{(R + 2)(R - 2)(R + 1)(R - 1)}, \dfrac{2R^2 - 8}{(R + 2)(R - 2)(R + 1)(R - 1)},$

$\dfrac{R^2 + 3R + 2}{(R + 2)(R - 2)(R + 1)(R - 1)}$

13. $\dfrac{3v + 3t}{(v - t)(v + t)}, \dfrac{sv - st}{(v - t)(v + t)}, \dfrac{t^2v + t^3}{(v - t)(v + t)}$

15. $\dfrac{5w^2 - 15w - 50}{(w - 2)(w + 1)(w + 2)(w - 5)}, \dfrac{2w^2 - 8w - 10}{(w - 2)(w + 1)(w + 2)(w - 5)},$

$\dfrac{w^3 - 2w^2 - 4w + 8}{(w - 2)(w + 1)(w + 2)(w - 5)}$

17. $\dfrac{21ER}{6IR(R - I)}, \dfrac{10I^2}{6IR(R - I)}$

19. r_1r_2

EXERCISES 7.4

1. $\dfrac{5a}{36}$

3. $\dfrac{5Ab + 3a^2}{30ab}$

5. $\dfrac{4y^2 - 3w^3}{14w^2y}$

7. $\dfrac{25\theta^2 - 21\phi^2}{90\phi^3\theta}$

9. $\dfrac{b^2 - 3b + 2}{b - 3}$

11. $\dfrac{-6B + 6}{(B + 1)(B - 2)}$

13. $\dfrac{q^2 + 5q}{(q + 3)(q - 3)}$

15. $\dfrac{4s^2 + 12s + 4}{(s + 3)(s - 3)(s + 2)}$

17. $\dfrac{2r^2 + 3i^2}{ir(i - r)}$

19. $\dfrac{-z + 6}{(z - 2)(z - 3)}$

21. $\dfrac{-R^3 + 4R^2 + 3R - 5}{(R + 1)(R - 1)(R + 2)(R - 2)}$

23. $\dfrac{(1 + i)^n - 1}{i(1 + i)^n}$

25. $p = \dfrac{2mL}{d^2 - L^2}$

EXERCISES 7.5

1. $x = \dfrac{14}{5}$

3. $R = \dfrac{9}{7}$

5. $p = -\dfrac{5}{2}$

7. $E = -\dfrac{1}{5}$

9. $r = 22$

11. $F = \dfrac{fQ + f}{Q}$

13. $r = \dfrac{Z - EZ}{E}$

15. $i = \dfrac{5}{2}$

17. $b = \dfrac{5}{9}$

19. $d = -\dfrac{19}{3}$

21. $A = -\dfrac{9}{7}$

23. $f = \dfrac{5}{7}$

25. $\dfrac{IR_s}{R_y + R_s}$

27. no solution

29. $D = \dfrac{100P + EP - 20}{H}$

31. $Q = 6$

33. $R_T = \dfrac{R_1R_2R_2}{R_2R_3 + R_1R_3 + R_1R_2}$

35. $s_a = \dfrac{pD - 2t_ss_s}{2t_a}$

UNIT 7 SELF-EVALUATION

1. $\dfrac{Q}{Q - 5}$

2. $\dfrac{3}{r_1 + 2r_2}$

3. $\dfrac{X - 4}{2X + 3}$

4. $\dfrac{5}{2q}$

5. $\dfrac{I + i}{3i}$

6. $\dfrac{3a^2}{2}$

7. $\dfrac{I + R}{IR}$

8. $\dfrac{7u - 2v}{(u - 2v)(u + 2v)(u + v)}$

9. $w = 8$

10. $r = -9$

Unit 8

EXERCISES 8.1

1. (3,2)
3. Inconsistent
5. (−4,3)
7. (−5,−1)
9. Inconsistent
11. (3,3)

EXERCISES 8.2

1. (9,2)

3. (1,1)

5. $\left(-\dfrac{1}{6},\dfrac{3}{2}\right)$

7. (2,4)

9. Inconsistent

11. (12,−20)

13. $\left(-\dfrac{2}{3},-\dfrac{19}{3}\right)$

15. (2,−1)

17. $\left(-\dfrac{23}{39},-\dfrac{24}{39}\right)$

19. Inconsistent

21. $\left(30,-\dfrac{20}{3}\right)$

23. (−2,−10)

25. $\left(\dfrac{84}{15},\dfrac{16}{15}\right)$

27. (6,1)

29. (−2,−2)

31. Dependent

33. (−26,−8)

35. Inconsistent

37. 165 and 75

39. 9 ft and 5 ft

41. 60 rpm and 40 rpm

43. (1,5,−1)

45. (3,−3,1)

EXERCISES 8.3

1. (2,5)
3. (0,−4)

5. (0,−4)

7. (3,−1)
9. Inconsistent
11. (−2,−3)
13. Dependent
15. $\left(\dfrac{1}{2},-\dfrac{3}{2}\right)$

17. (−2,0)
19. (−1,−1)
21. $\left(-\dfrac{5}{8},\dfrac{7}{8}\right)$
23. 17 and 20
25. 18 kg at $5.20
　　12 kg at $3.20
27. 11 kg, 17 kg

EXERCISES 8.4

1. 0
3. 378
5. 0
7. 9016
9. $(a_3 - a_2)(b_2{}^2 - b_1{}^2)$
11. (1,−1,2)
13. (4,−3,1)

15. (−2,2,−2)
17. (11,22,33)
19. (3,−3,6)
21. (1,−1,1)
23. 34 in., 40 in., 22 in.
25. 8 cm, 10 cm, 12 cm

UNIT 8 SELF-EVALUATION

1. (−3,−4)

2. (−1,−3)
3. (6,6)
4. $\left(\dfrac{3}{4},-\dfrac{7}{4}\right)$
5. (−2,3,−2)

6. $\left(\dfrac{2}{3},-\dfrac{1}{3}\right)$

7. Dependent
8. (.1,−.3)

9. Inconsistent

10. 11 and 20

Unit 9

EXERCISES 9.1

1. x^9
3. 10^7
5. 3^6
7. b^7

9. I^5

11. 10^2
13. A

15. $\dfrac{1}{w}$

17. $\dfrac{1}{10}$

19. $\dfrac{1}{n^6}$

21. $v^3 w^3$

23. $36T^2$

25. $\dfrac{Z^2}{R^2}$

27. $\dfrac{f_s^4}{81}$

29. Q^8

31. 10^{42}

33. 5^{12}

35. $25i^6$

37. $\theta^{12}\phi^4$

39. $\dfrac{G^8}{H^{12}}$

41. $\dfrac{729X_1^6}{X_2^9}$

43. 10^4

45. $i^3 e^4$

47. $\dfrac{q^2}{p^3}$

49. 100

51. $250{,}047\,\alpha^9\beta^{12}$

53. $\dfrac{1331\pi^{13}}{25p^6}$

55. 625×10^8

57. w^3

59. $2d\pi^2 v - 4d^2 f\pi v + 2d^3 f^2 v$

EXERCISES 9.2

1. 1

3. 1

5. 1

7. 2

9. 1

11. 10^4

13. $\dfrac{32p^5}{3}$

15. $\dfrac{1}{10^8}$

17. $\dfrac{5T^2}{2}$

19. R_1^5

21. 1

23. 1

25. 1

EXERCISES 9.3

1. $\dfrac{1}{g^2}$

3. $\dfrac{1}{10^4}$

5. f^3

7. 32

9. $\dfrac{4}{k}$

11. $\dfrac{1}{5r_0}$

13. $\dfrac{c^9}{8}$

15. $18\theta^2$

17. 1

19. 10^5

21. i

23. 10

25. $\dfrac{vw}{u^2}$

27. 10^6

29. $\dfrac{f}{Td^2}$

31. L^2

33. $\dfrac{1}{43^4 g^8}$

35. $\dfrac{z^6}{x^4}$

37. $\dfrac{5^3 E^3 c^3}{2^3}$

39. $\dfrac{-4J_x^8}{27J_y^9}$

41. $\dfrac{36}{f_1^8}$

43. $\dfrac{m^2}{v^2}$

45. $\dfrac{g^2}{200h}$

47. $\dfrac{R - I}{IR}$

49. $\dfrac{2\beta + 3\alpha^2}{\alpha^2\beta}$

51. $\dfrac{R_1 R_2}{R_1 + R_2}$

53. $\dfrac{1}{f} = \dfrac{1}{p} + \dfrac{1}{q}$

55. $\dfrac{\epsilon_0 v^2}{2d^2}$

57. $\dfrac{D^2}{d^2 x C_n}$

EXERCISES 9.4

1. $t^{1/2}$

3. $e^{3/5}$

5. $\dfrac{1}{R^{1/3}}$

7. $S_a^{1/4} S_b^{1/4}$

9. $\dfrac{2^{3/8}}{Z^{3/8}}$

11. $F^{1/4}$

13. $\dfrac{s^{9/4}}{r^{5/6}}$

15. $\dfrac{j}{h^2}$

17. $\dfrac{L_0(c^2 - v^2)^{1/2}}{c}$

19. $\dfrac{2I^{1/3}}{3K^{2/3}D}$

EXERCISES 9.5

1. 2.86×10^2
3. 1.8×10^4
5. 4.1×10^{-2}
7. 2.387×10
9. 8.42×10^{-4}
11. 2.3×10^6
13. 5.73×10^{-6}
15. $370,000$
17. 0.000285

19. 0.7
21. 3.02×10^5
23. 4.0×10^{-4}
25. 6.70×10^3
27. 4×10^{-4}
29. $602,472,000,000,000,000,000,000$
31. 3.3×10^2
33. 5.49×10^{-4}

EXERCISES 9.6

1. 7.6×10^{-6}
3. 2.0×10^{-10}
5. 7.5×10^{-2}

7. 7.5×10^{11}
9. 5.6×10^2
11. 5.0

13. 5.3×10
15. 1.143259423×10^4
17. $9.006562778 \times 10^{14}$

UNIT 9 SELF-EVALUATION

1. $\dfrac{1}{R^5}$

2. $-\dfrac{27}{s^3 t^3}$

3. 3

4. $\dfrac{1}{25x^2 y^4}$

5. $\dfrac{3y^2}{5xz^4}$

6. $\dfrac{108a^8 b}{c}$

7. $1 - \dfrac{1}{n^4}$

8. $B^{3/4}$

9. 3.52×10^5

10. 7.8×10^{-3}

11. 3.2×10^6

12. 3.5×10^2

Unit 10

EXERCISES 10.1

1. 7
3. 2
5. 3
7. -3
9. -5
11. 1
13. No real root
15. 2
17. 11
19. -10
21. -8
23. 1
25. p
27. D^3

29. n^2
31. c
33. $-T^5$
35. y^3
37. W^3
39. t^6
41. $4p$
43. $2d^2$
45. $ab^2 c^3$
47. $3BC^2$
49. $b^{1/2}$
51. $V^{3/2}$
53. $T^{2/3}$
55. $\sqrt[3]{r}$

57. $\sqrt[3]{W^2}$
59. $\sqrt[3]{2^2}$
61. $\sqrt{a}$
63. $\sqrt[4]{5}$
65. B
67. 27
69. $\dfrac{1}{4}$
71. 27
73. $\dfrac{27}{8}$
75. $\dfrac{9}{25}$

EXERCISES 10.2

1. $3\sqrt{2}$
3. $5\sqrt{2}$
5. $4\sqrt{3}$
7. $8\sqrt{3}$
9. $-6\sqrt{2}$
11. $12\sqrt{5}$
13. $p^3 \sqrt{p}$

15. $-x^4 \sqrt{x}$
17. $b\sqrt{ab}$
19. $2m\sqrt{2m}$
21. $3\sqrt{3}$
23. $-3\sqrt{2}$
25. $a^2 b\sqrt{ab}$
27. $d\sqrt[3]{d^2}$

29. Y
31. $E\sqrt[5]{D^2 E}$
33. $-4K^5 \sqrt{2}$
35. $\dfrac{F}{2}\sqrt[3]{F}$
37. $2X^2$
39. $LP\sqrt[3]{P^2}$
41. $T = .1$
43. $V = 1.4$
45. $Z = 8$

EXERCISES 10.3

1. $\sqrt{91}$
3. $\sqrt{22}$
5. $-\sqrt{15}$
7. $\sqrt[4]{24}$
9. $4\sqrt{3}$
11. $-5\sqrt{2ab}$
13. $\sqrt[3]{2a^2}$

15. $rs^2\sqrt{r}$
17. $3m\sqrt{2mn}$
19. $6\sqrt{42}$
21. $-20\sqrt{5}$
23. $6x^3\sqrt{35}$
25. $\sqrt{42}$

27. $b\sqrt[3]{a^2b}$
29. $48\sqrt{15}$
31. $6PQ\sqrt[6]{Q}$
33. $2\sqrt{3} - \sqrt{6}$
35. $6\sqrt{15} - 3\sqrt{6}$
37. 234.5

EXERCISES 10.4

1. $\sqrt{10}$
3. $\sqrt{3}$
5. $\sqrt[3]{11}$
7. $2\sqrt{2}$

9. $\sqrt[4]{2a}$
11. $a\sqrt{11c}$
13. $\dfrac{\sqrt{14}}{7}$
15. $\dfrac{\sqrt{10}}{6}$

17. $\dfrac{\sqrt{6}}{2}$
19. $\sqrt[3]{rs}$
21. $\dfrac{3\sqrt{7}}{7}$
23. $\sqrt{6}$

25. $\dfrac{\sqrt[4]{r^3}}{r}$
27. $m\sqrt{2}$
29. $\dfrac{2\sqrt{2p}}{p}$
31. 1.2

EXERCISES 10.5

1. $5\sqrt{7}$
3. $2\sqrt[3]{m}$
5. $\sqrt{2d}$
7. $6\sqrt{2} - 2\sqrt{3}$

9. $17\sqrt{p}$
11. $3\sqrt{2} - 5\sqrt{3}$
13. $3\sqrt{2}$
15. $\dfrac{13}{6}\sqrt{6}$

17. $.098\sqrt{x}$
19. $2b\sqrt{b} + 2d\sqrt{e}$
21. $4\sqrt[3]{2}$

EXERCISES 10.6

1. $-7 + 4\sqrt{5}$
3. $-18 + 8\sqrt{2}$
5. 78
7. $13 + \sqrt{15}$

9. $10 - 4\sqrt{6}$
11. $B + 3\sqrt{B} + 2$
13. $X - Y$

15. $6R - 11\sqrt{R} - 10$
17. $16P - M$
19. $\sqrt[3]{4} - \sqrt[3]{2} - 30$

EXERCISES 10.7

1. $\dfrac{5(\sqrt{3} + 1)}{2}$
3. $\dfrac{9(7 + \sqrt{2})}{47}$
5. $\dfrac{\sqrt{14} - 3\sqrt{2}}{-2}$
7. $-(49 - 20\sqrt{6})$
9. $\dfrac{9 + 3\sqrt{7} - 3\sqrt{2} - \sqrt{14}}{7}$

11. $\dfrac{2W^2 - \sqrt{5}W - 5}{4W^2 - 5}$
13. $\dfrac{63 - 13\sqrt{e} - 6e}{49 - 9e}$
15. $2 + \sqrt{3}$
17. $\dfrac{H + \sqrt{H}}{H - 1}$
19. $\dfrac{3R - 5\sqrt{6RS} + 8S}{3R - 2S}$

UNIT 10 SELF-EVALUATION

1. $8d$
2. $e\sqrt[5]{e^2}$
3. $a^{1/2}$
4. $\sqrt{y}$
5. $2\sqrt{3}$
6. $a^2\sqrt[3]{a^2}$

7. 15
8. $3\sqrt{2}$
9. $y\sqrt[4]{y}$
10. $\sqrt{15} - 2\sqrt{21}$
11. $-2 + \sqrt{14}$
12. $84 - 18\sqrt{3}$

13. $\sqrt[3]{3}$
14. $\sqrt[4]{a^3}$
15. $\dfrac{5\sqrt{5} - \sqrt{10}}{23}$
16. $2\sqrt{5}$
17. $24 - 20\sqrt{2}$
18. $-5\sqrt{p}$

Unit 11   **EXERCISES 11.1**

1. $x = 0,2$

3. $m = 5,-5$

5. $R = -2,-4$

7. $T = -2,8$

9. $q = 2,3$

11. $a = \dfrac{5}{3}, -\dfrac{7}{2}$

13. $y = \dfrac{5}{2}, -\dfrac{5}{2}$

15. $b = \dfrac{5}{4}$

17. $E = 0,\dfrac{1}{7}$

19. $n = -5,-7$

21. $h = -\dfrac{2}{5}, -\dfrac{7}{6}$

23. 10 and 16, or -10 and -16

25. $\dfrac{5}{2}$ or 3

27. 7 cm, 11 cm

29. $\dfrac{1}{3}$ or 8

31. 2 ft

33. 8

EXERCISES 11.2

1. $m = \pm 3$

3. $B = \pm \sqrt{7}$

5. $x = 8,-2$

7. $T = -13,11$

9. $C = 5 \pm \sqrt{6}$

11. $f = -10 \pm 4\sqrt{2}$

13. $d = 11 \pm \sqrt{11}$

15. $a = -\dfrac{5}{2}, -\dfrac{1}{2}$

17. $y = \dfrac{4 \pm \sqrt{5}}{3}$

19. $r = -2 \pm \sqrt{2}$

EXERCISES 11.3

1. $t = 1, -7$

3. $y = \dfrac{-1 \pm \sqrt{5}}{2}$

5. $E = \dfrac{1 \pm \sqrt{13}}{12}$

7. $m = \dfrac{5}{2}, -3$

9. $b = -\dfrac{5}{2}, \dfrac{1}{3}$

11. $D = \dfrac{-3 \pm \sqrt{129}}{10}$

EXERCISES 11.4

1. $E = 3 \pm \sqrt{13}$

3. $I = -4 \pm 2\sqrt{3}$

5. $r = \dfrac{-3 \pm \sqrt{17}}{4}$

7. No Real Solution

9. $U = 3,\dfrac{1}{3}$

11. No Real Solution

13. $W = \dfrac{-1 \pm \sqrt{15}}{2}$

15. $e = \dfrac{6 \pm \sqrt{61}}{5}$

17. $Z = \dfrac{3 \pm \sqrt{33}}{4}$

19. $n = \dfrac{-7 \pm \sqrt{1489}}{60}$

21. $S = \dfrac{3 \pm \sqrt{6}}{3}$

23. $D = \dfrac{9}{4}, -1$

25. $\dfrac{4 + \sqrt{22}}{3}$ ft

EXERCISES 11.5

1. Two unequal, real solutions

3. Two unequal, real solutions

5. No real solution

7. One real double solution

9. Two unequal, real solutions

11. Two unequal, real solutions

13. One real double solution

15.

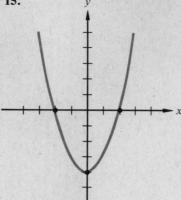

17.

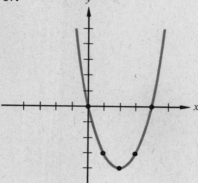

19.

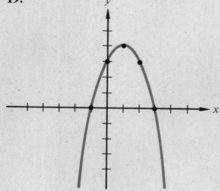

21.

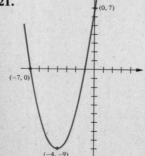

EXERCISES 11.6

1. $5j$
3. $44j$
5. $-9j$
7. $\sqrt{15}j$
9. $1 + 16j$
11. $x = 1 \pm \sqrt{2}j$
13. $T = \dfrac{1}{2} \pm \dfrac{\sqrt{39}}{2}j$
15. $a = \dfrac{3}{2} \pm \dfrac{\sqrt{3}}{2}j$
17. $R = -\dfrac{3}{8} \pm \dfrac{\sqrt{7}}{8}j$
19. $M = -\dfrac{1}{14} \pm \dfrac{\sqrt{83}}{14}j$

21. $7 - 7j$
23. 12
25. $10j$
27. $8 - 7j$
29. $8 - j$
31. $-26 - 18j$
33. 37
35. $\dfrac{21}{29} - \dfrac{20}{29}j$
37. $\dfrac{15}{26} + \dfrac{3}{26}j$
39. $24 + 28j$

UNIT 11 SELF-EVALUATION

1. $b = -7, 2$
2. $R = \dfrac{7}{2}, -\dfrac{5}{3}$
3. $f = 0, 10$
4. $W = -\dfrac{2}{7}$
5. $a = \pm 10$
6. $y = \pm\sqrt{19}$
7. $T = 8, 2$
8. $p = \dfrac{2 \pm 3\sqrt{2}}{7}$
9. $r = -6, 2$
10. $N = -\dfrac{5}{2}, \dfrac{1}{3}$
11. $x = \dfrac{1 \pm \sqrt{10}}{3}$

12. $E = \dfrac{-7 \pm \sqrt{57}}{4}$
13. $y = \dfrac{1 \pm \sqrt{73}}{12}$
14. $R = \dfrac{-2 \pm 3\sqrt{21}}{5}$
15. $n = \dfrac{7}{2} \pm \dfrac{\sqrt{23}}{2}j$
16. One real, double solution
17. Two unequal, real solutions
18. No real solution
19. 3 or $\dfrac{1}{3}$
20. 140 kg, 160 kg
21. 8 m by 26 m
22. 11,12

23.

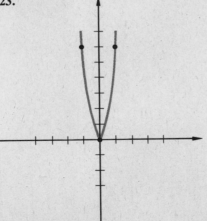

24.

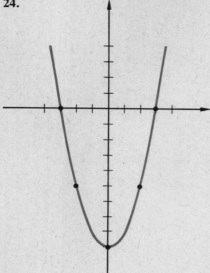

25.

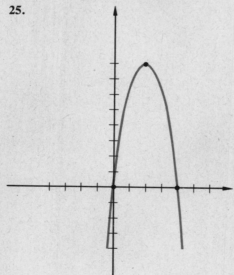

26. $32 - 16j$

27. $-2 - 9j$

28. $43 - 59j$

29. 73

30. $\dfrac{23}{13} + \dfrac{2}{13}j$

Unit 12   **EXERCISES 12.1**

1.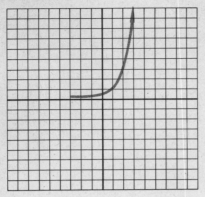

3.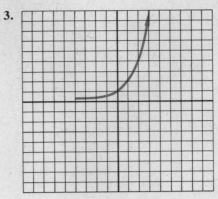

5.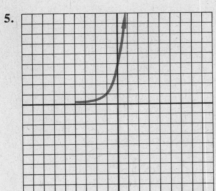

7.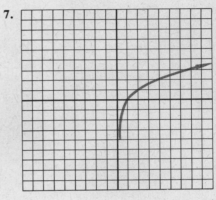

9.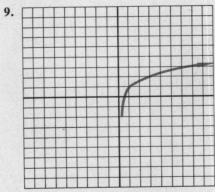

11.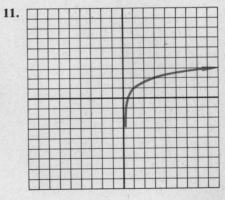

13.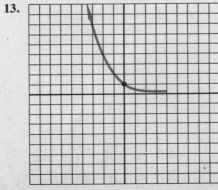

EXERCISES 12.2

1. $\log_4 16 = 2$

3. $\log_{10} 1000 = 3$

5. $\log_2 \frac{1}{2} = -1$

7. $\log_6 1 = 0$

9. $3^3 = 27$

11. $10^1 = 10$

13. $2^{-1} = \frac{1}{2}$

15. $22^0 = 1$

17. 2

19. $\frac{1}{3}$

21. 5

23. 1

25. 0

27. -3

29. 4

31. -3

33. -1

EXERCISES 12.3

1. 0.5922

3. 0.9138

5. 0.7782

7. 1.5670

9. $0.8415 - 2$

11. 3.7810

13. $0.8513 - 3$

15. 2.7482

17. 5.2553

EXERCISES 12.4

1. 3.61

3. 9.01

5. 1.04

7. 452

9. 0.329

11. 2630

13. 0.00526

15. 10.4

EXERCISES 12.5

1. $0.1616 + 2$

3. $0.4392 + 7$

5. $0.9530 - 1$

7. $0.1725 + 2$

9. $0.2680 + 6$

11. $0.0275 - 11$

13. $0.4144 - 1$

15. $0.9772 + 1$

17. 0.2376

19. $0.9727 + 1$

21. $0.9312 - 1$

23. 1.465

25. 2.730

27. 2.741

29. $x = 14$

31. $x = 9$

33. $x = 5$

35. $x = 2$

EXERCISES 12.6

1. 7.46×10^7

3. 2.61×10^{-5}

5. 7.01×10^{-1}

7. 9.25×10^{-3}

9. 2.31×10^{35}

11. 4.2×10^{-4}

13. 1.73

15. 2.6×10

17. 4.08×10^{-1}

19. 1.55×10

21. 1.15×10^3

23. 5.76×10^3

25. 1.76

27. 4×10

EXERCISES 12.7

1. 3.61

3. 6.35

5. 1.04

7. 3.80

9. -0.416

11. 2.8668

UNIT 12 SELF-EVALUATION

1.

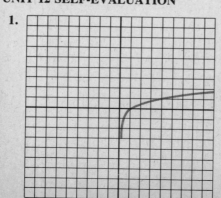

2. $x = 4$
3. $x = -4$
4. $\log_3 0.33 = -1$
5. $10^{-2} = 0.01$
6. 4
7. 3.5465
8. $0.5775 - 2$
9. 335
10. 0.00505
11. 1.7296
12. $x = 4$
13. 3.35×10^4
14. 0.922
15. 6.58

Unit 13   **EXERCISES 13.1**

1. $\dfrac{8}{17}, \dfrac{15}{17}, \dfrac{8}{15}, \dfrac{17}{8}, \dfrac{17}{15}, \dfrac{15}{8}$

3. $\dfrac{1}{2}, \dfrac{1}{2}, \dfrac{2\sqrt{3}}{3}, \sqrt{3}, \dfrac{2\sqrt{3}}{3}, \sqrt{3}$

5. $\dfrac{3}{5}, \dfrac{3}{4}$

7. $\dfrac{17}{15}, \dfrac{8}{17}$

9. $\dfrac{5}{4}$

11. $\dfrac{15}{17}$

13. $\dfrac{24}{7}$

15. $\dfrac{40}{9}$

EXERCISES 13.2

1. .4695
3. 2.577
5. .9877
7. .6316
9. 1.091

11. .2334
13. 15°
15. 46°
17. 26°30′
19. 46°10′

21. 37°20′
23. 72°20′
25. 7265 m kg
27. 3990 m kg

EXERCISES 13.3

1. $A = 36°50′$
   $B = 53°10′$
   $c = 5$
3. $B = 58°$
   $b = 3$
   $c = 4$

5. $A = 40°$
   $B = 50°$
   $c = 157$
7. $A = 41°20′$
   $b = 3{,}167$
   $a = 2{,}785$

9. $A = 58°40′$
   $B = 31°20′$
   $a = 143$
11. $A = 47°10′$
   $b = 8.2$
   $c = 12$

13. 11.9 m
15. 32°20′

17. 49 m
19. 120 m

21. 16°
23. 71 m

EXERCISES 13.4

1. $\dfrac{-1}{\sqrt{2}}, \dfrac{-1}{\sqrt{2}}, 1, -\sqrt{2}, -\sqrt{2}, 1$

3. $-\dfrac{1}{2}, -\dfrac{\sqrt{3}}{2}, \dfrac{1}{\sqrt{3}}, -2, \dfrac{-2}{\sqrt{3}}, \sqrt{3}$

5. $-\dfrac{\sqrt{3}}{2}, -\dfrac{1}{2}, \sqrt{3}, \dfrac{-2}{\sqrt{3}}, -2, \dfrac{1}{\sqrt{3}}$

7. $\dfrac{\sqrt{3}}{2}, -\dfrac{1}{2}, -\sqrt{3}, \dfrac{2}{\sqrt{3}}, -2, \dfrac{-1}{\sqrt{3}}$

9. $0, -1, 0, \text{undefined}, -1, \text{undefined}$

11. $-.9325$
13. $-.9455$
15. 6.314
17. 1.004
19. -17.17
21. .6734
23. 334°40′
25. 119°50′
27. 321°20′
29. 358°
31. 310°40′

EXERCISES 13.5

1. (.278,1.98)
3. (−85.6,180)
5. (−5.33,−102)
7. (19,198°30′)
9. (22.2,53°40′)

11. (6.6,104°)
13. $V_x = 48.4$ km/hr
   $V_y = 194$ km/hr
15. 190 km/hr
17. 129,43°20′

19. 21°30′,279 kg
21. 2500
23. 8 kg, 5 kg
25. 514 kg, 613 kg

EXERCISES 13.6

1.

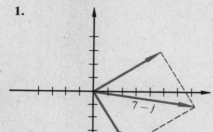

3.

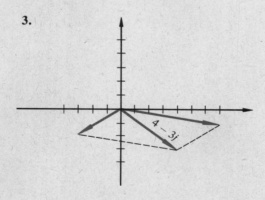

5.

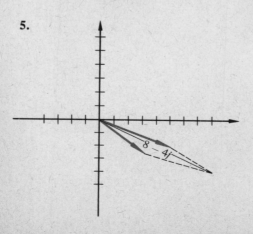

7.

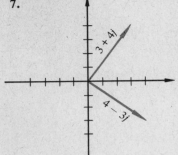

9.

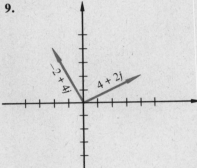

11.

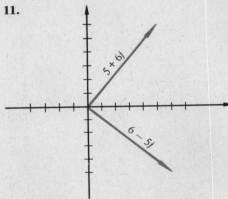

13. $(-15.8, 9.86), (.223, .016)$

15. $(13, 12), (.52, .24)$

17. $7.62e^{1.17j}$

19. $7.81e^{5.40j}$

21. 5 cis $159°20', (-4.68, 1.76)$

23. 2.6 cis $174°10', (-2.59, 264)$

EXERCISES 13.7

1. $C = 74°10'$
$b = 35$
$c = 45$

3. $a = 1.39$
$c = 4.18$
$A = 14°20'$

5. $A = 47°$
$a = 235$
$c = 289$

7. $B = 24°$
$C = 125°$
$c = 14$

9. $B = 38°50'$ $B' = 141°10'$
$C = 119°50'$ $C' = 17°30'$
$c = 10$ $c' = 3.5$

11. $B = 34°30'$
$C = 8°20'$
$c = 5$

13. 297 ft

EXERCISES 13.8

1. $A = 106°40'$
   $B = 48°10'$
   $C = 25°10'$
3. $c = 5.1$
   $A = 21°10'$
   $B = 37°10'$

5. $b = 7.69$
   $A = 34°40'$
   $C = 19°20'$
7. $A = 34°50'$
   $B = 99°40'$
   $C = 45°30'$

9. $A = 23°20'$
   $B = 33°20'$
   $C = 123°20'$
11. $B = 41°50'$
   $C = 8°$
   $a = 7.1$

EXERCISES 13.9

1.

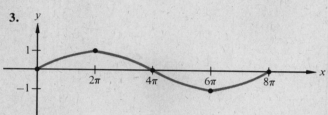

3.

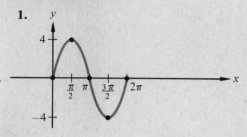

5.

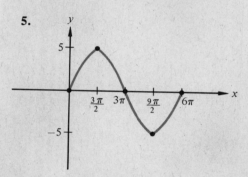

7.

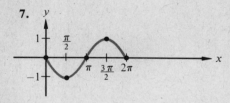

9.

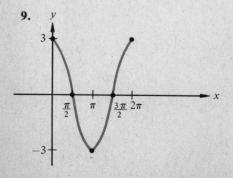

11.

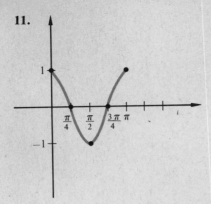

13.

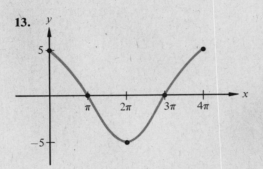

15.

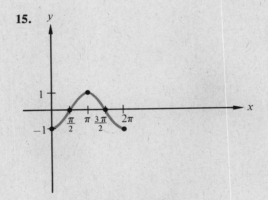

17.

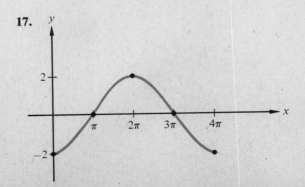

19.

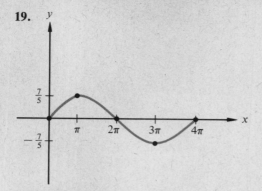

EXERCISES 13.10

1.

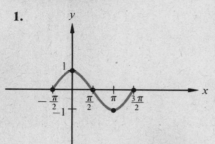

3.

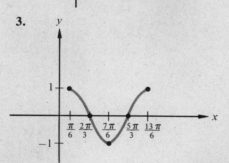

5.

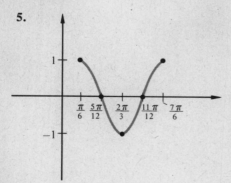

7.

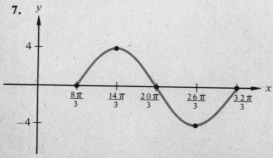

9.

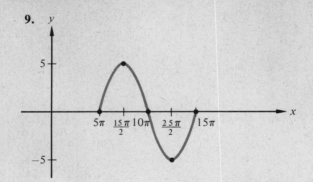

11.

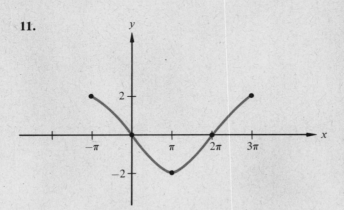

EXERCISES 13.11

1.

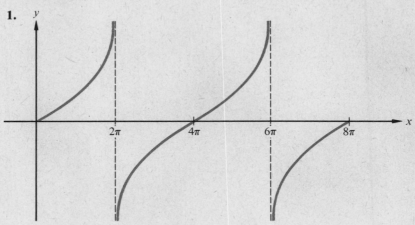

3.

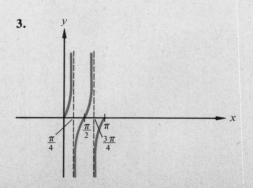

5.

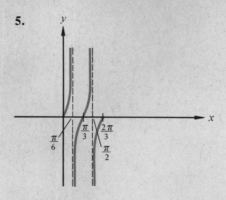

7.

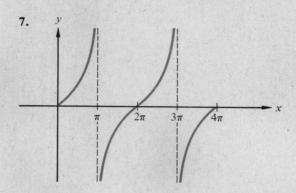

EXERCISES 13.12

1. $\dfrac{\pi}{4}, \dfrac{3\pi}{4}$

3. $\dfrac{\pi}{3}, \dfrac{2\pi}{3}, \dfrac{4\pi}{3}, \dfrac{5\pi}{3}$

5. $\dfrac{\pi}{3}, \dfrac{2\pi}{3}, \dfrac{4\pi}{3}, \dfrac{5\pi}{3}$

7. 90°, 270°, 45°, 225°

9. $\dfrac{\pi}{2}, \dfrac{7\pi}{6}, \dfrac{11\pi}{6}$

11. 18°30′, 198°30′, 26°30′, 206°30′

13. $\dfrac{\pi}{6}, \dfrac{5\pi}{6}$

15. 199°30′, 340°30′

17. No Solution

19. 38°10′, 141°50′

UNIT 13 SELF-EVALUATION

1. $-\sqrt{3}$

2. -2

3. $-\dfrac{1}{2}$

4. 160°

5. 252°

6. $-\dfrac{3}{5}, -\dfrac{4}{5}, \dfrac{3}{4}, -\dfrac{5}{3}, -\dfrac{5}{4}, \dfrac{4}{3}$

7a. .6428

7b. .6412

7c. -2.947

8. 153 kg

9. $A = 76°40′$
   $B = 13°20′$
   $a = 38$

10. 6.2 m

11. a.

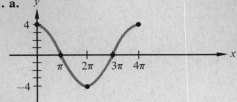

b.

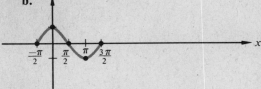

c.

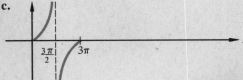

12. (4,3)

13. (13,22°40′)

14. $A = 38°$
$B = 99°$
$C = 43°$

15. 119 kg

16.

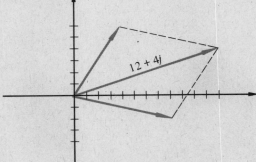

17.

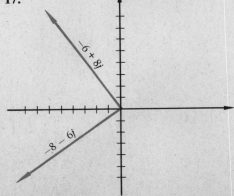

18. $B = 41°40'$ $B' = 138°20'$
      $C = 101°$ $C' = 4°40'$
      $c = 29.7$ $c' = 2.72$

19. $(-6.4, -4.3)$

20. $\theta = 33°40', 213°40', 111°50', 291°50'$

Index

Geometry Formulas

Plane Figures

		Perimeter	Area
Rectangle		$P = 2(a + b)$	$A = ab$
Square		$P = 4b$	$A = b^2$
Parallelogram		$P = 2(a + b)$	$A = bh$
Rhombus		$P = 4b$	$A = bh$
Trapezoid		$P = a + b + c + d$	$A = \dfrac{a + b}{2}h$
Triangle		$P = a + b + c$	$A = \dfrac{1}{2}bh$

		Circumference	Area
Circle		$C = 2\pi r$ $C = \pi d$	$A = \pi r^2$